*This manual is dedicated to Deputy Chief Steve Storment,
and the members of the City of Phoenix Fire Department
Hazardous Incident Response Team,
who insisted we write down the
"gee whiz stuff".*

Air Monitoring Instrumentation

A Manual for Emergency, Investigatory, and Remedial Responders

Carol J. Maslansky
and
Steven P. Maslansky

 Van Nostrand Reinhold
New York

Cover: Rescue party members prepare to enter a mine after an explosion in early
20th Century Great Britain. A caged canary is used to test the atmosphere in the
mine (The Bettmann Archive, Inc.).
Members of the Westchester County, NY Hazardous Materials Response Team
carrying modern-day air monitoring instruments (Photo by Steve Napolitano).

I(T)P   Van Nostrand Reinhold is a division of International Thomson Publishing.
         ITP logo is a trademark under license.

Printed in the United States of America.

Van Nostrand Reinhold
115 Fifth Avenue
New York, New York 10003

International Thomson Publishing
Berkshire House
168-173 High Holborn
London, WC1V7AA, England

Thomas Nelson Australia
102 Dodds Street
South Melbourne 3205
Victoria, Australia

Nelson Canada
1120 Birchmount Road
Scarborough, Ontario
M1K 5G4, Canada

16 15 14 13 12 11 10 9 8 7 6 5 4 3 2 1

Library of Congress Cataloging-in-Publication Data

Maslansky, Carol J.
    Air monitoring instrumentation: a manual for emergency,
investigatory, and remedial responders / Carol J. Maslansky,
Steven P. Maslansky.
        p.   cm.
    Includes index.
    ISBN 0-442-00973-9
    1. Air—Pollution—Measurement—Handbooks, manuals, etc. 2. Air
sampling apparatus—Handbooks, manuals, etc. I. Maslansky, Steven
P. II. Title.
TD890, M37 1993   92-10410
628.5'3' 0287—dc20   CIP

Contents

10. Field Applications   171

11. Emergency Response   197

Appendices   219

Index   301

Preface

Air monitoring instruments are used every day to make decisions regarding health and safety risks to hazardous waste site workers and emergency responders, as well as workers at fixed facilities, the general public, and the environment. This manual is designed to assist you in understanding how to operate air monitoring instruments. It discusses instruments which detect organic and inorganic vapors and gases, oxygen deficiency and excess, and combustible gases. Topics include theory of operation, use, what the readings actually mean, limitations, calibration, and trouble-shooting. Although oriented toward emergency response and hazardous waste work, we hope this manual will be useful for other types of instrument users as well.

This is NOT a health and safety manual for hazardous waste workers or emergency responders. It is beyond the scope of this manual to discuss hazard recognition, levels of protection, operations, containment and control, sampling techniques and strategies, and decontamination of personnel and equipment. Numerous texts and manuals are available on these topics for emergency responders[1-9] as well as hazardous materials investigatory and remediation workers[10-14].

Similarly, we do not wish to suggest that individuals completing this manual can replace trained health and safety professionals. In many cases, air monitoring programs are designed by professionals for specific sites and hazards. Health and safety plans (HASPs) for hazardous materials sites also discuss air monitoring in detail. A well-written HASP should also contain explicit guidance in terms of what worker activities can be undertaken and the type of protective equipment that must be utilized, based, in part, on instrument readings. It is always best to consult a trained professional when using air monitoring instrumentation as part of a health and safety program.

This manual should also be helpful to emergency response incident commanders and supervisors of investigatory or remediation operations, as well as those who interpret readings or make decisions based on air monitoring readings. A lack of a basic understanding of how monitoring instruments operate and their inherent limitations can result in misinterpretation of collected information, which may jeopardize the health and safety of all concerned.

If you are wondering what you have gotten yourself into—relax! Sit down and enjoy yourself! This is not a traditional textbook—too many text books require nothing of their readers except sufficient attention to simply read (and then all too often forget) the material. A few authors try to impress their readers with seemingly important information that has nothing to do directly with understanding the topic at hand; some authors talk down to their audience by using technical jargon; others write about subjects that they know only from other books or reference materials.

We will do none of that to you. This is a straightforward manual designed to assist you in understanding how air monitoring instruments work, and how to use instrument responses in making intelligent decisions. We both have practical, hands-on and field experience with nearly all the instruments described. We have been teaching air monitoring instrumentation skills for quite a few years; this manual is a result of our field and teaching experiences.

While it may be helpful to have previous experience in emergency response or hazardous materials site work, you do not need an advanced degree in industrial hygiene, engineering, chemistry, or physics to understand and complete this manual. Although previous experience with the instruments described may also be of value, it is not essential for your understanding; we have attempted to provide pictures and diagrams of the equipment described. We do urge you to take out any instruments you have available and follow along with the text whenever possible.

We will only briefly discuss radiation detectors and will not cover instruments which sample and measure dusts, particulates, and aerosols. These topics have been thoroughly discussed in several valuable references[15-17]; we heartily recommend them.

Most of the examples given are derived from our own experiences; they emphasize the use of air monitoring instruments as tools to assist the user in evaluating the potential risk to personnel from some types of air contamination. Our emphasis on instrumentation does not in any way suggest that other hazards should be ignored. Instrument response is only one of many parameters that must be integrated into the hazard and risk evaluation process.

Each chapter introduces specific types of instruments and discusses a variety of aspects including limitations, accessories, and interpretation of meter response. Later chapters discuss selection and utilization of meters during field investigations or emergency response. Some readers may become impatient with this format, however, since most situations require the use of multiple instruments, an understanding of all the tools that are available is necessary before they can be put to intelligent use. There will be times when you do not agree with our assessment of an instrument; you may not agree with some of our recommendations. Some readers may consider our approach too simple or too much like a cookbook. To that we say–good for you! Remember, you are supposed to think and make your own decisions even as you complete this manual. Your decisions must be based on your own experiences and the equipment you have at hand. Besides, even the best recipe in the best cookbook can be improved; just remember that a good cook always reads the original recipe first and knows what each of the ingredients are.

Throughout this manual we employ some basic arithmetic calculations. You will notice that we round off numbers whenever possible, and attempt to show how to reach the same approximate answer without the use of a calculator, or even a pencil and paper. We do this deliberately because you may not always have ready access to pen, paper and calculator.

This manual is organized in such a way that concepts are introduced and followed by questions and problems with answers; we employ this approach to ensure your understanding of the material. Review questions and problem sets are available at the end of each chapter to enable you to assess your own understanding and progress. These are learning tools and we encourage you to use them. Plenty of space has been provided for calculations and answers. It is not the end of the world if you answer incorrectly! Mistakes are why there are erasers on pencils. To make a mistake, recognize the error, and then correct it is the best way to learn and remember. Answers and discussions of odd-numbered problems can be found in Appendix E. Course instructors may obtain answers to even-numbered problems by writing to the publisher on official letterhead. If you cannot understand a concept or answer, or need to discuss some aspect of this manual with us, please do not hesitate to call or write to us. We encourage feedback from our readers.

It is probably evident to you by now that to successfully complete this manual you will have to do more than read. You will have to think and actually participate by working out problems and answering questions. So, before we get any further–stop reading and get a pencil and an eraser!

1. Andrews, L.P. (Ed). 1992. Emergency Responder Training Manual for the Hazardous Materials Technician. New York: Van Nostrand Reinhold.

2. Noll, G.G., Hildebrand, M.S. and Yvorra, J.G. 1988. Hazardous Materials–Managing the Incident. Stillwater, OK: Fire Protection Publications.

3. International Fire Service Training Association (IFSTA). 1988. Hazardous Materials for First Responders. Stillwater, OK: Oklahoma State University.

4. Noll, G.G. and Hildebrand, M.S. 1992. Gasoline Tank Truck Emergencies–Guidelines and Procedures. Stillwater, OK: Fire Protection Publications.

5. Tolke, G. (Ed). 1993. Hazardous Materials Response Handbook. Quincy, MA: National Fire Protection Association.

6. International Association of Fire Fighters (IAFF). 1991. Training for Hazardous Materials Team Members. Washington, DC: IAFF.

7. York, K.J. and Grey, G.L. 1989. Hazardous Materials/Waste Handling for the Emergency Responder. New York: Fire Engineering.

8. International Fire Service Training Association (IFSTA). 1984. Haz Mat Response Team Leak & Spill Guide. Stillwater, OK: Fire Protection Publications.

9. American Society for Testing and Materials (ASTM). 1983. A Guide to the Safe Handling of Hazardous Materials Incidents. Special Technical Publication 825. Philadelphia: ASTM.

10. Andrews, L.P. (Ed). 1990. Worker Protection During Hazardous Waste Remediation. New York: Van Nostrand Reinhold.

11. Martin, W.F., Lippitt, J.M. and Prothero, T.G. 1987. Hazardous Waste Handbook for Health and Safety. Stoneham, MA: Butterworth Publishers.

12. Levine, S.P. and Martin, W.F. (Eds). 1985. Protecting Personnel at Hazardous Waste Sites. Stoneham, MA: Butterworth Publishers.

13. Simmons, M.S. (Ed). 1991. Hazardous Waste Measurements. Chelsea, MI: Lewis Publishers, Inc.

14. U.S. Environmental Protection Agency. 1987. A Compendium of Superfund Field Operations Methods. Washington, DC: USEPA Office of Emergency and Remedial Response.

15. Hering, S.V. (Ed.) 1989. Air Sampling Instruments for Evaluation of Atmospheric Contaminants, 7th edition. Cincinnati, OH: American Conference of Governmental Industrial Hygienists.

16. Ness, S.A. 1991. Air Monitoring for Toxic Exposures. New York: Van Nostrand Reinhold.

17. Shapiro, J. 1990. Radiation Protection–A Guide for Scientists and Physicians, 2nd edition. Cambridge, MA: Harvard University Press.

Acknowledgments

We would like to thank our publisher, Van Nostrand Reinhold, for their patience, and Mr. Thomas Bock for text design, page format, and initial typesetting. Special thanks to Ms. Lisa Zaccharia of LZ Graphics for preparing all artwork, graphic figures, and appendices, and for her invaluable suggestions and assistance during final typesetting, page layout, and proofing. We also thank Mr. Steve Napolitano for his photographic talents and for his patience. We also appreciate the assistance of Ms. Pam Markelz of SEC-Donohue, Mr. Steve Vanburen of the New York Power Authority, and Mr. Mario Manfredi of AAA Emergency Supply. We also thank the many manufacturers who generously provided photographs and other materials. Special kudos go to Mine Safety Appliances Company for providing archival photographs, as well as Photovac, Inc. and The Foxboro Company for generously contributing instruction manuals and technical information. Additional thanks to Photovac, Inc. for their valuable discussions and hospitality at their Deer Park, NY facility. Special thanks to the manufacturers who allowed us to use instrument-specific information in the Appendices: Mine Safety Appliances Company, National Draeger, Inc., Scott Aviation, and Gas Tech, Inc. We particularly wish to thank Mr. Rod Turpin of the USEPA Environmental Response Team in Edison, NJ, for reviewing this manual and for his encouragement and suggestions. We also thank the other reviewers solicited by VNR for their valuable comments. Finally, we thank Mr. Lawrence Conklin for his superb proofreading.

1

Why Monitor?

LEARNING OBJECTIVES

1. Define the reasons why air monitoring is conducted.
2. List the factors to be considered when selecting an instrument.
3. Describe the difference between Class I, Class II and Class III materials.
4. Describe the differences between Division 1 and Division 2 locations.
5. List some of the gases and vapors which belong in Groups A through D.
6. Describe the difference between intrinsically safe and non-incendive instruments.
7. Explain the difference between sensitivity and selectivity.
8. Explain the difference between factory calibration and field calibration checks.
9. Describe initial meter start-up and check-out procedures.

Air monitoring is performed for different reasons. The purpose of monitoring will vary with the situation as well as the actual, potential, or anticipated hazards. Obviously, air monitoring procedures at a factory, refinery, or other fixed facility that has a limited number of known hazards will be very different than at an uncharacterized hazardous waste site or an uncontrolled hazardous materials release.

In most situations, however, the basic impetus to perform air monitoring remains the same: to provide information regarding the type and relative quantity of hazards present. This information can, in turn, be used to assess the risk to workers, responders, the general public, and the environment. Monitoring information can be utilized to select the appropriate personal protective equipment; it can also be used to delineate areas where protective equipment or evacuation is required. Other reasons to monitor are to assure that a facility or procedure is complying with applicable regulations, to evaluate the effectiveness of mitigation techniques, to assess the efficiency of emissions control, and to determine the need for specific medical monitoring for exposed personnel.

The way air monitoring is performed will depend on the actual or perceived hazards present. For example, at a fixed facility such as an oil refinery, the hazards are known; worker overexposure and perhaps flammable atmospheres will be daily concerns that require constant or periodic monitoring. Entry to confined spaces trigger other concerns, including oxygen deficiency and the presence of toxic vapors such as carbon monoxide and hydrogen sulfide.

At an uncontrolled hazardous materials release, emergency responders must initially identify the hazards present using recognition clues such as the location of the release, the type and configuration of containers, labels or placards, or other sources of information including shipping papers, material safety data sheets (MSDS), or information from site workers, or bystanders.

Figure 1-1: Workers are familiar with hazards at this facility and periodic monitoring is used to detect potentially hazardous conditions. (Author photo.)

Air monitoring can be used to confirm suspicions regarding the type of hazards present as well as to rule out other potential hazards.

Uncharacterized hazardous materials storage or waste sites present unique problems. These sites may offer few clues regarding the type of hazards present. Initial site characterization procedures are now mandated and specified by the Occupational Safety and Health Administration (OSHA) in 29 CFR 1910.120[1]. This standard, initially promulgated as an interim rule on December 19, 1986[2], and published as final rule on March 6, 1989[3], applies to hazardous waste clean-up operations and initial investigations at Superfund

Figure 1-2: Uncontrolled or uncharacterized sites may contain unknown hazards and require continuous air monitoring strategies. (Author photo.)

(CERCLA) sites, landfills, and those sites designated by State or local governments, corrective actions at RCRA sites, and emergency response sites. The Environmental Protection Agency (EPA), National Institute for Occupational Safety and Health (NIOSH), Coast Guard, and OSHA have jointly or individually published guidelines for hazardous waste site activities[4-7].

Primary site characterization at a hazardous waste site must include monitoring for ionizing radiation, flammable or explosive atmospheres, oxygen deficiency, and toxic substances. Results of initial air monitoring are used to determine the type of personal protective equipment (PPE), including respiratory protection, that is appropriate for that particular site. An ongoing air monitoring program must be established after initial characterization to assure continued proper selection of PPE, and to provide a means to assess contamination controls and workplace practices. As a minimum, periodic monitoring must be conducted when work begins at a different portion of the site, a different type of activity is initiated (e.g. sampling drums as opposed to well drilling), another type of contamination than that originally discovered is handled, or workers are in an area with obvious contamination (e.g. leaking drums, lagoons or holding ponds, spill areas).

It should also be recognized that valid air monitoring data can eliminate questions or allegations of worker overexposure, as well as provide documentation of adherence to environmental and occupational standards. This requires, however, that monitoring be conducted by trained personnel in a manner that will withstand legal challenge.

BASIC INSTRUMENT DESIGN

Most instrument designs are sufficiently similar to allow a generic description of basic components and operational features. Manufacturers, however, may use different terms to describe the same components.

The instrument itself is often called a **meter**, and is usually contained within a box or **case**. The **readout display** indicates the relative amount of material present; the display may be digital, analog, or represented as a bargraph. The **sensor** or **detector** is the component which responds to the contaminants present in air. **Controls** or **function switches** include the on/off switch, zero adjust, range or scale selector, span adjust, battery check, and alarm on/off/adjust. Controls may be configured as knobs, toggles, switches, buttons, screws, or keypads; one switch may control more than one function. **Alarms** are activated at a pre-set response level and may be audible, visual, or a combination of both.

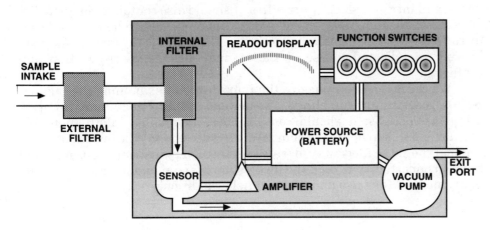

Figure 1-3: Schematic of a basic meter equipped with a vacuum pump.

The **battery** provides electrical power to the entire instrument; batteries may be re-chargeable or disposable. Electronic components are contained within the meter box and are relatively inaccessible. Sample air may enter the sensor by passive diffusion or be drawn into the instrument by a **vacuum pump** or **aspirator bulb** via a **sample line** or **hose** which conducts sample air into the meter. A **probe** is an inflexible extension to the end of a sample line. The sample line may be fitted with a **filter** to exclude particulates or a **liquid trap** to prevent liquid from entering. Sample air enters through an **inlet** or **intake**, and **exits** the meter through the **exit port** or **discharge**.

USING AND CHOOSING AN INSTRUMENT

The Manufacturer

This may seem obvious, but it is **very** important to consider the manufacturer of an air monitoring instrument prior to purchase. A reputable company, with a history of designing, manufacturing, servicing and repairing instruments, is the one to choose. The manufacturer should also have a record of improving designs and offering more than one model to meet different needs. It is a gamble to buy an instrument from a fledgling company; what will you do if the company goes out of business? It will be impossible to replace sensors, or to get your instrument serviced or repaired. Remember a multi-year guarantee is worthless if the guarantor is no longer around!

The Instruction Manual

One way to determine the quality of the manufacturer is to examine the instruction manual which accompanies the instrument. Most manufacturers will gladly provide a manual prior to purchase; in some cases you may be required to purchase the manual, which is usually less than $50. The small price paid to purchase a manual is a wise investment, especially if you discover that an instrument costing thousands of dollars will not meet your needs.

The quality of instrument workmanship is usually reflected in its manual. The manual should be well organized, and easy to read and understand; manuals translated from foreign languages can be especially difficult to comprehend.

A good instruction manual should contain clear, easy to understand sections on theory of operation, limitations, troubleshooting, calibrating, maintenance, cleaning and servicing, replacing sensors or other components, and intrinsic safety warranties; it should also contain a complete list of accessories and replacement parts. Buyer beware when there is an introductory section explaining how to use the manual; the instrument may be very complicated to use or the manual may be poorly written and confusing.

We cannot over-emphasize the importance of examining the manual **before instrument purchase**. The manual contains more than just operating instructions; it should contain other invaluable information. We recommend that a copy of the manual always accompany the instrument; the original should be filed away for safe keeping. Every instrument user should have an opportunity to read the instruction manual and practice using the instrument **before** use under actual field conditions.

A good instruction manual should contain the following information in an easy-to-find format:

- table of contents
- limitations, warnings, and operating precautions
- summary of general specifications
- a description of the meter (this should include a diagram of controls as well as a photograph of the meter itself)
- intrinsic safety approvals
- initial inspection and set-up procedures; this often includes directions on how to insert the sensor
- daily start-up and operating procedures
- theory of operation
- calibration and field check procedures
- response curves or conversion factors
- troubleshooting
- general and corrective maintenance (this should include information on battery charging, as well as sensor, filter, and battery replacement)
- calibration procedures (this should include a photograph or diagram of internal span adjustment controls)
- alarm set point as well as directions how to change it
- descriptions of accessories and replacement parts, including order numbers
- electrical or schematic drawings
- address and telephone number of authorized service center

The general specification section summarizes the performance, operation, and phyisical characteristics of the instrument. This section is the best place to start when reviewing the manual, and usually contains the following information:

- type of sensor(s)
- detection range(s)
- accuracy (± a percentage of the reading)
- response time
- operating temperature range
- operating oxygen range (if appropriate)
- sample air flow rate
- power source
- battery life or operating time on full battery charge
- overall dimensions and weight
- intrinsic safety approvals and warranties

The instruction manual should also contain an adequate number of pictures, figures, diagrams to assist the user in understanding the operation and use of the instrument. In addition, the manual should contain a description of the instrument with technical specifications, as well as at least one picture or diagram indicating all instrument components and controls. One "manual" we examined had been reproduced on a few sheets of paper without even a figure or picture of what the instrument looked like when fully assembled!

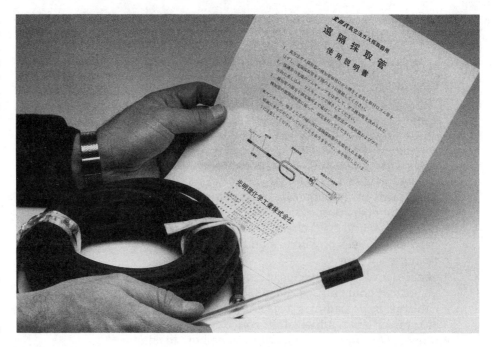

Figure 1-4: Make sure foreign manufacturers supply adequate translations; the Japanese instructions for this equipment was of little value. The manufacturer did not have an English translation. (Photo by Steve Napolitano.)

Material	Group
Acetaldehyde	C
Acetone	D
Acetylene	A
Ammonia	D
Benzene	D
1,3-Butadiene	B
Butane	D
Carbon Disulfide	–
Carbon Monoxide	C
Cyclohexane	D
Diethyl Ether	C
Ethyl Acetate	D
Ethylene	C
Ethylene Oxide	B
Formaldehyde (Gas)	B
Gasoline	D
Hexane	D
Hydrogen	B
Hydrogen Sulfide	C
LPG	D
Methane	D
Methanol	D
Methyl Ethyl Ketone	D
Methyl Mercaptan	C
Naphtha (Petroleum)	D
Propane	D
Propylene Oxide	B
Styrene	D
Toluene	D
Turpentines	D
Xylene	D

Table 1-1: Group classification of selected flammable gases and vapors.

Inherent Safety

Instruments which require battery power to operate may be a source of ignition in the presence of flammable gases or vapors. Any instrument operated in an area where flammable conditions are anticipated should be certified as **intrinsically safe** or **explosion proof**. Explosion proof instruments allow flammable gases to enter but have an enclosure expressly designed to contain and withstand any explosion so it does not spread. Intrinsically safe devices are designed to be incapable of releasing sufficient thermal or electrical energy to cause ignition of specific categories of hazardous atmospheres.

Minimum standards for inherent safety in hazardous atmospheres have been defined by the National Fire Protection Association (NFPA) in its National Electrical Code (NEC)[8]. Categories of hazardous atmospheres are defined by Class, Group, and Division.

Class describes the type of flammable material that produces the hazardous atmosphere. Class I represents flammable gases and vapors such as methane, hydrogen, gasoline, and hydrogen sulfide. Class II consists of combustible dusts such as flour, grain, or coal dust. Class III denotes ignitable fibers.

Class I flammable gases and vapors are classified into **Groups A through D**. Group A contains only one gas, acetylene. Group B contains gases and vapors such as butadiene, ethylene oxide, formaldehyde, hydrogen, and propylene oxide which have wide flammable ranges. Groups C and D encompass a large number of gases and vapors. Group designation involves examining the autoignition temperature of the material, as well as explosion pressures and maximum safe clearance between parts of a clamped joint under different conditions. Group assignments may be made as a result of actual testing or by chemical analogy or similarity to chemicals previously tested[9]. Many Group C and D materials have not been tested; such group classification based on chemical analogy may be incorrect.

Some chemicals have attributes that require safeguards beyond those required for Groups A through D. Such a material is carbon disulfide which has an extremely low autoignition temperature.

The term **Division** is used to describe the location of the hazardous atmosphere. A Division 1 location is any area where under normal conditions, there may be the generation or release of flammable gases or vapors on a continuous, intermittent, or periodic basis, into an open or unconfined area. Thus a gasoline terminal, propane filling station, or drum filling facility which handles flammable liquids would all be classified as Class I, Division 1 locations.

Division 2 describes locations where the generation and release of flammable gases or vapors are anticipated only from ruptures, leaks, or other failures from closed systems or containers. A manufacturing plant which uses flammable liquids which are normally confined within a closed system is a Division 2 location. If the plant suffers an explosion and flammable liquids and vapors are released, it is now a Division 1 location. Similarly, an abandoned waste site containing intact drums of flammable liquids is a Division 2 location. If the site contains leaking drums of flammable liquids it is a Division 1 location.

Figure 1-5: The back of a Scott Alert Model S109 diffusion meter which detects carbon monoxide, oxygen, and combustible gases. Intrinsic safety approvals are clearly marked. Note the five set screw controls are accessible but cannot be inadvertently changed. (Photo by Steve Napolitano.)

An instrument is certified as explosion proof or intrinsically safe for a given Class, Division, and Group. **Intrinsically safe** instruments are approved for Division 1 locations while **non-incendive** instruments are approved only for Division 2 locations. Certification is awarded only for specific atmospheres; it is not certified for use in atmospheres other than those indicated. Division 1 locations are more hazardous than Division 2; test protocols for Division 1 devices are therefore more stringent. An instrument certified for Division 1 locations is therefore also approved for Division 2. Instruments with the highest level of inherent safety for flammable gases and vapors will be certified for Class I, Division 1, Groups A,B,C,D.

All certified devices must be clearly and permanently marked to show Class, Division, and Group; it is insufficient for the instrument to be labeled as intrinsically safe. Remember that NFPA defines standards for inherent safety but it does not certify instruments. Testing and certification is performed by Factory Mutual Research Corp. (FM) or Underwriter's Laboratory, Inc. (UL).

Older instruments which were FM or UL certified at the time of manufacture may no longer be eligible for such approvals. Inherent safety approvals are renewed periodically, in order to ensure instrument design meets current specifications. Manufacturers do not broadcast information regarding the loss of certification; it is up to the user to determine if the make and model in use is still approved.

Imported and domestic instruments may have European certification similar to that provided by UL or FM. European standards are defined by the European Committee for Electrotechnical Standardization (CENELEC)[10]. CENELEC requirements for construction and testing of electrical equipment for use in potentially hazardous atmospheres is similar to NEC standards. Each country which adopts the European standards publishes them in their own language. The British Standards Institution publishes the English-language version. Certification that an instrument meets CENELEC standards may be awarded by a variety of European testing laboratories; certified instruments must carry an approval plate. The symbol **EEx** indicates the instrument conforms to one or more CENELEC standards; other symbols are added to define which standards and under what conditions.

Figure 1-6: Although the handle indicates this instrument is intrinsically safe, there is no FM or UL approval plate. (Photo by Steve Napolitano.)

Material	Subdivision
Acetaldehyde	A
Acetone	A
Acetylene	C
Ammonia	A
Benzene	A
1,3-Butadiene	B
Butane	A
Carbon Disulfide	C
Carbon Monoxide	A
Cyclohexane	A
Diethyl Ether	B
Ethyl Acetate	B
Ethylene	B
Ethylene Oxide	B
Gasoline	A
Hydrogen	C
LPG	A
Methane	A
Methanol	A
Methyl Ethyl Ketone	A
Methyl Mercaptan	A
Petroleum Naphtha	A
Octane	A
Propane	A
Propylene Oxide	B
Styrene	A
Toluene	A
Turpentines	A
Xylene	A

Table 1-2: Subdivision classifications of selected flammable gases and vapors under CENELEC standards.

Instruments may be certified as meeting requirements for **increased safety**, **flameproof enclosure**, or **intrinsic safety** in potentially explosive atmospheres. Increased safety instruments are designated on the approval plate as "e", flameproof enclosures as "d", and intrinsic safety as "i". **Group I** instruments are designed for use in mines susceptible to firedamp (methane). **Group II** instruments are for use in locations containing gases and vapors other than methane (or in addition to methane). Group II gases and vapors are separated into **Subdivisions** A, B, or C; Subdivision C is reserved for the highest hazard gases, such as hydrogen and acetylene. Table 1-2 lists some representative gases for each Subdivision; note that most materials classified under Group D and C by NEC/NFPA standards (Table 1-1) are placed in Subdivision A under CENELEC.

CENELEC defines an intrinsically safe instrument as incapable of causing ignition in normal operation with a specified number (1 or 2) of faults or malfunctions. Instruments labelled "ia" are intrinsically safe with two faults, while "ib" equipment qualified with a single fault. Maximum acceptable temperature for the instrument housing are indicated by temperature classes T1 through T6. The highest rating is T1 (450°C); ratings at or above T4 (135°C) should be considered acceptable.

An instrument comparable to NEC/NFPA Class I, Division 1, Groups A, B, C, D (the highest level of intrinsic safety) would be rated as EEx ia IIC T6. The presence of an explosion-proof enclosure in a combustible gas indicator is designated by the letter "d" (i.e. EEx d ia IIC T6).

Instruments certified only for use in atmospheres containing methane and/or coal dust are tested and certified by the Mine Safety and Health Administration (MSHA). Group I instruments under CENELEC standards are also only for methane-containing atmospheres (i.e. EEx ib I T4). These instruments should not be used in the presence of other gases and vapors.

Instrument Design and Operation

Field monitoring instruments should be lightweight and portable, sturdy, compact, weather and temperature resistant, and relatively simple to operate and maintain. It should have a convenient handle as well as a shoulder strap; leather straps cannot be decontaminated and should be replaced. Any instrument used for emergency response should have a short warm-up interval.

The instrument should have a readout display that is easy to see and read under a variety of lighting conditions. Digital displays are often very difficult to see in the glare of sunlight. LED displays literally disappear under bright light. Analog displays without an illumination option will be difficult to read under low light conditions.

Most instruments are subject to slight fluctuations in readings when sampling zero or background concentrations; this is often interpreted as electronic drift and is most apparent during warm-up. While such fluctuations are more obvious in digital display instruments, they also occur in analog display models. Many users incorrectly assume that analog display instruments are more reliable because there are less overt vacillations in readings. Meter response to slight concentration changes, however, are more readily observed on digital instruments.

Controls should be conveniently placed, adequately labeled, and large enough to adjust or reset while wearing gloves. Controls should have locking hubs or be positioned so that the user cannot inadvertently change settings.

Most instruments are equipped with visible and audible alarms that are activated at a level preset at the factory. The alarm settings should be easily adjusted; if possible get a meter with an audible alarm shut-off. There will be situations when you do not want the alarm. Some instruments automatically reset when the concentration present falls below the alarm level. Such instruments should never be left unattended, since they will not indicate if an alarm condition has occurred and then resolved itself. A safer alternative is to use a meter that requires manual resetting after an alarm condition has occurred.

Many monitoring instruments are equipped with a battery-powered pump or a hand aspirator that draws sample air into the instrument where the sensors are located; the alternative is the diffusion instrument with an internal or external sensor that must be physically in the atmosphere it is sampling. That is, for diffusion instruments, the sensor must get to the sample, while for sample draw instruments, the sample can be brought to the sensor. Small personal monitors have sensors placed at the surface of the instrument which are in direct continuous contact with ambient air; these monitors can be adapted for remote sampling by using a sample draw adaptor and bulb aspirator or battery-powered pump. Other diffusion instruments have external diffusion sensors attached to the instrument by a long cable.

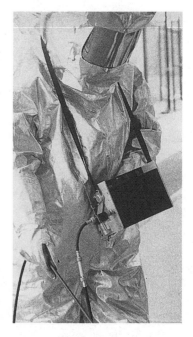

Figure 1-7: Bulky suits or big stomachs not only make it difficult to read an instrument display, they may also bump into and change meter control settings. (Author photo.)

Diffusion meters have several advantages. There is no pump or moving parts, and the sensor can provide immediate information about the monitored environment. Diffusion instruments, however, are more susceptible to damage and inadvertent contamination because the sensor is more exposed. Diffusion meter readings may be more affected by excessive moisture, particulates, and temperature extremes.

One drawback often overlooked with external diffusion sensors is the accessibility of the atmosphere that you wish to sample. A meter with a large external sensor the size of a package of cigarettes will be of little value when sampling for combustible gases through the holes in a manhole cover. Diffusion meters will not provide reliable results in atmospheres where there are fluctuating contaminant concentrations or inordinate air turbulence; in these situations, a sample draw instrument should be used.

Figure 1-8: The Scott Alert S109 is a diffusion instrument designed to be worn as a personal monitor; openings at the top front of the instrument allow sample air to enter by diffusion. A bulb aspirator adaptor can convert it to a sample draw instrument (right) for testing of confined spaces prior to entry. (Photos by Steve Napolitano.)

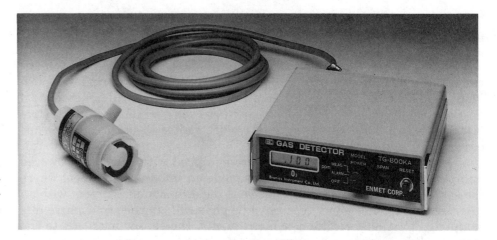

Figure 1-9: An ozone gas detector with an external diffusion sensor attached by a long cable. (Courtesy Enmet Corp.)

Sample draw meters with pumps or hand aspirators are less susceptible to external contamination, since in many cases only the sample hose or probe contacts the suspect atmosphere. If the sample hose is grossly contaminated it can simply be discarded and replaced. Running the meter in clean air is usually all that is necessary for internal decontamination of the meter and hose that were exposed only to gases and vapors. Other advantages of sample draw meters is that the sample hose or extendible probe can access small or remote locations and the sensor is less likely to be damaged since it is protected within the instrument. The primary drawback of this design is sample dilution; any leak in the sample system will result in a false negative or an incorrect low reading. Other disadvantages include a lag in response time which can be significant when the sample must travel a considerable distance through the sample hose, and pump malfunction or inadequate sample flow.

Where applicable, sensor changeover should be simple and quick. In most cases, sensor replacement should involve simply unplugging the old sensor and replacing it with a new, factory-calibrated sensor from the manufacturer. After replacement, the user should then determine the instrument and sensor are working properly by performing a calibration check. Avoid instruments with sensors that require monthly or periodic readjustment, or that must be returned to the manufacturer for sensor replacement. Check the instrument manual for a diagram which depicts where the sensors are housed. Choose a meter that has readily accessible sensors, otherwise you may have to disassemble the entire instrument just to get at the sensor.

If the unit is powered by disposable batteries, changeover should be easy and quick. If the battery is rechargeable, it should have a long life. Usable battery life is diminished in nickel-cadmium (NiCad) batteries that are recharged before they are completely drained. Acid batteries are usually designed to be continuously plugged into a trickle charger; when stored off the charger they can become rapidly depleted.

There should be a battery check setting as well as a visible and an audible alarm when the battery power is inadequate. Meters with rechargeable batteries that are designed to be stored in the trickle charge mode usually do not have a battery check. The user is informed of battery status only when the low battery alarm is triggered.

Finally, check what optional accessories are available. Many meters can be outfitted with multiple lengths of hose as well as a variety of particulate or activated charcoal filters; dilution assemblies, and traps to prevent entry of liquid or aerosols may also be available.

Instrument Response

Each instrument is designed to respond to a particular type of hazard within a specific concentration or **detection or operating range**. For instance, combustible gas indicators (CGIs) may detect combustible gases in % gas by volume, % LEL, or ppm equivalents; instruments which measure oxygen (O_2) concentrations read in % O_2 by volume in air; meters which detect toxic gases such as hydrogen sulfide (H_2S) and carbon monoxide (CO) are sensitive to ppm concentrations in air. Other instruments will simply sound an alarm in the presence of low concentrations of certain gases and vapors but are not designed to give an actual reading of the amount present.

The lower detection limit is the lowest concentration that the meter will accurately and reliably respond to; the upper limit is the concentration which saturates the sensor or detector and elicits the maximum meter response.

Whenever possible, choose an instrument with a relatively **short response time**. Response time is the interval required to obtain a sample and attain approximately 90% of the final reading or response. Direct reading instruments typically have a response time of several seconds to several minutes. Response time is not the same as **lag time**; lag time is the interval between the introduction of a material and the first observable meter response. Lag time and response time can depend on a variety of factors including how the sample reaches the sensor or detector (passive diffusion or drawn in by a pump), sampling rate, length of sampling hose, use of accessories such as prefilters and liquid traps, temperature, and the type of sensor or detector used. **Recovery time** is the time required for the instrument to return to zero or to a normal ambient reading after removal from the sample atmosphere. Recovery time can be influenced by the air sampling rate, length of sample hose, contaminant concentration, as well as temperature and humidity.

Instruments used for initial response to emergencies should have a short **warm-up time**. We define warm up time as the interval required before the instrument can be calibrated. Some manufacturers define warm-up time as the interval before the instrument can be used; in many cases, this is simply the time it takes to turn the instrument on, which is not correct. Insufficient warm-up can result in significant drift in readings as the sensor stabilizes, making it difficult to obtain meaningful information.

Decide how sensitive the instrument must be to meet your needs. **Sensitivity** is defined by the detection limit, or how small a concentration can be reliably detected. Sensitivity is important when very small changes in concentration may be hazardous. Most highly sensitive instruments tend to require a longer warm-up, more careful handling, and more frequent calibration. Remember that a meter with multiple operating ranges is usually limited by its most sensitive range.

The **selectivity** of the meter should also be examined. Selectivity defines what type of materials will be detected, and which other interfering

INACCURATE PRECISE

INACCURATE IMPRECISE

ACCURATE PRECISE

Figure 1-10: The difference between accuracy and precision can be envisioned using the bullseye.

compounds will affect meter response. Most air monitoring instruments are not very selective; meter response will usually require at least some interpretation by the operator. Read the limitations section of the instruction manual carefully. Beware of a manufacturer that lists no limitations or interfering substances; just because none are listed does not mean that none exist.

Precision is a measure of the reproduciblity and reliability of a meter response to a known concentration. A meter which repeatedly produces the same response or gives meter readings which are very close to each other is considered precise. Precision has nothing to do with **accuracy**. Meter A may consistently give a reading of 50 ppm in response to a sample known to contain 100 ppm; such a meter would be considered inaccurate but precise. On the other hand, Meter B may give inconsistent meter readings such as 70, 90, 120, 80, and 100 ppm. Meter B is not precise, although each individual reading is more accurate than that provided by Meter A.

Instrument Calibration and Relative Response

Air monitoring instruments are calibrated by the manufacturer to respond **accurately** to one particular vapor or gas. Calibration is a process of adjusting the instrument response until the reading corresponds to the actual concentration present. Initial calibration is performed at the factory, using multiple concentrations of calibrant gas. After factory calibration the instrument will respond accurately only to the calibrant gas within its detection range. The meter will not respond accurately to other gases or vapors, but will rather give a **relative response** reading that may be higher or lower than the actual concentration present. This does not suggest that the meter is useless except for situations when the calibrant is present; it simply means that the meter responds to other gases or vapors using the calibrant as a reference. When detecting gases other than the calibrant gas, the concentration read is often expressed as calibration gas equivalents, response equivalents, equivalent units, or simply units.

Depending on the type of instrument, the manufacturer may provide **relative response curves** or **conversion factors** for gases or vapors other than the calibrant gas. The number of curves or factors available can be useful in assessing how thorough the manufacturer has been in developing the instrument. Check to ensure that the same calibrant has been used for each detection range; for instance, some combustible gas indicator manufacturers offer meters with the %LEL range calibrated to methane while the ppm range is calibrated to another gas, such as hexane.

More than likely you will not actually be calibrating your meter but rather checking to ensure that the meter still maintains its factory calibration. The **calibration check** is performed with a single concentration of calibrant gas. For example, a carbon monoxide detector may be factory calibrated to accurately respond to carbon monoxide concentrations between 1 ppm and 200 ppm. A calibration check can be performed by the operator using a single concentration of calibrant gas, such as 25 ppm carbon monoxide. If the instrument responds correctly to 25 ppm, the user can be reasonably assured it will accurately respond to other concentrations within the detection range of 1-200 ppm. **The calibration check is the only means of demonstrating that the instrument is working properly**; if the meter does not respond appropriately to the calibrant gas, it should be adjusted until an accurate reading is obtained. The calibration check is often called the **field calibration check** when it is performed on site under the temperature conditions of anticipated use.

Calibration should be checked prior to instrument use; in emergency response situations, this may not always be feasible. In this case, instruments should be checked at regular intervals, such as every week. Make sure the calibration check procedure for the instrument is relatively easy and quick. The manufacturer should offer a calibration kit as an accessory item. Examine the manual to determine how difficult it is to make minor instrument adjustments. Knobs or set screws should be clearly marked and easy to reach.

Other Considerations

Actual or anticipated **environmental and site conditions** can affect the operation of many instruments. These include temperature, humidity, elevation and barometric pressure, particulates, static electricity, electromagnetic fields, oxygen concentrations, ambient lighting, and the presence of interfering compounds. It is important that field calibration checks be performed under the actual conditions of use; this is especially important when operating instruments at extreme temperatures.

Information about the **nature of the hazard** is also important when selecting an appropriate instrument. For a material known to be present, this may include flammability, vapor pressure, vapor density, and ionization potential. Toxicity and exposure limit information will assist in assessing the potential health effects and the need for personal protective equipment.

Cost is another factor that must be considered. Do not fall into the trap of purchasing the most sensitive instrument with many options when all you might need is the standard, basic meter which costs considerably less. Remember that price does not always reflect utility. Multiple sensor instruments are more expensive to operate and maintain; sensors usually last about a year and then must be replaced. All meters should be considered expendable; buy with the idea that you may have to replace it every few years.

Finally, the **personnel** available to operate and maintain the air monitoring instrument must also be considered. Someone should be responsible for instrument use and maintenance. All personnel, however, should be able to use the meter, as well as to interpret readings, perform calibration checks, and troubleshoot basic problems in the field. It is a dangerous mistake to entrust these functions to only one or two individuals; anyone whose activities are affected as a result of monitoring information should have a basic level of instrument literacy. We cannot recommend the common (and potentially disastrous) practice of having a responder or worker report meter responses to someone else for interpretation. If personnel do not receive formal instrument training, then get an instrument that is simple to operate and interpret, and that has a manual that is easy to understand.

START-UP PROCEDURES

Start-up procedures should always be conducted in a non-contaminated area. Turn the instrument on and verify that the meter responded appropriately to turn-on. Depending on the meter, such a response may be needle deflection, illumination of digital readout display, audible and/or visible alarms, or operation of the battery-operated pump. This information can be found in the operating procedures section of the manual. If necessary, reset the alarm functions.

Next, check the battery. It is imperative that there is sufficient battery power to operate the sensor. If there is an inadequate or low battery condition, shut the meter off and recharge or replace the batteries, as appropriate. **Never use a meter with inadequate battery supply**; it will not give a valid reading and can endanger the user as well as other personnel. It is not always possible to tell how depleted a battery is simply by listening to the pump or watching how quickly the meters turns on. Remember that battery life can be significantly shortened by cold temperatures.

Some meters do not have a battery check option; these meters are usually designed to be continually plugged into a trickle battery charger. The meter is unplugged from the charger just prior to use. Often, however, the meter must be removed from the charger hours or days prior to use. In these cases, it is not possible to assess the percentage of maximum charge left in the battery pack. These instruments give no indication of battery life until an alarm, accompanied by a readout display or warning light, indicates a low battery condition. The alarm on these instruments should never be turned off or disabled, since it may be the only indication of insufficient battery power.

Next, make sure the alarm is working properly. Turn the zero adjust control until the meter display reaches the alarm set point; an audible and/ or visible alarm should be activated. Disregard this step if the meter has no alarm, or if the zero adjust is relatively inaccessible.

Return the meter reading to background readings that would be normally encountered in a non-contaminated atmosphere (i.e. 0% LEL, 0 ppm hydrogen sulfide) and reset the alarm. Allow sufficient time for the meter to warm up and stabilize; this is especially important for low concentration meters such as ionization detectors. Failure to allow adequate warm-up time will result in variable instrument response as the readings fluctuate until the sensor is fully operational. Do not attempt to calibrate until the instrument has completed the minimum warm-up interval.

A calibration check can be performed when meter readings are stable. We always advocate conducting calibration checks as recommended by the manufacturer. This is can be very important in situations which ultimately involve litigation; a successful calibration check eliminates any question regarding initial instrument performance. Remember that the calibration check is the **only** acceptable method to ensure the meter is functioning properly.

We are often asked if we advocate calibrating the meter with all accessories attached. In most cases, our answer is no. The reason for this is that often the accessories, especially long sample hoses, increase the volume of calibrant gas required. The calibration check is used to verify that the instrument is working properly; it should not be used to detect leaks in the sampling system. Such leaks often go unnoticed when the gas sample is introduced under pressure at optimal sample flow for the instrument.

After completing the calibration check, appropriate attachments should be added to the instrument. Instruments with battery-operated or bulb-type aspirator pumps should always be operated with a liquid trap (if available) and particulate filter. Attach an additional filter holder onto the sample hose in case a charcoal or other type of filter is required. Check all fittings and ensure they are snug. Make sure the liquid trap is properly oriented; if necessary, use a strap or tape to keep it in place.

When all accessories are properly positioned, place your finger over the sample probe inlet; if the probe has multiple inlets, kink the sample hose to prevent sample air from entering. There should be a definite interruption in pump function in meters with battery-powered pumps; in some meters an alarm may also sound. If the meter has a bulb-type hand aspirator, depress the bulb completely; if no leaks are present the bulb should remain deflated. This type of leak check is usually advocated by the manufacturer; however, some vacuum pumps may be susceptible to eventual sealant rupture. Always consult the instruction manual before adding the leak-check to the standard start-up protocol.

REFERENCES

1. 29 Code of Federal Regulations, Part 1910. Hazardous Waste Operations and Emergency Response.
2. Federal Register, Volume 51, Number 244, Friday, December 19, 1986, pages 45654-45675.
3. Federal Register, Volume 54, Number 42, Monday, March 6, 1989, pages 9294-9336.
4. U.S. Environmental Protection Agency. 1992. Standard Operating Safety Guidelines. Office of Emergency and Remedial Response, Hazardous Response Support Division, Environmental Response Team.
5. U.S. Department of Labor, Occupational Safety and Health Administration. 1986. Hazardous Waste Inspections Reference Manual.
6. U.S. Department of Health and Human Services, Public Health Service, Centers for Disease Control, National Institute for Occupational Safety and Health. 1982. Hazardous Waste Sites and Hazardous Substances Emergencies. NIOSH Worker Bulletin.
7. National Institute for Occupational Safety and Health, Occupational Safety and Health Administration, U.S. Coast Guard, U.S. Environmental Protection Agency. 1985. Occupational Safety and Health Guidance Manual for Hazardous Waste Site Activities.
8. National Fire Protection Association. 1986. Classification of Gases, Vapors and Dusts for Electrical Equipment in Hazardous (Classified) Locations. Batterymarch Park, MD.
9. National Advisory Board, Commission on Engineering and Technical Systems, National Research Council. 1982. Rationale for Classification of Combustible Gases, Vapors, and Dusts with Reference to the National Electric Code, Publication NMAB 253-6.
10. European Committee for Electrochemical Standardization (CENELEC). 1977. Electrical Apparatus for Potentially Explosive Atmospheres (BS5501 Parts 1-7, English Version). London: British Standards Institution.

CHAPTER REVIEW

1. List three reasons why air monitoring is conducted. _____
   _____
   _____ .

2. List three types of hazardous conditions that can be detected by using air monitoring instruments.
   _____
   _____

3. An instrument with a Class I, Division ____ approval should be used at a response involving a leaking gasoline tanker.

4. Acetylene is the only material in listed in Group ____.

5. List several Group B gases _____ .

6. Carbon disulfide is a Group _____ gas.

7. List five factors that may influence the selection of an instrument: _____

   _____

   _____

8. List five components of a typical air monitoring instrument. _____

   _____

   _____

9. List some factors which affect lag time and response time._____

   _____

10. The only reliable method to verify that an instrument is working properly is to perform
    a _____.

11. True or false? A Class I, Division 1 instrument is safe to use at an incident involving flammable dusts.

12. A drum storage facility containing intact drums of flammable solvents is a Division____ location. If
    drums of flammable solvent are leaking it is a Division ____ location.

13. True or false? An instrument which consistently gives a reading that is one-half the correct value can
    be considered inaccurate and precise.

14. The time required for an instrument to return to zero or normal background readings after responding
    to a sample atmosphere is the _____ time.

15. List several advantages and disadvantages of passive diffusion sensor instruments.

    _____

    _____

16. List several advantages and disadvantages of sample-draw instruments.

    _____

    _____

17. True or false? When sampling warm gases and vapors, the instrument operator must consider the
    potential for condensation of sampled gases and vapors within the cooler sample hose.

18. Where and when should initial instrument turn-on, warm-up, and field calibration be peformed?

    _____

19. An operator has difficulty setting an instrument to normal ambient background readings. Should the
    operator use the instrument?_____

20. Most air monitoring instruments are designed to respond accurately to only one gas or vapor. This
    vapor or gas is the _____.

2 Properties of Hazardous Materials

LEARNING OBJECTIVES

1. Describe the different types of hazards that may be encountered at a hazardous materials site.
2. Discuss various physical and chemical properties of chemicals and their importance.
3. Define the criteria used to determine the flammability of a material.
4. Define the routes of exposure and some effects of chemical overexposure.
5. Identify and define exposure limits and the organizations responsible for their existence.
6. Discuss the difference between hazard and risk.
7. Understand what is meant by relative concentration.

The natural inclination of many readers is to skip this chapter, and rush ahead to later chapters that seem more interesting. We urge you to resist! A sound understanding of the chemical/physical properties of chemicals is necessary when determining when and how to use air monitoring instruments. It is also critical when attempting to control and contain hazardous materials.

It is easy to pigeonhole hazardous materials into a single specific category, such as flammable/ignitable, toxic, corrosive, or reactive. Single-category classifications, such as the placard and label systems developed by the DOT and UN, are valuable tools. Such systems, however, indicate only the **primary** hazard of the material. While it is extremely important to understand the nature of the hazard described, it is very unusual for a hazardous material to have only one hazard associated with it. **Most hazardous materials have multiple hazards** associated with their manufacture, transport, storage, use, or disposal.

The hazards associated with a material will often change relative to the amount of material present. For example, the chlorinated solvent trichloroethylene or TCE is usually considered a toxic material and is designated a carcinogen by several nationally recognized organizations. Ground water containing very low concentrations of TCE is considered contaminated and no longer potable; TCE exposure in the workplace is regulated. Since workplace concentrations are usually kept relatively low, it is not generally recognized that TCE, with a flash point of 90 °F, is also a Class IC Flammable Liquid.

It is important then, to understand how and why chemicals act the way they do under certain conditions. It is not necessary to draw chemical structures, to have a college degree, or understand molecular orbital theory. We will not discuss concepts dealing with ideal gases and liquids, since they are not encountered at hazardous materials sites (our apologies to Messrs. Boyle, Charles, Raoult, and especially to Mr. Henry for not including his law!).

All that is required is a willingness to practice **thinking** about how chemical/physical properties affect the behavior of chemicals; that will be easily accomplished if you thoroughly read this chapter and then complete the exercise at the end.

WHAT IS A HAZARD?

Look up the term **hazard** in a standard dictionary and it will be defined as a danger, or something that causes danger or difficulty. Hazard as a term is incomplete, for it does not indicate exactly what the hazard is. The concept of hazard differs between individuals and their experiences. In addition, what is hazardous is often defined by the situation in which it is encountered. For instance, a honey bee may be considered a hazard to a person who is extremely allergic to bee stings, while to another it may be an insignificant nuisance and ignored.

Hazard may also be defined as a lack of predictability, or an uncertainty. This second definition also aptly fits conditions often found at hazardous materials sites. At hazardous materials responses, every effort must be made to expect the unexpected. Remember that while your primary focus may be on the chemical hazard, there are other types of hazards present that must also be recognized. Hazard recognition is the first step in risk assessment and mitigation process.

HAZARD RECOGNITION

Hazardous materials responders and site workers encounter many real or potential hazards. They may also be called upon to perform tasks or put themselves in situations that are potentially dangerous. The goal is to minimize the risk by understanding the nature of **all** hazards involved. The use of protective equipment increases risk by decreasing hearing, vision, and agility, and encouraging temperature stress; these effects limit the ability of the responder to recognize newly encountered hazards. Remember, **health and safety are not necessarily mutually inclusive**; decreasing health risks often increases safety risks.

Electrical hazards include downed or intact electrical wires, buried cables, electrified third rails of subways and trains, generators, and electrical equipment. Lightning is also an electrical hazard, especially when working around or wearing metal equipment. The presence of water at the scene increases the hazard.

Physical hazards are often referred to as general safety hazards. Natural physical hazards such as uneven terrain, unstable slopes, steep grades, holes, ditches, falling debris, sharp or jagged edges, and wet, muddy or slippery surfaces fall into this category. Other physical hazards may be associated with structures on the site, such as open manholes, ladders, catwalks, and damaged supports as well as ropes, wires, and hoses which can trip or entangle site responders. Cylinders under pressure, regardless of their content also present a potential physical hazard.

Mechanical hazards are caused by moving parts which may cause crushing injuries or entrap the responder. These include heavy equipment, tools, hoists, augers, drilling equipment, and vehicles. Mechanical and physical hazards are responsible for the majority of injuries which occur at hazardous waste sites.

Biological hazards or agents are living organisms or their products that can cause illness or death to the individual exposed. Biological hazards are found in hospital or laboratory wastes that contain infectious materials. Carriers such as ticks, mosquitoes, and fleas can transmit diseases to human hosts. Plants and insects may produce allergic skin reactions in sensitive individuals. Wild animals such as bear, wolves, wild dogs, rodents, and snakes, may carry disease or present other hazards.

Radiation hazards may also be present. Radiation is commonly found in industry, medical and research laboratories, and hospitals. Low levels of radiation are often found at landfills. Alpha particles are large, slow moving, and have limited capacity for penetration; alpha radiation is usually stopped by clothing and the outer skin layers. Beta radiation consists of small, fast-moving particles which can cause skin burns and damage to tissue below the skin. Alpha and beta particles can be inhaled and are potential respiratory hazards. Gamma rays easily pass through clothing and tissue and can cause serious damage.

Noise is a hazard that is often overlooked. Noise interferes with normal communication between workers and can be distracting. Noise may also cause physical damage that can result in temporary or permanent hearing loss.

Temperature stress includes adverse effects on the body as a result of exposure to hot or cold temperature conditions. Heat stress is caused by overheating the body and loss of fluids through sweating. Heat stress is a major hazard for workers wearing personal protective equipment which limits dissipation of body heat and prevents evaporation of moisture. Cold stress or hypothermia is caused by the inability to maintain normal body temperature. Frostbite occurs when there is local freezing of tissue. Cold stress is increased under damp, wet, or windy conditions.

Confined space hazards can exist in areas which have restricted means of entry and egress. Confined spaces are not designed for continuous occupancy and may be oxygen deficient; these areas also have the potential for the accumulation of toxic or flammable gases.

Fatigue is usually not considered a hazard, but rather something that "goes with the territory". We do not agree. Work at hazardous materials sites typically involves tasks that are performed in other industries, but with the added burden of personal protective equipment that produces inordinate physical stress. Emergency responses to hazardous materials incidents often require prolonged activity in high levels of protective equipment. Fatigue adversely affects performance, decreases alertness, and increases carelessness. Poor physical condition, obesity, substance abuse, and the presence of pre-existing medical conditions increase the hazard from fatigue.

Chemical hazards often have more than one hazard associated with them. For example, the substance morpholine, an organic amine, is toxic, flammable, and corrosive. A good, informal definition of a hazardous substance that certainly applies to chemical hazards, is:

Any substance that jumps out of its container at you when something goes wrong and hurts or harms the things it touches.

(Ludwig Benner, in Recognizing and Identifying Hazardous Materials, National Fire Academy, Emmitsburg, MD, 1985.)

PROPERTIES OF HAZARDOUS MATERIALS

The physical properties of a chemical substance will affect how the material will act upon release from its container. That is, will it sit there as a solid, quickly spread as a liquid, evaporate into a vapor, or immediately escape as a gas? Will the vapor or gas rise and dissipate or settle into low lying areas? Will the liquid or solid material float, sink, or dissolve in water? Will the material produce sufficient vapors to present a flammable hazard?

Physical State

A chemical may exist as a solid, liquid, gas, vapor, or aerosol. Temperature, pressure, and concentration may affect the physical state of a material. Most chemicals can exist in more than one state. For instance, the liquid chemical, water, can also exist as a solid (ice), mist (rain), and vapor (steam).

A **solid** is a material that has definite size and shape. A **liquid** has definite volume, but no shape. Liquids take on the shape of their container. **Gases** are materials that are not liquids or soilds. Gases have no shape or volume, and are completely airborne at room temperature. Gases can be changed to a liquid or solid, but only when subjected to increased pressure and/or decreased temperature; **liquified gases** such as propane and butane are maintained under these conditions. **Vapors** are the evaporation products of materials that are liquid under normal temperature; liquids such as water, methylene chloride, ethyl alcohol give off vapors. Solid materials can also produce vapors; for example, naphthalene and para-dichlorobenzene are used in moth balls because they slowly release vapors which repell moths.

An **aerosol** is a mixture of solid or liquid particles dispersed in a gas. Aerosols found in households are usually generated by spray cans or bottles, and include spray paints, bug killers, deodorants, cleaners and disinfectants.

Dusts, smoke, particulates, mists, fumes, and fog are all different types of aerosols. **Dusts** are derived from solids by mechanical means such as crushing or grinding stone. Dusts are also generated from organic materials such as grain and cotton dust, and flour. Dusts are usually spherical in shape and differ from fibers which are not spherical. **Mists** are small liquid droplets suspended in air or another gas. Mists may be formed mechanically by spraying or splashing, or by condensation of a vapor into the liquid state. A highly visible mist is often called **fog**.

Fumes are very small particles formed from solid materials by evaporation and subsequent condensation from the gaseous state, generally after heating or a chemical reaction. When heated, solid materials such as metals and waxes form fumes. The most commonly encountered fumes are formed by the evaporation of metals during welding; solid metal particles are formed as the gaseous phase cools. **Smoke** is a complex mixture of particles, vapors, and gases produced as a result of incomplete combustion of organic materials. **Smog** is a combination of smoke and fog.

Physical Properties

Physical properties of chemicals are determined under standardized, laboratory conditions and are published in reference texts or manuals. Most physical properties are defined for normal temperatures and pressures.

Vapor pressure (VP) defines the ability of a liquid or solid to evaporate into the air. Vapor pressure is dependent upon temperature; vapor pressure increases dramatically as the temperature of the material increases. Vapor pressures for liquids and solids are usually measured at 68°F (20°C) and defined in millimeters of mercury (mm Hg). Chemical compounds that exist as a gas at room temperature have vapor pressures defined in pounds per square inch (psi) or atmospheres (atm). Normal atmospheric pressure is 760 mm Hg, 14.7 psi, or 1 atmosphere. Materials that are gases or vapors at room temperature will have vapor pressures greater than these values.

Another way to look at vapor pressure is to consider it a measure of **how much the material wants to become airborne**. Liquids or solids with high vapor pressures are highly volatile and quickly evaporate into the air. The higher the vapor pressure, the higher the vapor concentration will be immediately above the surface of the material, or in the **headspace** when the material is in a closed container. Visualize a closed container half full with a liquid; a certain amount of liquid will evaporate into the headspace of the container. Eventually the vapor in the headspace will exert so much pressure on the walls of the container and the liquid at the bottom that some vapor will be forced back into the liquid. When this happens the container and its contents are considered to be in equilibrium. The maximum pressure exerted by the vapor against the walls of the container and the liquid is its vapor pressure.

Most mobile liquids (i.e. liquids that readily pour at room temperature) have vapor pressures of at least 1 mm Hg. Compounds with vapor pressures greater than 100 mm Hg are highly volatile; materials with vapor pressures between 1 and 100 mm Hg will release significant concentrations into the atmosphere. Do not assume that viscous liquids and solids have negligible vapor pressures! Organic materials that are solid at room temperature may have vapor pressures of 0.1 or more. For example, p-dichlorobenzene and naphthalene are both solids at room temperature and are found in the home as mothballs or moth flakes; their vapor pressures are 0.44 and 0.08 mm Hg, respectively.

We can roughly calculate how much vapor will be given off by a material based on its vapor pressure. There is a simple formula used to estimate the concentration of vapors in ppm that will collect above the liquid or solid when it is confined in a closed container at room temperature. We call this calculation the "1300 rule".

Chemical	Vapor Pressure (mm Hg)
Acetone	180
Benzene	75
Chloroform	160
Chlorine	6.3 atm
Ethylene oxide	1.4 atm
Ethyl ether	440
Fuel oil (#4)	2
Isopropyl alcohol	33
LPG	8.6 atm
Methyl alcohol	92
PCBs	0.001
Toluene	20
Trichloroethylene	60
Vinyl chloride	3.4 atm
Water	25
Xylenes	9

Table 2-1: Vapor pressures of common materials at 68°F in mm Hg or atmospheres (atm).

Vapor Pressure x 1300 = Headspace concentration in ppm

For instance, the vapor pressure of 1,1,1-trichloroethane (methyl chloroform) is 100 mm Hg. The concentration in the headspace of a container above the liquid can be calculated:

100 mm Hg x 1300 = 130,000 ppm

This is an **estimation** (plus or minus 1-2% of the calculated value) of the actual concentration inside a closed container. Temperature, for example, will affect vapor pressure and therefore the actual concentration present. Increasing temperature will dramatically increase vapor pressure. We can appreciate this by thinking about how water evaporates. On a hot day, water quickly evaporates; on a cool day, it takes much longer.

Even though the actual vapor pressure of the material is not identical to the value measured at 68°F, the 1300 rule can still give you an idea of the potential concentration that may be present in a container or confined space, and will often guide you when deciding what type of instrument to use.

The 1300 rule cannot be used for materials that have vapor pressures greater than 760 mm Hg which exist as a gas or vapor at room temperature. Gases with vapor pressures greater than atmospheric are already completely airborne. The rule must be used only with materials that are normally liquids or solids at room temperature.

What maximum concentration of vapors would you anticipate in a closed container containing moth flakes composed of 100% naphthalene (VP 0.08 mm Hg)?

0.08 x 1300 = 104 ppm (or approximately 100 ppm)

What is the approximate maximum headspace concentration for acetone at room temperature (VP 180 mmHg)?

180 x 1300 = 234,000 ppm

Vapor density (VD) is the weight of a vapor or gas relative to the weight of an equal volume of air at comparable temperature and pressure. The vapor density of air is equal to 1. Materials with vapor densities less than 1 are lighter than air and if released, will tend to rise and dissipate. Chemicals with vapor densities greater than 1 are heavier than air and will sink or settle.

Some chemical reference manuals, usually emergency response guides, give vapor density values. Other guides may offer the specific gravity of the vapor or gas relative to dry air; this value, usually designated **specific gravity (gas)**, and vapor density are one and the same and are used interchangeably. Other reference manuals oriented toward professional chemists do not offer vapor density measurements but rather give the actual weight of the gas in grams per liter (gm/l). This value is of little use unless you know how much air weighs. At 0°C and normal atmospheric pressure, air weighs 1.29 gm/l. The vapor density of air is therefore calculated as 1.0 (1.29/1.29=1.0). Vapor densities of other materials are determined by dividing the weight of the gas or vapor in gm/l by 1.29. For instance, one liter of propane at 0°C weighs 2.02 gm. The vapor density of propane is therefore 2.02/1.29 = 1.56.

What is the vapor density of hydrogen sulfide (H_2S, 1.54 gm/l)? Is H_2S heavier or lighter than air?

Vapor density = 1.54 / 1.29 = 1.19 or about 1.2. H_2S has a vapor density that is greater than 1.0, therefore it is heavier than air.

Vapors from materials that are liquid or solid at ambient temperature and pressure will have vapor densities greater than 1.0. The vapor density of gases is dependent on their chemical composition. Common gases that are lighter than air are methane, ammonia, helium, neon, and hydrogen. Gases that have vapor densities approximately the same as air include nitrogen, carbon monoxide, and ethane. Representative gases that are heavier than air include oxygen, propane, butane, sulfur dioxide, nitrogen dioxide, chlorine, bromine, and butadiene.

If the reference sources available do not offer information regarding vapor density or the weight of the vapor in gm/l, it is still possible to estimate the vapor density. This can be done by finding the molecular weight (MW) of the chemical or product of interest, and dividing it with the approximate molecular weight of air, which is 29. For instance, the approximate vapor density of ammonia gas (NH_3), which has a molecular weight of 17.03, is about 0.6 (17/29 = 0.59).

What is the approximate vapor density of hydrogen sulfide (H_2S, MW 34.08)?

Vapor density = 34 / 29 = 1.17 or about 1.2. Note how close this value is to the vapor density calculated earlier using the weight of the vapor in gm / l.

Vapor density is somewhat dependent on the temperature of the vapor or gas. Vapor density and vapor pressure decrease as temperatures decrease. Remember that lighter than air compressed gases are much cooler than ambient temperatures when initially released from pressure; these gases often initially behave like heavier than air gases until they warm up to ambient temperatures. Likewise, heavier than air gases or vapors that have been heated may rise on hot air thermals.

Vapor density is important in situations where high concentrations are known or anticipated to occur, as when a propane tank ruptures, a large volume of a volatile chemical is released, or in confined spaces where vapors can stratify. Vapor density is less important when dealing with ppm concentrations, when the predominant gas present is air, with a vapor density of 1.0.

Boiling point (BP) is the temperature at which the vapor pressure equals atmospheric pressure. At their boiling point, materials will "boil" away or rapidly evaporate. Most materials that are liquids at room temperature have boiling points greater than room temperature, otherwise, they would quickly evaporate into the air. Liquids with relatively low boiling points, such as ethyl ether (BP 94°F) are often kept refrigerated to reduce evaporation. Chemicals that exist as gases at normal temperature (68°F) and pressure (760 mm Hg) have, by definition, low boiling points. Liquified gases with boiling points less than room temperature such as propane (BP - 44°F), butane (BP 32°F), and ethylene oxide (BP 51°F) are maintained under pressure as compressed liquids.

Specific gravity is the ratio of the density or weight of a substance to that of a reference material. When dealing with liquids and solids, the density of the material in grams per cubic centimeter (gm/cc) is compared to water. Since the specific gravity of water is 1.0 gm/cc, the specific gravity of the material is actually equal to its density. Materials with densities less than 1.0 are lighter than water and will float on water, materials which have densities greater than 1.0 will tend to sink in water.

The specific gravity or density varies with temperature. The temperature of the water reference and the material are usually indicated along with the density or specific gravity value. The density of acetone is listed in a chemical reference as 0.789 at 20/4; 20/4 pertains to the temperature in °C. The top number is the temperature of acetone, the lower number refers to the temperature of the reference material, water. Specific gravity of liquids and solids must not be confused with vapor density or specific gravity of gases! When specific gravity values are determined for gases they will normally be presented in such a fashion to prevent inadvertent mistakes.

Water solubility is the quantity of a material that will mix or dissolve completely in water. Materials that are **miscible** in water will freely mix regardless of the quantities present. Some references consider a material to be **insoluble** or have **negligible solubility** in water if its solubility is less than 1 part in 1000 parts water, or 0.1%. In some references, the term **practically insoluble** may be used to indicate the same thing. While these solubilities may be negligible by some standards, they may be indeed significant in terms of potential environmental or ground-water contamination! If a reference indicates an organic material is insoluble or practically insoluble, determine what criteria were used. Some references simply copy qualitative solubility data from other sources and do not provide any criteria.

For non-miscible materials, water solubility is typically measured by the weight or volume of material that will dissolve in a volume of water. Solubility may be stated as 1 part material in so many parts water, or grams of material per gram of water, or grams of material per milliliter of water, or gram of material per cubic centimeter (cc) of water. Do not let this confuse you! The good news is that all of these are the same thing. One cc or ml of water weighs 1 gram. For instance, the solubility of table salt or sodium chloride (NaCl) is 1.0 gm salt in 2.8 milliliters (ml) water at 25°C. This is comparable to 1 gm salt in 2.8 gm water, or 1 part salt in 2.8 parts water.

Solubility is often given as a percentage. A material with a solubility of 1 part in 100 has a solubility of 1%. The percentage is determined by dividing the parts of material by the number of parts of water. For example, the aromatic solvent benzene is soluble in 1430 parts water and has a solubility of approximately 0.07% (1/1430 = 0.000699 x 100 = 0.0699%).

What is the percent solubility of salt, if 1 part salt is soluble in 2.8 parts water?

Percent solubility = 1/2.8 = 0.357 x 100 = 35.7%

A material that is miscible will mix with water, while other materials may sink, float, and/or mix with water to the limits of their solubility. Water solubility is obviously important when predicting the behavior of materials in water. Water solubility in the field often differs from that observed in the laboratory, however. In many cases, "field solubility" is higher than anticipated; this is often caused by adsorption or dispersal of the material in the presence of other dissolved and solid organic compounds in water.

Flammability

Flash point (FP) is a parameter that is defined for liquids and solids. Flash point is the temperature that a liquid or solid must be raised to in order to emit sufficient vapors to form an ignitable mixture with air. At the flash point, the concentration of vapors present will ignite or flash over in the presence of a source of ignition such as a lighted match or cigarette. The flammability hazard of a substance is directly related to its flash point. The lower the flash point, the greater the hazard.

Until recently, a liquid or solid with a flash point less than 100°F was considered flammable by USDOT, while combustible materials had a flash point between 100°F and 200°F (49 CFR 173.117-119). New DOT regulations[1] now define a flammable material as a substance with a flash point of less than 140°F while a combustible material has a flash point at or above 140°F and less than 200°F. However, for domestic shipments (i.e. not transported by vessel or aircraft), **a material with a flash point at or**

greater than 100°F which does not meet the definition of another hazard class **can be reclassified as combustible**. The domestic exception allows the continued use of the previous DOT definitions of flammable and combustible materials. The EPA and OSHA definition of an ignitable material is similar to that of USDOT, an **ignitable material** has a flash point of less than 140°F; solids that can spontaneously combust are also considered ignitable by EPA.

Remember that the temperature of the material must be raised to the flash point temperature before ignition or flash over can occur. Liquids and solids do not burn, rather the vapors released actually burn. In order to release sufficient vapors, the substance must be heated and release vapors. Think about your last birthday cake—remember the candles? (Never mind about how many there were.) Initially the flame was very small, then the wax became soft and began to melt. As the wax melted, wax (paraffin) vapors were released and burned; the flame then became brighter and higher.

Substances that are gases at ambient temperature and pressure do not have flash points. Flash point is the temperature that a **liquid** or **solid** must be heated to in order to release sufficient vapors to burn. No heat is necessary to release gases. Gases are already airborne!

Unless the substance is already subjected to an external heat source (as in our candle and flame analogy), ambient temperature will be an important factor to consider when appraising the potential fire hazard. For instance, if the weather conditions are cloudy, and the air temperature is 70°F, there is little likelihood that a material with a flash point of 90°F will release sufficient vapors to flash over unless the vapors are confined. On the other hand, if it is a sunny day, with an ambient temperature of 80°F, and the material is spilled on a hot, exposed asphalt surface, there is a good chance for a flash over if a source of ignition is available. Substances with flash points less than ambient temperature are always fire hazards. For example, gasoline poses a significant fire hazard; it has a flash point of –45°F.

The **flammable range** is the concentration range of a vapor or gas within which combustion will occur. Within the flammable range, the vapor or gas mixes with oxygen in the air and burns. Outside the flammable range, the mixture is either too lean or too rich to burn. The flammable range is defined in terms of the percent by volume of gas or vapor in air. Every chemical capable of burning has a flammable range.

The **lower explosive limit** (LEL) or lower flammable limit (LFL) is the minimum concentration of gas or vapor in air that will ignite in the presence of a source of ignition. Concentrations less than the LEL are too lean and will not ignite. The **upper explosive limit** (UEL) is the maximum concentration of gas or vapor in air that will ignite; higher concentrations are too rich to burn. The concentration range between the LEL and UEL is the flammable or explosive range.

Traditionally, the terms used for upper and lower flammable limits (LFL-UFL) were reserved for liquids; explosive limits (LEL-UEL) were used only for materials that were gases at normal temperature and pressure. This distinction was considered unnecessary; today, the terms are used interchangeably.

Chemicals vary in their flammable range. Some substances have narrow flammable range while others have a very wide range. In general, the wider the flammable range, the greater the fire hazard. It is a dangerous mistake, however, to assume there are no other potential hazards when operating outside the flammable range. Concentrations outside the flammable range may still pose a toxic or asphyxiation hazard to unprotected personnel. In addition, vapors that are initially too rich or too lean can be diluted or concentrated and eventually reach the flammable range.

The flash points (FP) and flammable ranges (LEL - UEL) for some commonly encountered chemicals are presented in Table 2-2. LEL and UEL values are presented as percent gas or vapor by volume in air.

Chemical	FP (°F)	LEL (%)	UEL (%)
Acetone	– 4	2.5	13.0
Acetylene	Gas	2.5	100.0
Carbon Monoxide	Gas	12.5	74.0
Ethyl Alcohol	55	3.3	19.0
Gasoline	– 45	1.4	7.6
Kerosene	100-162	0.7	5.0
Methane	Gas	5.0	15.0
Methyl Alcohol	52	6.0	36.0
Naphthalene	174	0.9	5.9
Propane	Gas	2.1	9.5
Toluene	40	1.2	7.1
Trichloroethylene	90	8.0	10.5

Table 2-2: Flash points and flammable ranges for commonly encountered materials.

Human Exposure and Toxic Effects

In order for a chemical to produce a toxic or adverse health effect, it must come in contact with, or gain entry into, the body. Exposure to chemical hazards may occur through different **routes of exposure**: inhalation, skin and eye contact or absorption, ingestion, or injection. Inhalation of airborne contaminants is the most common route of exposure.

Water solubility of a gas or vapor is important in determining how much will actually reach the lungs. Water soluble gases such as ammonia and hydrogen chloride dissolve readily in the nasal passages and upper respiratory tract; these materials are upper respiratory irritants. Less soluble gases and vapors such as nitrogen dioxide and phosgene can penetrate deeper into the lower respiratory tract and lung tissue where they can produce significant damage.

Although the skin is a marvelous organ and an excellent barrier against its external enviroment, many chemicals are readily absorbed through the skin. Absorption is more rapid when skin has been compromised or damaged by abrasion, laceration, inflammation, dehydration, or sunburn. Organic solvents such as gasoline, xylene, and acetone remove oils from the skin and facilitate skin absorption by increasing the permeability of the skin. Chemicals that can be absorbed through the skin can also be absorbed through the eye.

Ingestion or swallowing a chemical is a common route of exposure to chemicals encountered every day. Chemical additives, flavor enhancers, antioxidants, food dyes, and artificial sweeteners are a regular part of our diet. Ingestion of chemicals at a hazardous material site, however, should be an unlikely route of exposure. Chemicals may be inadvertently ingested however, if there is hand-to-mouth activity on site prior to decontamination or washing. These activities include smoking, chewing tobacco or gum, eating, drinking, or biting fingernails.

Injection is the least common route of exposure. Concerns about contracting hepatitis and immune deficiency diseases have increased awareness of biological injection hazards. Injection may also occur if an open wound is exposed to a chemical.

Factors that influence toxicity include the extent and duration of exposure, the route of exposure, and the inherent susceptibility of the individual. Age, sex, diet, inherited traits, physical fitness and health, use of alcohol, drugs, prescribed medications, and pre-existing medical conditions all affect individual susceptibility.

Acute toxicity refers to the ability of a chemical to produce adverse health effects as a result of a single exposure of short duration. **Chronic toxicity** results when a chemical produces damage after repeated exposure. Some chronic effects, such as cancer, pulmonary fibrosis, emphysema, and organ damage have long latency intervals, develop slowly, or require years of repeated exposure before they are manifested.

Systemic toxins produce damage to specific organs or organ systems. It is not unusual for a chemical to produce multiple organ effects; chlorinated hydrocarbons, for example, may affect the liver, kidney, nervous system, and heart. **Irritants** produce pain, swelling, redness, and inflammation of exposed tissues. Extreme irritants are considered corrosive. **Sensitizers** are substances which induce an allergic reactions in sensitive individuals. Reactions may include skin rashes with itching and blister formation, watery, itching eyes, nasal discharge and sneezing, or asthma-like symptoms which make breathing difficult.

Carcinogens are substances that cause cancer. **Mutagens** cause mutations; a mutation is a change in our genetic material or genes. **Teratogens** produce birth defects by damaging the fetus while it develops in the mother's womb. Some chemicals produce lethal damage and cause the fetus to be aborted. In excessively large quantities, alcohol, aspirin, and vitamin A can act as human teratogens.

Exposure Limits

Standards and guidelines have been developed by several organizations to prevent chemical overexposure of workers, responders, and community receptors. Each limit defines or recommends the maximum concentration in air that is acceptable for a **specified period of time**. The **duration** of exposure is an important factor in defining **overexposure**. It is well recognized that an individual can be safely exposed to a relatively high concentration for a very short duration, while exposure over a long period of time can result in significant toxicity. Exposure limits are established only after careful examination of the effects of low level human exposure, as well as acute and chronic animal toxicity test data.

In order to use these values properly, it must be recognized that human response to exposures vary. Even though safety factors are incorporated into workplace exposure limit values, they should not be expected to protect everyone. Sensitive individuals exist in all populations who may find odors objectionable or experience slight effects such as irritation or headache. Workplace exposure limits are based on the anticipated response of an average young male worker, who is in good health and physical condition. Many factors are not considered and may affect an individual response to exposure including exposure to other chemicals, age, sex, physical fitness, weight, allergies and other pre-existing medical conditions, pregnancy, and the use of tobacco, drugs, alcohol, and prescribed or over-the-counter medications.

Permissible exposure limits or **PELs** were established by the Occupational Safety and Health Administration (OSHA) in 1971. PELs are defined in 29 CFR 1910.1000 for nearly 600 materials; there are individual standards for another 24 substances. The PEL is the maximum average concentration that a worker may be exposed to for 8 hours a day, 40 hours per week. Notice that although only one concentration in air is ascribed to each PEL, it is an **average** concentration for an 8-hour work day. A worker may be exposed to higher or lower amounts throughout the work day as long as the 8-hour average does not exceed the PEL. The **time weighted average** or TWA for 8-hours or another time interval is basis for most occupational exposure limits.

A **ceiling** is used when the concentration in air cannot exceed a specific concentration at any time during exposure. Ceiling limits are often used for respiratory irritants. The ceiling limit is the **maximum** concentration that an individual should be exposed to during the work day.

A **short term exposure limit** or **STEL** is the peak concentration of a substance that a worker may be exposed to for a short period of time. Exposure to concentrations above the STEL may result in irritation, sleepiness, difficulty breathing, or other symptoms which may affect worker efficiency or safety. STEL exposures should not exceed 15 minutes each and should be followed by a 60 minute interval without exposure. There should not be more than 4 STEL exposures per day.

The American Conference of Governmental Industrial Hygienists (ACGIH), founded in 1938, recommends **threshold limit values** or **TLVs**. While ACGIH TLVs are recommended guidelines, compliance with OSHA exposure limits are required by law. TLVs are reviewed, updated, and published on a yearly basis by ACGIH. New chemicals are added and pre-existing exposure limits are adjusted in response to new information. OSHA exposure limits can be changed only by amending the original law passed by Congress. Not surprisingly, ACGIH TLVs are usually more current and are often lower than the OSHA limits. When evaluating exposure limits, examine both OSHA and ACGIH, then select the lower value.

ACGIH TLVs are recommended exposure limits that are similar to OSHA limits. The **TLV-TWA** is analogous to the OSHA PEL and refers to the maximum average concentration that is acceptable for an 8 hour day, 40 hour work week. ACGIH also recommends STELs and Ceiling limits. ACGIH is currently phasing out many STEL values and recommends the use of an **excursion limit**, which is a multiple of the TWA. The excursion limit is the peak concentration that workers should be exposed for no more than a total of 30 minutes within any work day. Although three times the TWA is recommended, the maximum excursion limit concentration should not exceed five times the 8-hour TWA.

An **action level** is a specific concentration or level in air at which a particular action is required. OSHA action levels are typically one-half the 8-hour TWA; at this level, OSHA regulations (29 CFR 1910.1001-1047) to protect employees take effect and typically require employee training, medical monitoring, and medical record keeping. These and other protective measures are already required for hazardous materials workers and responders under the OSHA Hazardous Waste Operations and Emergency Response (HAZWOPER) rule (29 CFR 1010.120).

Workplace environmental exposure limits or **WEELs** have been developed by the American Industrial Hygiene Association (AIHA) for substances for which no OSHA or ACGIH exposure limit exists. WEELs have been established for 61 substances; WEELs for nearly 40 additional workplace substances or processes are currently being prepared.

The National Institute of Occupational Safety and Health (NIOSH) is an agency within the Department of Health and Human Services with a responsibility to identify workplace hazards, conduct research and field investigations within the workplace, develop and publish air sampling and analysis methods for chemicals, and recommend workplace exposure limit standards. Each **NIOSH recommended exposure limit** or **REL** is initially published as part of a criteria document; NIOSH RELs may be found in the NIOSH Pocket Guide to Chemical Hazards. In most cases, the REL is based on a 10-hour TWA; NIOSH RELs are recommendations to OSHA. It is up to OSHA to act on those recommendations and propose realistic exposure limits.

NIOSH has also established **IDLH** or **Immediately Dangerous to Life and Health** values for some hazardous materials. The IDLH is the concentration in air of any toxic, corrosive, or asphyxiant substance that poses an immediate threat to life or would cause irreversible or delayed adverse health effects. The IDLH also represents the maximum concentration in air from which one could escape within 30 minutes without suffering any escape-impairing symptoms, such as dizziness, or eye or respiratory tract irritation.

At and above IDLH concentrations air-purifying respirators are not permitted; positive pressure supplied air must be used. IDLH atmospheres must not be entered unless supplied air is already in use. If there is a potential for IDLH conditions, and air-purifying respirators are being used, air monitoring must be conducted continuously.

AIHA has established **Emergency Response Planning Guidelines** or **ERPGs** for about 20 hazardous materials. The ERPGs provide estimated concentration ranges and anticipated adverse health effects as a result of a one hour exposure to three different concentration ranges. The ERPG-1 value represents the least hazard and results in only mild effects; the ERPG-3 value represents the highest concentration and the greatest hazard.

RISK VERSUS HAZARD

A substance is hazardous because it has one or more inherent characteristics that are considered dangerous such as flammable, toxic, reactive, radioactive, corrosive. The **degree of hazard** is a function of the specific hazardous properties of the material and more importantly, its concentration. For instance, glacial acetic acid is a corrosive, combustible liquid and should be considered hazardous, yet few would consider the vinegar in our salad dressing a significant hazard. The difference is that the concentration of acetic acid in vinegar is very low.

The **degree of risk** is the potential harm that may occur as a result of exposure to the hazard. This is a function of the hazards present as well as the likelihood that exposure can or will occur. There is no risk if exposure to the material is not possible. For example, drums of benzene are buried at a site 10 feet below ground. Benzene is a known human carcinogen and a flammable liquid. What is the risk of benzene exposure to site workers? There is no risk until the drums are dug up and opened. Once the drums are encountered, the hazard can be defined in terms of concentration and the risk of flammability or toxic overexposure determined.

What the Numbers Mean

Throughout this chapter we have been defining concentrations using different terms. Although most readers understand these terms, a quick review is appropriate. The most commonly used term used to describe concentration is **parts per million** or **ppm**. A ppm is an incredibly small amount; as a volume, it is equivalent to one drop of vermouth in approximately 22 gallons or 350 glasses of gin (now that's a dry martini!). A part per billion (ppb) is equivalent to a drop of gin in 22,000 gallons of gin.

Many types of air monitoring equipment are sensitive to concentrations less than one ppm. Percent by volume concentrations can be compared to ppm concentrations because 100% by volume is equal to a million parts per million or 1,000,000 ppm.

It follows then, that:

50% by volume	=	500,000 ppm
25%	=	250,000 ppm
10%	=	100,000 ppm
5%	=	50,000 ppm
1%	=	10,000 ppm
0.5%	=	5,000 ppm
0.1%	=	1,000 ppm
0.01%	=	100 ppm

Using this information, the concentration of material present in air can be easily determined. For example:

The LEL of gasoline is 1.4% by volume in air; how many ppm is that and how does it compare to a TWA of 300 ppm and a STEL of 500 ppm?

1.4% by volume in air	*=*	*? ppm*
1% by volume in air	*=*	*10,000 ppm*
1.4 x 10,000	*=*	*14,000 ppm*

At the LEL for gasoline, the concentration present is about fifty times the TWA (14,000/300) and thirty times the STEL (14,000/500).

The same type of calculation can be used when dealing with substances in another medium, such as water. For example:

The solubility of material CXP is 1 part in 500. With everything else equal, what is the maximum concentration that you would expect to find in a water sample containing this material at maximum solubility?

1/100	*=*	*1%*
1/1000	*=*	*0.1%*
1/500	*=*	*2/1000 or 0.2%*
1%	*=*	*10,000 ppm*
0.2%	*=*	*0.2 x10,000 2,000 ppm*

The solubility of CXP is 1/500 or 0.2%; at maximum solubility, 2000 ppm of CXP will be dissolved in water.

REFERENCES

1. Federal Register, Volume 55, Number 246, Friday, December 21, 1990, pages 52606-52671.

CHAPTER REVIEW

1. Normal atmospheric pressure is _____ mm Hg.

2. Gases are materials that have vapor pressures greater than _____ mm Hg or _____ psi at room temperature.

3. 1300 x _____ = the approximate saturated headspace concentration in ppm.

4. A storage tank contains the solvent ethyl benzene (VP 10 mm Hg); the tank is approximately half full. The ambient temperature is between 70 and 80 °F. Estimate the approximate concentration of vapor in the tank _____.

5. Vapor density is the weight of a gas relative to the weight of an equal volume of _____.

6. At 0°C and 760 mm Hg normal pressure, air weighs _____ gm/l.

7. The vapor density of air is _____.

8. When the actual weight of the gas in gm/l is known, the vapor density can be approximated by dividing the weight of the gas by _____.

9. One liter of ethylene gas at 0°C and 760 mm Hg weighs 1.26 grams. The approximate vapor density of ethylene is _____.

10. The molecular weight (MW) of air is _____.

11. If the molecular weight is known, vapor density can be estimated by dividing the MW of the substance by _____.

12. Acetone has a MW of 58. The approximate density of acetone vapors is _____. Will acetone vapors tend to rise or settle? _____.

13. Liquids with boiling points less than _____ are frequently refrigerated to retard evaporation.

14. Specific gravity of a liquid or solid is determined by the ratio of the density of the substance relative to _____.

15. The density of water at 4°C is _____.

16. Tetrachloroethylene has a specific gravity of 1.62; this material is _____ than water.

17. A material that mixes completely with water regardless of the concentration is said to be _____ with water.

18. Sugar has a solubility of 1 part in 0.5 parts water. This is equivalent to _____ gm in _____ L water, or _____ gm in _____ ml water.

19. The percent solubility of ethyl benzene is 0.01%. This is equivalent to 1 part in _____ parts water.

20. Propylene oxide has a flash point of –35°F. This material is considered _____ by USDOT.

21. A-1 Kerosene has a flash point of 135°F. This material would be classified as _____ under USEPA guidelines.

22. Liquids and solids have flash points. Substances that are _____ at normal temperature and pressure do not have flash points.

23. The concentration range between the LEL and UEL is the _____.

24. Four routes of exposure for chemicals to enter the body are _____

25. At IDLH concentrations, what type of respiratory protection is NOT allowed? _____
_____.

26. What type of respiratory protection should be used when IDLH concentrations are present?
_____.

27. 1% by volume x _____ = concentration in ppm.

28. A material has an LEL of 1.9%. What is the LEL concentration in ppm? _____.

29. A tank containing n-hexane is about one-tenth full. The VP of n-hexane is 150 mm Hg. What is the approximate vapor concentration within the tank? _____.

30. Is the vapor concentration of n-hexane in the tank within the flammable range? The flash point of hexane is −7°F, the LEL is 1.1%, the UEL 7.5%. _____.

3

Detector Tubes And Dosimeters

LEARNING OBJECTIVES

1. Discuss different types of colorimetric detector tubes and how they work.
2. Describe different types of aspirating or vacuum pumps and how they work.
3. Describe the proper use of pumps.
4. List the steps to follow when using a detector tube.
5. Recognize the limitations of detector tubes.
6. Given a situation, select the appropriate detector tube.
7. Interpret the results obtained from various types of detector tubes.
8. Understand and describe the use of different types of dosimeters and badges.

The first detector tube was developed in 1917 at Harvard University and patented in 1919 for the detection of carbon monoxide[1]. A Bureau of Mines report on the use of a detector tube to estimate low concentrations of hydrogen sulfide was published in 1935[2]. Detection of aromatic hydrocarbons in air was described in 1950[3]. Since that time, **detector tubes**, also called **colorimetric indicator tubes**, have become an important tool for rapid detection and measurement of contaminants in air.

A detector tube is an hermetically sealed glass tube containing an inert solid or granular material such as silica gel, alumina, resin, pumice, or ground glass. The inert material is impregnated with or mixed with one or more reagents which change color when specific types of air contaminants are introduced. The length of the color change or stain, or the intensity of color change as compared to a comparative standards indicates the amount of material present.

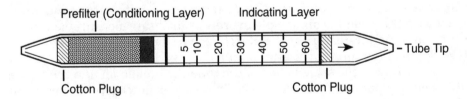

Figure 3-1: A colorimetric indicator tube.

An air sample is drawn into and through the tube by a pump, which may be a bellows, piston, or bulb type. Unless the manufacturer indicates otherwise, **one pump stroke pulls 100 cc of sample air through the tube**. When a pump stroke is performed, 100 cc of air is pushed or pulled out of the body of the pump through one or more exhaust ports; a valve prevents air from re-entering the pump through the exhaust ports. Sample air is then drawn through the tube and into the pump until the 100cc vacuum is replaced.

33

The color change or stain is caused by a chemical reaction between the reagents in the tube and contaminants in the sample air. A color change thus indicates the presence of air contaminants as well as the approximate relative concentration present. The reagents within each tube are usually specific for a particular type or class of chemical. The reagent mix to indicate the same material, however, often varies between different manufacturers. For instance, ammonia detector tubes typically contain a reaction mixture consisting of an acid and a pH indicator. Ammonia reacts with the acid to form an ammonium salt; this reaction is accompanied by a change in pH which causes the indicator to change color. Dräger and Mine Safety Appliances (MSA) utilize bromophenol blue as a pH indicator while the Sensidyne uses cresol red. In the presence of ammonia, the Dräger and MSA tubes change from gold to dark blue while the Sensidyne tube changes from purple to yellow.

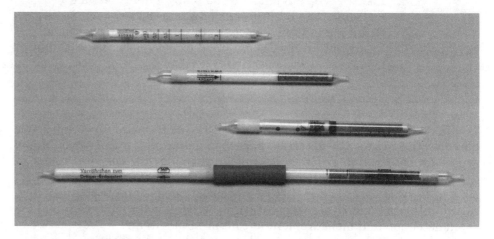

Figure 3-2: Types of detector tubes (from top to bottom): calibration scale; indicating-comparison layer; tube with sealed reagent ampule; separate reagent and indicator tube combination. Liquid in the sealed ampule is released by pressing down firmly on the dots. (Photo by Steve Napolitano.)

Most detector tubes contain a cotton or glass wool filter at each end to prevent particulates and aerosols from entering the tube. The tube may also contain a prefilter or conditioning layer to remove humidity or interfering gases and vapors. Some tubes, however, are a bit more complex, and may contain sealed ampules that must be broken and activated prior to sampling. Other detector tubes require the use of a **pre-tube** or **reagent tube** in concert with the colorimetric detector tube. The reagent tube is used to remove contaminants that are known to interfere with measurement, or react with the incoming air sample in order to facilitate measurement of the contaminate of interest. The tube that changes color and is used to determine the concentration present is always called the **indicator tube**.

Prior to use, the sealed ends of the tube are broken off and the tube is attached to a pump. Most tubes have an arrow or flow direction indicator. **The arrow always points toward the pump**. If a sample is taken with the arrow in the wrong direction, results may be invalid since the air did not pass through the prefilter or conditioning layer; another sample should be taken with a fresh, unused tube.

The amount of time it takes to complete one pump stroke, or put another way, for 100cc of sample air to pass through the detector tube, is called the **pump stroke interval**. The interval varies between tubes and is a function of tube resistance to air flow and the sampling rate of the pump. Pump stroke intervals may be as short as 5-10 seconds and as long as several minutes. A general rule of thumb is the larger the inner diameter of the tube, the shorter the pump stroke interval.

TYPES OF ASPIRATING OR VACUUM PUMPS

There are a number of detector tube and pump manufacturers. These manufacturers design their tubes to be used only with their pumps. Pumps offered include piston, bellows, or bulb type. The pump is designed to pull a known volume of air into and through the detector tube. While the design and function may appear simple and interchangeable, they are not. Sampling rates and vacuum pressures for pumps, and flow resistance and reaction rates of tubes, vary between manufacturers. Interchanging tubes and pumps of different manufacturers will produce significant measurement errors[4]. **Tubes and pumps of different manufacturers should NOT be interchanged**.

When selecting a manufacturer, it is important to examine the ranges and types of detector tubes offered as well as the types of pumps required to use the tubes. It is not unusual that more than one type of pump will be needed to efficiently utilize all the various types of tubes offered by one manufacturer.

Piston Pumps

A piston pump is composed of a cylindrical body, a piston connected to a shaft that runs through the cylinder, and a handle attached to the shaft. When using a piston pump, look for an index mark on the cylinder and on the handle. The index mark may be a triangle, dot, line, or some other character. The index mark on the handle must be aligned with the mark on the cylinder. Make sure the handle is pushed all the way into the cylinder, then pull the handle out all the way until the shaft is fully exposed and the handle locks into place. Do not twist the handle while pulling back. (A few piston pump

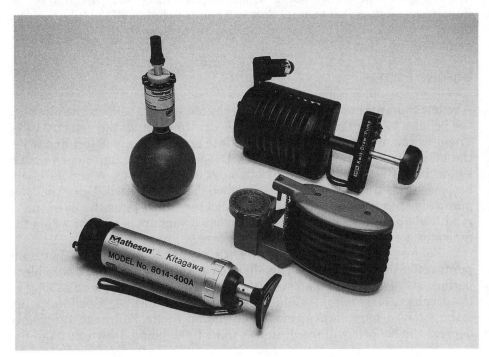

Figure 3-3: Types of pumps (clockwise from upper left): MSA Thumb Pump™, MSA bellows, Dräger bellows, Matheson-Kitagawa piston pump. (Photo by Steve Napolitano.)

Break off tube tips.

Insert tube into pump.

Pull handle straight out.

Air sample is drawn through tube during pump stroke interval

Turn handle and push back into cylinder.

Ready for another pump stroke, or remove tube and read results.

Figures 3-4: How to use a piston pump. (Photos by Steve Napolitano.)

models require the user to lock the handle in place with a quarter turn after the handle is pulled back completely.) At the end of the pump stroke, unlock the shaft by turning the handle a quarter turn and push the handle into the cylinder; realign the index marks and the pump is then ready for another pump stroke.

The piston pump shaft usually has at least two stops where the pump can be locked. These correspond to one-half (50 cc) and one full (100 cc) pump stroke volume. Some pumps have four stops, which allows the user to select 25, 50, 75, or 100 cc volumes. Others pumps have multiple orifices which control pump stroke volume.

Piston pumps usually have a tube tip breaker at the top of the cylinder, near the tube holder or orifice; a few manufacturers have added a broken tube tip reservoir at the top of the cylinder. When the tip is broken off it falls into the reservoir and is held there until a rubber plug is removed and the glass shaken out. Such an option may sound handy, however, the fine ground glass accelerates degradation of seals and the rubber plug is often lost after a short time. Some manufacturers also offer stroke counters on the body of the pump adjacent to the handle; no matter how the sales brochure may read, piston pump stroke counters are not automatic. The user must reset the counter prior to each use, and advance the number by one after each pump stroke.

Instructions accompanying detector tubes that are used with piston pumps include the minimum time interval required to complete a pump stroke. This is to ensure that a full 100 cc of sample air passes through the tube. Since air flow resistance and reaction rates vary between tubes, the pump stroke interval also varies between tubes. Some manufacturers offer piston pumps with pump stroke finish indicators. These are usually situated adjacent to the tube connector and are virtually impossible to read if the pump body is held in the normal position of use. A major drawback of the piston pump, then, is the need to actually time each pump stroke; this can become tedious when multiple pump strokes are required.

Another purported disadvantage of the piston pump is that both hands must be used to perform a pump stroke. While this is true, we fail to see its importance. Remember that it takes two hands to break off the tube tips and insert the tube into the tube connector, regardless of the type of pump employed.

Bellows Pumps

A bellows pump is operated by pushing the air out of the bellows and allowing the bellows to re-inflate as sample air is drawn through the detector tube. The user can observe the bellows re-inflate and can easily determine when the pump stroke interval is over. The Dräger bellows pump comes with a small chain that is connected to both sides of the bellows. When the chain becomes taut, the pump stroke is finished. The MSA Kwik Draw® bellows pump also has a pump stroke finish indicator which faces the user and is readily visible. The indicator is yellow when a vacuum is present; when the indicator goes black the pump stroke is over.

Bellows-type pump manufacturers are quick to remind you that their pumps offer the convenience of one-handed operation. It is indeed possible to perform a one-handed pump stroke with a Dräger bellows pump as long as the user has a large hand. A significant drawback of the Dräger bellows pump is the potential to perform an inadequate pump stroke. Failure to completely depress the bellows will result in less than 100cc being drawn through the tube. The bellows must be completely flattened by applying pressure to both ends of the pump. When wearing gloves, or if the user has small hands, it may be necessary to use both hands to ensure that an adequate pump stroke is taken. Another common problem encountered when using the Dräger pump is in the release. After the bellows is completely depressed, it must be released quickly in order to seat the exhaust valve properly. If the pump is released slowly, air can re-enter the bellows through the exhaust valve instead of through the tube. Quick release can be more difficult when the one-handed technique is used; many users prefer the two-handed method.

Break off tube tips.

The MSA bellows pump handle is designed to be used with one hand. The pump handle is pushed in and then released. The user must push the handle in completely before releasing in order to achieve a complete pump stroke of 100 cc.

Insert tube into pump.

Both the Dräger and MSA pumps offer automatic pump stroke counters; complete compression of the bellows pushes the counter down and advances it by one stroke. Failure to completely depress the bellows will result in an inaccurate count. The counter must be reset back to zero prior to each new tube.

Pull bellows up and release quickly.

Thumb Pump

The MSA Thumb Pump™ is an aspirator bulb type pump which is operated by squeezing the bulb lengthwise with the thumb to force air out of the pump. A compression ring and internal stop allows 100 cc sample air to be drawn through the tube. The pump has a counter that must be advanced by the user after each pump stroke.

Proper hand position for two handed pump stroke; push down rather than pull up.

Pump Testing and Maintenance

Piston pumps which do not have a broken tip reservoir require the least maintenance and offer the greatest service life. The bellows type pump offers a shorter service life and requires more maintenance. All piston pumps should be disassembled at regular intervals and the interior of the cylinder examined and cleaned. Vacuum grease should be applied at least every six months to piston pumps; O-rings or gaskets should be inspected and replaced at the same time.

Allow bellows to completely re-inflate before taking another stroke or removing tube.

Pumps should be tested for leaks prior to each day's use. A quick leak test can be performed by inserting an unopened tube into the pump's tube connector. Take one full pump stroke and wait at least two minutes (some pump manufacturers suggest waiting even longer). At the end of two minutes, if a leak is not present, the handle of a piston pump should rebound sharply toward the pump to the starting position when released. The thumb and bellows pumps should still be completely deflated.

Figure 3-5: How to use a Dräger bellows pump. (Photos by Steve Napolitano.)

Another leak-test method is to fill a sample bag with 100 cc air using a 50 or 100 cc syringe. Attach the pump to the bag and take one full pump stroke. Wait no more than one minute, then close the sample bag valve and remove the pump. Attach a 5 cc syringe and attempt to remove any residual air from the bag. If no leaks are present, the bag should be completely empty and less than 1 or 2 cc air can be collected. Repeat this process at least twice more, then average the results. Leakage of 3% per minute (i.e. 3 cc can be collected from the bag after a pump stroke), should be considered excessive and unacceptable.

If there is an indication of a leak, the pump should not be used until it is serviced. Piston pumps usually require replacement of an O-ring and greasing with vacuum grease, or replacement of the tube connector. Bellows pumps may require replacement of tube connector, screen, or valves; a pin-hole leak in the bellows may also be the culprit. Check the manufacturer's instructions for more specific information on field leakage tests, maintenance, and repair.

The pump volume and flow rate should also be checked periodically. Pump volume should be within 2% of the target capacity of 100 cc. Pump flow rate should be within 10% of the flow rate indicated by the manufacturer. A simple procedure using a bubble flow meter to determine pump volume and flow rate has been described[1].

TYPES OF DETECTOR TUBES

Most manufacturers use more than one type of measurement method, which often varies with the concentration range as well as the air contaminant detected. There are four basic types of detector tubes: direct reading with a concentration scale, direct reading with a concentration conversion chart or scale, color intensity, and color comparison.

Direct Reading

The easiest detector tubes to use are those with a calibration scale marked directly on the tube; the length of color change corresponds to the actual concentration on the scale after a specified volume of air is pulled through the tube. In some cases, the concentration read off the scale must be multiplied by a factor; nevertheless, the approximate actual concentration present can be determined directly off the tube without referring to a conversion chart in the instructions.

Figure 3-6: A direct-reading tube; the concentration is read directly from the scale on the tube.

A variation of the direct-read tube is the concentration-pump stroke tube. The scale is read off the tube and divided by the number of pump strokes (ps) taken to give the concentration present. For example, if the reading on the tube is 1000 ppm-ps, and 10 pump strokes were used to achieve the reading, then

$$\frac{1000 \ ppm\text{-}ps}{10 \ ps \ taken} = 100 \ ppm \ present$$

If 1/4 ps was used, then the actual concentration present would be 4000 ppm; this type of tube can be quite useful when a wide detection range is required.

Concentration Conversion

Another direct reading tube is the type that has a millimeter (mm) scale; the length of color stain is measured, and a conversion table is used to convert mm stain to concentration. A different type of conversion method is the concentration chart. The detector tube is placed on the chart and the

CALIBRATION CHART

Number of Strokes	Light Figures Below Indicate Length of Stain in Millimeters				Bold Figures Indicate Concentration in Parts Per Million				
1/2	0 mm	4	6.5	9	10	13	16	20	24.5
	0	25	50	75	100	150	200	300	400
1	0 mm	4.5	8	11.5	16.5	23	28		
	0	20	40	60	100	150	200		
2	0 mm	4	9.5	17.5	25.5	32.5			
	0	10	25	50	75	100			

Figure 3-7: Example of a millimeter to ppm conversion chart. What concentration is present in ppm if one pump stroke were used and the tube has a 23 mm stain? (Answer 150 ppm.)

concentration is read off the chart. Conversion tubes are also easy to use, however, the actual concentration present cannot be determined without referring to the conversion chart or table.

Color Intensity

Some manufacturers use the intensity of color change to determine the approximate actual concentration. The deeper or more intense the color, the greater the approximate concentration present. The color of the tube is compared to standard color indicator tubes or a color intensity chart provided by the manufacturer. Color intensity tubes are useful when the chemical reaction used to detect the material does not produce a distinct border between reacted and non-reacted portions of the tube. In some cases, the color change must develop over a period of time after sampling.

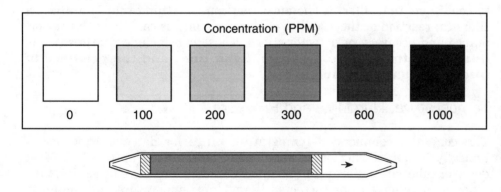

Concentration (PPM)

| 0 | 100 | 200 | 300 | 600 | 1000 |

Figure 3-8: A color intensity tube and chart. When using color charts, the color intensity is proportional to concentration. A waiting time or development time is usually required. This detector tube shows a reading of approximately 300 ppm.

Color Comparison

A variant of the color intensity tube is the color comparison detector tube. When using this type of tube, the number of pump strokes required for the indicating layer to match a comparison layer in the tube is counted; the approximate concentration is then determined by referring to a pump stroke conversion chart in the instructions.

Figure 3-9: The Dräger Hydrocarbons 0.1%/b detector tube for liquid hydrocarbons is an example of a color comparison detector tube. The top tube is unused; the indicating layer of the lower tube matches the comparison layer. If the gas present is propane, and 11 pump strokes are required to match the comparison layer, the approximate concentration present is 0.8% or 8,000 ppm.

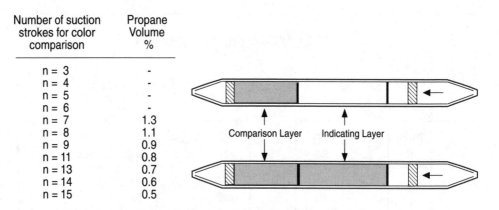

Number of suction strokes for color comparison	Propane Volume %
n = 3	-
n = 4	-
n = 5	-
n = 6	-
n = 7	1.3
n = 8	1.1
n = 9	0.9
n = 11	0.8
n = 13	0.7
n = 14	0.6
n = 15	0.5

Qualitative Indicator Tubes

All manufacturers offer tubes that are used as yes-no tubes to alert the user to the presence of air contamination. These tubes usually have no scale or concentration indicators; the presence of air contamination is indicated by a color change. In most cases, the relative concentration of contaminants present cannot be determined on the basis of stain length or intensity. A common yes-no tube is the so-called "poly" tube or "polytest" tube designed to detect a wide spectrum of organic and inorganic gases and vapors. Matheson offers separate qualitative detector tubes for organic and inorganic materials. Each tube is divided into four or five sections; the pattern of discoloration may assist the user in determining which tube to use next. Qualitative tubes should never be used, however, to determine the identity or approximate concentration of contaminants in air.

LIMITATIONS OF DETECTOR TUBES

Detector tubes, because they are filled with reactive chemicals, are sensitive to conditions that affect a chemical reaction. Anything that can affect a chemical reaction or the stability of stored chemicals can affect the performance and accuracy of detector tubes. These can include **temperature, humidity, atmospheric pressure, light, time, and the presence of interfering gases or vapors**.

Temperature, Humidity, and Pressure

The general recommended temperature range for detector tube use is typically 0 to 40°C, although some tubes can be used up to 50°C. Most detector tubes are calibrated at 20 to 25°C and 50% relative humidity; tube readings may have to be corrected if ambient temperature and humidity differ significantly from the calibration conditions. When a correction is required, the manufacturer supplies this information in the instructions. Compensating for temperature and humidity conditions may be accomplished by use of a correction table, conversion factors, or an increase or decrease in the number of pump strokes required.

Detector tubes which are calibrated at pressures other than ambient (i.e. at pressures other than 760 mm Hg or 1013 mbar) may require correction. The conversion factor or equation will be noted in the manufacturer's instructions. Typically, the conversion is as follows:

$$\text{Corrected reading} = \text{tube reading} \times \frac{\text{calibration pressure}}{\text{sampling pressure}}$$

Light

Detector tubes come in sealed boxes that are usually light-tight which can be resealed or securely closed after opening. Remember that light can affect stored chemicals; that is why so many chemicals, including prescriptions, over-the-counter medication, fine wines, even some cooking oils, are supplied in amber or colored glass containers. Light accelerates chemical decomposition or degradation; it is important then to protect detector tubes from light until ready for use. Never leave tubes out in the sun, the combination of rapid heating and ultraviolet light can accelerate decomposition of tube reagents to the point where such tubes can be can rendered useless within a few hours.

Time and Storage Conditions

The reagents stored within each detector tube are chemically reactive, and are often incompatible with each other. Such chemical reactivity is necessary to elicit an immediate response. Unfortunately, chemical reactivity has its drawbacks; one is that light and heat can accelerate degradation. Another is that even under optimal conditions, the reagent mixture will begin to degrade over time. All detector tubes come with expiration dates after which the tube should not be used for field sampling. Most tubes have a shelf life of two or three years; a few tubes have limited storage life and must be used within one year or less of manufacture.

The expiration date found on each box of tubes represents the maximum shelf life under **optimal** storage conditions. Appropriate storage temperatures are recommended by the manufacturer and can be found in the instructions. If storage conditions are not specified by the manufacturer, be sure to store tubes in a cool place away from sunlight and heat at a temperature of no more than 25°C. Refrigeration is often mentioned to optimize shelf life, however, it is not recommended if boxes of tubes are repeatedly removed and then re-refrigerated. Repeated cycles of warming and cooling can accelerate degradation of some reagents. Improper storage can result in reagent deterioration, poor performance, and incorrect results, even in unexpired tubes.

Interfering Gases and Cross Specificity

Detector tubes were originally designed to monitor workplace air for one, specific, **known** contaminant such as carbon monoxide or hydrogen sulfide. These detector tubes were calibrated with the contaminant of concern and were intended to measure only that particular gas or vapor. As detector tubes were developed for more gases and vapors, however, it became clear that there was only a limited number of possible chemical reactions that could be used to indicate the presence of air contamination. It also became evident that tubes could react not only to one material, but also to many other air contaminants.

Despite what you may have heard or read, there is **not** a unique and specific detector tube for each chemical. Nearly all tubes will react with more than one material, not just the chemical that the tube is calibrated to. For instance, an acetone tube calibrated with acetone will give a positive response when acetone is present within a specific concentration range. Other organic chemicals which are structurally similar to acetone, or which react with the reagent mixture in a manner similar to acetone may also be detected. A typical acetone tube may be expected to react to other ketones such as methyl ethyl ketone (MEK) and methyl isobutyl ketone (MIBK); other organic materials such as acrolein, methyl acrylate, isopropyl acetate, and acetaldehyde may also be anticipated to produce a positive response. Since the acetone detector tube is calibrated only to acetone, the actual concentration of contaminants other than acetone cannot be accurately measured.

Interfering gases thus cause a positive response, even when the material for which the tube is calibrated is not present; this is called a **false positive** result. This is a very important concept; in the presence of **unknown** or **multiple** air contaminants, a positive response does not always indicate the presence of a one specific material. In such cases, a positive response only indicates the presence of contamination; the response may be caused by the vapor or gas for which the tube is calibrated, or it may be due to the presence of one or more interfering gases, or a combination of both.

Interfering gases which cross-react and produce a positive color change identical to that produced by the calibrant gas or vapor are listed by the manufacturer in the instructions, usually under **cross specificity** or **cross sensitivity.**

Other interfering gases can affect the reading by inhibiting the chemical reaction; in the presence of these gases, the tube will give an inaccurate result by indicating a lower concentration than is actually present. Other interfering gases or interferants will alter the chemical reaction and produce a color change **different** from the change produced by the calibrant. Such a color change informs the user that an interfering gas is present, however, the presence of other materials, such as the calibrant or cross-specific gases or vapors, cannot be confirmed or denied.

SPECIAL TUBES, KITS AND ACCESSORIES

Detector Tubes for Ions

Ion detector tubes are used to measure the concentration of specific ions in water not air. Tubes are available to measure chloride, sulfide, cyanide, iron, copper, residual chlorine, and salinity. Water is introduced into the tube by immersing the tube into the water sample, injecting water directly into the tube, or by drawing water up into the tube with an aspirator bulb.

Hazmat Kits

Several manufacturers have developed Hazmat Kits to detect the presence of certain classes of air contaminants and facilitate rapid interpretation of tube results. Dräger and Sensidyne kits both have a decision matrix; the user determines which tube to draw next, based on previous tube results. The MSA kit is divided into three sets of four tubes each. Sample air is drawn into the four tubes simultaneously by means of an aspirator bulb or Kwik-Draw bellows pump. All three sets are drawn and then the results are tabulated. A similar design is available for inorganic contaminants from Dräger. Two sets of five detector tubes are pre-positioned in a holder; air is drawn through all five tubes simultaneously using a bellows pump.

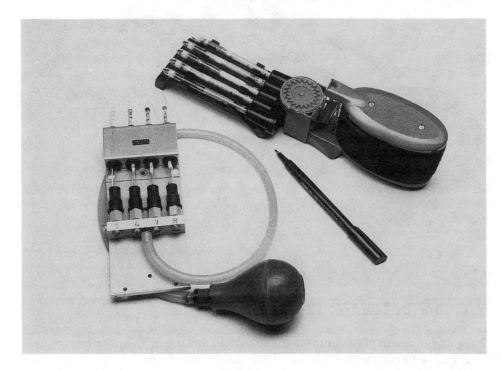

Figure 3-10: MSA Hazmat Kit tube holder with aspirator bulb (left) and Dräger Simultaneous Test Set with bellows pump (right). (A Steve Napolitano photo.)

Air Analysis Kits

Nearly all detector tube manufacturers offer analysis kits to measure the amount of carbon monoxide, carbon dioxide, oil mist, and water vapor in air that is to be used for breathing. An adaptor which holds the tube in place is usually placed in line with the flow indicator and regulator coming off the air compressor.

Accessories

All manufacturers offer **extension hoses** to facilitate remote sampling of tanks, conduits, sewers, and other areas prior to entry. Hose lengths usually vary between 3 and 5 meters (10 to 15 feet); lengths up to 50 feet have been used. The tube holder is always at the end of the hose, distal to the pump. MSA offers a solvent-resistant 25 foot extension hose on a reel; to use, simply place the tube in the holder, connect the pump, and reel out the required length of hose. A plastic tube holder protects the glass tube from breakage.

A **hot probe** is required if hot stack emissions or furnace gases are sampled. Heated sample air is drawn into the hot probe where it is cooled before entering the detector tube. A **pyrolyzer** is used to detect certain air contaminants such as freons by heating incoming sample air, thus forming thermal decomposition products that are then measured with a detector tube.

The Dräger Quantimeter is a battery powered pump that is used when many strokes are required to complete the test protocol. Once the required number of strokes is programed and the tube positioned, the Quantimeter is turned on and the pump stroke protocol is performed automatically.

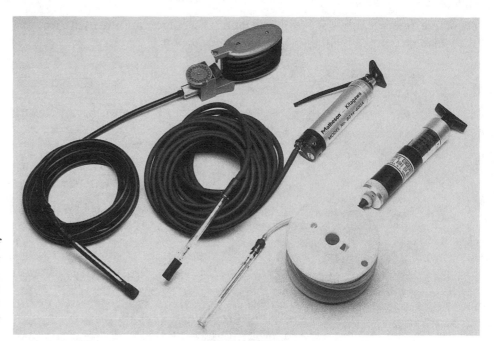

Figure 3-11: Some examples of extension hoses from left to right: Dräger, Matheson Kitagawa, MSA sampling reel. Note that the tubes are placed at the end of the hose distal to the pump. (Photo by Steve Napolitano.)

INTERPRETING DETECTOR TUBE RESPONSE

Detector tubes can give valuable information, as long as the user recognizes the limitations of detector tubes and does not over-interpret results. Remember that information on tube limitations are contained in the instructions that accompany each box of tubes. It is important to **follow instructions** regarding the number of pump strokes, pump stroke interval, and correction factors when determining the approximate concentration present. Some tubes contain an ampule which holds a liquid or solid reagent. The ampule must be broken at the correct time, either before or after drawing sample air through the tube. A few tubes require time for the indicating layer to change color. Also check instructions for advice concerning interfering materials which can affect results. Note what color change to expect; interfering materials may alter the chemical reaction and affect the final color change.

It is important that the tube fits securely in the pump; there should be no leaks through the inlet around the tube, or around the exhaust port, valves or gaskets. If air leaks into the pump, there will be reduced flow through the tube, and an inaccurate reading will be obtained. Remember that the **arrow always points toward the pump**, since it indicates the direction of air flow. If the tube has no arrow or other direction indicator then the tube can be placed either way in the pump.

Whenever possible, the tube should be watched for a response during each pump stroke. Let's say you are sampling an atmosphere for ammonia with a Dräger Ammonia 5/a tube; the instructions call for 10 pump strokes, after which concentrations between 5 and 70 ppm can be read directly off the tube. Suppose you took the 10 pump strokes and then looked at the tube, which was completely saturated (i.e. the color stain extended the entire length of tube well past the 70 ppm mark). How many ppm are present? With the information available, all you can say is that more than 70 ppm are present. Now let's say you watched the tube, and after the second pump stroke the color stain was at the 70 ppm mark; how many ppm are present? Now you have sufficient information to make the following calculation:

$$\frac{ppm\ tube\ reading\ x\ 10\ pump\ strokes}{number\ of\ pump\ strokes\ taken} = ppm\ ammonia\ present$$

$$\frac{70\ ppm\ x\ 10}{2} = \frac{700\ ppm}{2} = approximately\ 350\ ppm\ ammonia$$

It is essential, however, that you **do not end the pump stroke protocol prematurely**. Watch and keep a mental note of the stain progress down the tube with each pump stroke. Stain progression down the length of tube is not always additive; the first pump stroke may induce a color change of significant length while remaining pump strokes may cause the stain to advance only a small distance. Pump stroke protocols should be ended only if the stain actually reaches the end of the measurement range.

In many cases, **a negative response is more informative than a positive response**; this is especially true when dealing with unknowns. A negative response allows the user to exclude a substantial number of potential contaminants. For example, the lack of a color change on a MSA Petroleum Hydrocarbon tube indicates that **significant concentrations** of gasoline, kerosene, benzene, toluene, xylene, hydrogen sulfide, cyclohexane, propane, butane, pentane, hexane, heptane, octane, and nonane are **not** present. A positive response, on the other hand, indicates that one or all of these materials are present; without additional information, it would not be possible to determine which product is present, and how much is present.

The lack of a color change is often used to rule out the presence of contaminants. A negative response, however, does not mean that those contaminants are completely absent; it simply indicates that they are not detected within the measurement range of the tube. The MSA Petroleum Hydrocarbon tube reads in mm stain which is converted to ppm by means of a conversion chart. The lowest measured concentration on the chart is 200 ppm; a negative result thus indicates that if any of the petroleum hydrocarbons listed are present as contaminants, the total concentration in air is less than 200 ppm.

It is not unusual for a detector tube to give no color change but become quite warm, or even hot, during the pump stroke protocol. Some tubes may react to form a fume which is visible as it exits the pump when another stroke is taken; a color change does not always accompany this effect. These are signs that a non-detectable contaminant is present. Before making this interpretation, however, check the instructions to ensure that water vapor does not produce similar results.

It is not unusual that the end of the color change is uneven. In some instances, however, one side of tube shows significantly more stain than the other. This occurs when the packed reagents shift in the tube, producing an area or channel with a decreased resistance to air flow. This so-called channeling effect is often enhanced when tubes are handled roughly during shipment or use. If a tube demonstrates a jagged or uneven stain edge, the

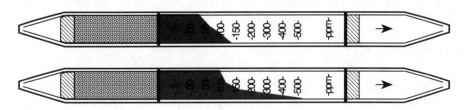

Figure 3-12: A detector tube with uneven stain (top) and tube with severe channeling effect; the lower tube results should be disregarded and a fresh tube sampled.

minimum and maximum reading should be recorded; in most cases it is appropriate to take the middle value as the concentration. If the tube shows a significant channeling effect, however, another reading should be taken with a fresh tube.

After pulling a number of tubes, or at the end of day, several pump strokes of clean air should be drawn into the body of the pump to remove residual reagents that may still be present. If a fume or smoke is formed during a pump stroke protocol, clean the pump with fresh air as soon as possible to prevent corrosion.

USING AND CHOOSING TUBES

Detector tubes are most commonly employed to determine the presence and extent of air contamination at the **time of sampling**. Detector tubes take discrete grab samples and give information that is valid only for the time and location of the sampling event. Air, and the contaminants it may contain, is not static. Contamination concentrations fluctuate because **air moves**. Change the location of sampling by a few feet and your results may differ significantly. Other variables such as temperature, humidity, and the presence of other vapors and gases also affect detector tube response.

It should be obvious, then, that detector tubes should not be used in place of continuous air monitoring, but rather to augment, enhance, or confirm the results obtained from other instruments. In some cases, tubes may be the only source of information; in these situations, detector tube readings should be taken frequently and at various locations throughout the potential exposure area. Remember that the accuracy of the reading may vary between different tubes and manufacturers. Some manufacturers supply the relative standard deviation which can be up to ± 35%; this information is found in the instructions and/or the detector tube handbook. When interpreting results, the deviation should also be considered. For example, a tube reading of 100 ppm with a standard deviation of 25% should be reported as 100 ± 25 or as a concentration range of 75–125 ppm.

Preparing To Use A Tube

When preparing to use a detector tube, the first thing to do is check the box and determine that the tubes are calibrated to the contaminant of interest; box labels usually also indicate the measurement range. Next verify that the tubes still have a valid expiration date; also note and record the lot number of the tubes. Look at the box for signs of mistreatment; make a record of your observations if the box is shredded, dented, crushed, melted, discolored, chewed, or otherwise abused.

Open the box and check the conditions of the tubes. If the box is in poor condition, examine the tubes very carefully for hairline cracks, shifted reagent mixture, or discoloration. If the tube contains an ampule, make sure it is still intact. Do not use tubes that are damaged. Tubes that are damaged, subjected to abuse, or stored improperly can give inaccurate results.

Information In The Instructions

Remove the instruction sheet from the box. Some manufacturers provide a sheet that seems to have been designed to rip or fall apart after a single inspection, others are impossible to refold properly. Avoid future problems with lost or unreadable instructions by making a copy of each instruction sheet and keep them in a safe place. Another option is to place all instructions in plastic holders and store them in a binder.

Make sure the instructions are for the tubes in the box! It is not unheard of that instruction sheets get mixed up when multiple types of tubes are used at the same time. Do not rely on the summary instructions on the box; these instructions do not supply conversion factors, limitations, anticipated color change, and other information required to properly use a detector tube.

Now look at the instructions. Hopefully you have familiarized yourself with each detector tube, and its instructions, before going out on site. The time to learn how to use a tube is **before** using it in the field! The instructions should contain the following information, not necessarily in this order:

- description of the tube (often with a diagram)
- measurement range
- testing procedure
- detection principle and anticipated color change
- limitations and corrections
- interfering gases/vapors (cross specificity)
- storage procedures and shelf life
- other information

Look at one of the tubes in the box. It should match the description given in the instructions. If the color of the tube does not match the instructions **do not use the tube**. The prelayer and reagent layer may change color when tubes are improperly stored or subjected to temperature extremes.

Check the measurement range and determine if the tube selected is appropriate for use. For example, if concentrations between 1 and 100 ppm are anticipated, it would be inappropriate to use a tube calibrated to detect percent by volume concentrations. Alternatively, if large concentrations are known to be present, a low concentration tube could be used to detect the presence of the material, but would be not useful to determine the actual concentration present. The measurement range can be extended many-fold for some tubes, this information is supplied in the instructions.

Make sure the tube will not create a hazard under the anticipated conditions of use! Some tubes are incompatible with flammable gases or high oxygen concentrations. For example, the Dräger Oxygen 5%/B detector tube generates sufficient heat in oxygen enriched atmospheres to be a potential flammability hazard for materials with low ignition temperatures.

Next, examine the section which lists the interfering gases and vapors; this section may also be called the specificity or cross-sensitivity section. If a mixture of different gases or vapors is known or suspected to be present, determine if any of these materials will affect the performance of the tube. If so, it may be necessary to select another tube with a different detection principle that is not affected. Oftentimes, a tube calibrated to the same material that relies on a different chemical reaction does not have the same interferants.

Now check the limitations section for temperature and humidity correction factors that will be required, based on current conditions. Note that the pump stroke protocol may vary with temperature and/or humidity.

Refer to the test procedure section for the number of pump strokes, duration of pump stroke, and anticipated color change. Determine how long the entire pump stroke protocol will take and if the time interval is acceptable. For example, the Dräger Sulfuric Acid 1/a tube requires 100 pump strokes; each pump stroke takes approximately 1 minute to complete. The entire protocol therefore requires 100 minutes. This tube would not be feasible for use with a bellows hand pump at an emergency response by personnel using self-contained breathing apparatus with an air supply limit of no more than 60 minutes. This tube could, however, be used with a battery-operated Quantimeter, which performs the requisite number of pump strokes automatically.

Finally, determine if the color change will be easily seen under test conditions. The user should already be familiar with the anticipated color change; this is usually accomplished by testing at least a few tubes from each lot. Many color changes are very difficult to detect in even the best of circumstances; these include tan to pale yellow, off-white to gray, yellow to pale green. Consider if they can be detected in bright sunlight, under artificial lighting, or in poor light conditions; remember that it is more difficult to detect a color change when wearing a full-face respirator. Comparing the sampled tube with an unreacted, sealed tube is often helpful in determining the extent of a color change.

When using the tube, remember to watch for a color change and end the pump stroke protocol when the tube is saturated. Keep a mental note of the stain progress down the tube with each pump stroke. Stain progress down the tube is not always additive; end the protocol prematurely only when the stain reaches the end of the measurement range.

It is not uncommon to get no color change after completing the test procedure. In some cases, a tube which shows no response can be reused; this will be clearly stated in the instructions. Do not reuse a tube unless the manufacturer clearly specifies that this is permissible.

When a color change is noted, initially mark the edge of the color stain with a marker at the end of the test procedure; if the stain is uneven or jagged, mark the limits of the stain. Do not attempt to store tubes and read them at a later time; the stain may fade, change color, or bleed into initially unreacted portions of the tube. Some tubes which rely on intensity of color change require a specified interval for color development. Other tubes require time for the color change to deepen and become more readable. For these tubes, mark the limits of the stain after sampling, and check again after several minutes to determine if the color change is easier to read. After the stain length has been marked on the tube, read the length of stain, then use the instructions to convert to actual concentration.

Used tubes should be retained until the end of the day; a cardboard or plastic 8-inch mailing tube is a handy carrier. Do not leave used tubes laying about. They contain reactive and corrosive chemicals; the broken ends can cut or puncture the skin. If the skin is punctured with an open tube, treat as for skin exposure with a corrosive and flush the wound with copious amounts of water.

Collect tubes for disposal in a 1-gallon plastic container (such as for liquid laundry detergent) until nearly full. Add tap water to cover the tubes, and let soak for at least 24 hours. Gently mix the contents and soak for another 24 hours. If the pH is between 5-10, water can be decanted into the sanitary sewer; the tubes are sealed in the plastic container and disposed of as solid household waste.

A final word regarding requirements for the users of detector tubes. It is apparent that the user must have sufficient reading and comprehension skills to understand the instructions. Basic arithmetic skills such as simple multiplication and division are also required. It should also be obvious that the user must have full spectrum color vision; individuals who are color blind or who cannot discern part of the visible color spectrum should not use detector tubes. On a similar note, sunglasses should not be worn when using detector tubes.

Certification

Detector tubes and pumps were at one time certified by NIOSH to meet specific performance standards. These standards are described in 42 CFR 84 and include specifications regarding accuracy at four concentrations, standard deviation of tube readings, and variation of stain length within tubes. Although the NIOSH certification program has been discontinued, the Safety Equipment Institute (SEI) provides voluntary, third-party certification and publishes a certified product list of detector tubes and pumps which have been tested and found to meet NIOSH criteria[6]. In addition to using NIOSH criteria, SEI also requires periodic quality assurance audits to ensure adherence to SEI standards.

DOSIMETER TUBES AND BADGES

When information is required concerning the concentration of a contaminant in air over a relatively long interval (e.g. during an 8-hour workday), long-term tubes, passive dosimeters, or badges can be used. These are usually attached to workers' clothing and positioned within the breathing zone. They may also be used as area samplers to assess worksite contaminant concentrations by placing them at strategic locations.

Figure 3-13: Long-term tubes connected to the pump by a hose. Tubes are positioned within the breathing zone of the wearer. (Photo courtesy MSA.)

Long Term Tubes

Long term tubes are detector tubes that are calibrated to specific workplace contaminants such as benzene, chlorine, carbon monoxide, or ammonia. The tube is usually clipped onto clothing and positioned within the breathing zone of the wearer. The tube is then connected to a small peristalic or diaphragm pump; the tube may be placed directly into the pump or connected by a small-diameter hose. The pump continuously pulls air through the tube at a set sample rate. Long-term tubes may also be positioned throughout the worksite to determine the concentration in a specific area. The length of color stain indicates the concentration of contaminant present during the sampling interval.

The length of stain is usually indicated in uL (microliters); this is transformed into an exposure value for the sampling interval based upon the sampling rate of the pump and the maximum sample volume recommended by the manufacturer.

For example, the Dräger Carbon Monoxide 10/a-L long term tube has a maximum sample volume of 4 liters. That means that no more than 4 liters or 4000 cc of air should be drawn through the tube over a period of 4 hours. This protocol therefore requires a sampling pump drawing 16-17 cc per minute, which is approximately 1000 cc or 1 liter per hour.

Suppose the CO tube reads 100 uL after 4 hours, using a flow rate of 17 cc per minute. The average exposure over 4 hours can be calculated by dividing 100 by 4 for a TWA (time weighted exposure) of 25 ppm per hour. In this case, 100 uL is equivalent to ppm, since the sample volume per hour was 1000 cc (1 uL in 1000 cc = 1 ppm).

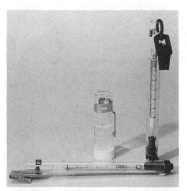

Figure 3-14: Passive bubbler for formaldehyde (middle) and dosimeter tubes for ammonia. Tubes are placed inside plastic holders to minimize fluctuations in air currents reaching the tube. (Photo by Steve Napolitano.)

It is important, then, that the pump flow rate and sampling interval be closely monitored. Pump rates can be determined using a bubble buret, electronic bubble meter, or rotameter. Sampling pumps should be regularly calibrated; pump calibration methods have been thoroughly described[7,8]. Specific calibration procedures are provided by the manufacturer in the instruction manual.

Long duration tubes have sampling rates that vary between 1 and 8 hours and require very low flow rates; the sampling pump should maintain rates between 10 and 20 cc per minute. Tubes should be checked periodically throughout the sampling interval to ensure that the capacity of the indicating layer is not overwhelmed. Remember that the calculated exposure value is valid only for the individual or immediate location sampled.

Long term tubes can also be used to detect very low concentrations by extending the sampling interval up to 24 hours. In this case however, results should be used only as a yes/no indication, rather than attempting to determine an actual concentration. Extended duration tubes are often difficult to read since the color stain may be lighter at low concentrations, or the stain may gradually change during the sampling interval.

Passive Dosimeter Tubes

Passive dosimeters do not require a sampling pump but rely on air currents and diffusion to move contaminants into the tube. The dosimeter is simply hung in place for a specified length of time; at the end of the interval the length of stain is read and the exposure value calculated. Passive dosimeters may be a length of stain detector tube, or a strip of paper within a glass or plastic tube. In some cases, the ppm per hour dose is read directly from the tube. Other tubes require the use of a calibration curve provided with the tube; still others use color intensity charts.

Passive dosimeters are convenient, relatively inexpensive, easy to use, and do not require a calibrated sampling pump. Wearer compliance is usually good, since they do not interfere with activity. Since passive dosimeters are not used with a battery-operated pump, they are safe for use in flammable atmospheres.

Dosimeter tubes are designed for long duration, low air flow use. These tubes rely on the diffusion principle to move contaminants into the tube. Diffusion tubes are designed to be used inside a plastic holder; the holder serves as a draft shield to prevent high velocity air from entering the tube and skewing results. In situations where air currents are variable or affected by worker activity, dosimeter results may over- or underestimate the actual cumulative exposure to the contaminate present.

Passive Badges and Tapes

Figure 3-15: Sure-Spot™ dosimeter badge in reusable plastic holder. (Photo courtesy GMD Systems.)

Passive badges rely on permeation of gases or vapors through a membrane onto a collection medium or chemically impregnated tape designed to change color when a minimum accumulated concentration is reached. Badge color is then compared to a chart to estimate the average concentration present.

Like dosimeter tubes, badges are easy to use; interpretation can be more difficult, especially when color changes are uneven or accompanied by flecking or speckling. Some passive badges are used with electronic readers to eliminate color interpretation errors. Badges and tapes are susceptible to drying, clogging with dust, and fading from sunlight. Unless approved by the manufacturer, unprotected badges should not be worn in the rain.

USING DOSIMETERS

Limitations

Dosimeters experience the same limitations as discussed for detector tubes. Remember that temperature and humidity conditions change over time, and may exert a greater effect, since the dosimeter is exposed for a prolonged interval. Most importantly, the presence of other gases and vapors can also affect dosimeter results.

The standard deviation for dosimeter tube results is typically at least 20-25%. Results should be given as a range rather than one number. For example, an ammonia dosimeter tube gives a result of 50 ppm ± 25%; the concentration that should be reported is therefore 50 ppm ± 12.5 ppm or 38 - 63 ppm.

Packaging and Instructions

Tube and badge dosimeters should be examined carefully to ensure they are not damaged. Take special care when handling badges. Do not touch the surface of the badge; skin oils can clog the pores and produce a splotchy result. Do not bend, crimp, or puncture the badge surface. Look at the color of the unused dosimeter and make sure it has not been inadvertently exposed.

Read the instructions! Although dosimeter tubes may look like colorimetric detector tubes, the instructions are very different. For example, most passive dosimeter tubes require that only one end of the tube be opened, not at both ends.

REFERENCES CITED

1. Direct Reading Colorimetric Indicator Tubes Manual, 1st Edition. 1976. Akron, OH: American Industrial Hygiene Association.
2. Littlefield, J.B., Yant, W.P., and Berger, L.B. 1935. A detector for quantitative estimation of low concentration of hydrogen sulfide. U.S. Bureau of Mines Report, Vol. 3276. Department of Interior, Washington, DC. 1935.
3. Hubbard, B.R. and Silverman, L. 1950. Rapid method for the determination of aromatic hydrocarbons in air. Arch. Ind. Hyg. Occup. Med. 2: 49-55.
4. Colen, F.H. 1972. A Study of the Interchangeability of Gas Detector Tubes and Pumps. Report No. TR-71. Morgantown, WV: National Institute for Occupational Safety and Health.
5. Kusnetz, H.L. 1960. Calibration of direct reading colorimetric gas detecting devices. Am. Ind. Hyg. Assoc. J. 21: 340-341.
6. Safety Equipment Institute. 1990. Certified Product List. Arlington, VA.
7. Ness, S.A. 1991. Air Monitoring for Toxic Exposures. New York: Van Nostrand Reinhold.
8. Hering, S.V. (Ed). 1989. Air Sampling Instruments for Evaluation of Atmospheric Contaminants, 7th Edition. Cincinnati, OH.

CHAPTER REVIEW

1. When using detector tubes, the intensity or length of _____ is used to indicate the amount of material present.

2. A pump is used to draw sample air through the tube; the volume of air in one pump stroke is _____ cc.

3. The arrow always points _____ the pump.

4. True or False? It is permissible to interchange tubes and pumps of different manufacturers. _____ .

5. Which pump design allows the user to take partial pump strokes? _____ .

6. The recommended temperature range for most tubes is _____ .

7. Name three other limitations of detector tubes _____ and_____ .

8. Gases which can produce false positive responses, or produce an unexpected color change are called _____ gases.

9. When using an extension hose for remote sample, the detector tube is always at the _____ of the hose.

10. If the end of the color stain is uneven, the minimum and maximum reading should be recorded, and the _____ used as the approximate concentration present.

11. A particular type of tube has a detection range of 0-1000 ppm, with a reported standard deviation of $\pm$ 15%. The tube shows a response of 100 ppm; the approximate concentration present is _____ ppm to _____ ppm.

12. A tube protocol calls for 10 pump strokes, after which the concentration can be read directly off the tube. The range of the tube is 10 to 500 ppm. After 5 pump strokes, the color stain is just to 500 ppm. Approximately how many ppm are present? _____ .

13. A tube protocol calls for 20 pump strokes, after which the concentration is read directly off the tube. The range of the tube is 100 to 1000 ppm. After 10 pump strokes, the color stain is just to 500 ppm. What should you do?_____ .

14. Tubes or badges used to determine the concentration of air contaminants in the breathing zone over a 4 or 8 hour period during the work day are called _____ .

15. How should the lack of a color change be reported? _____ .
    What does it suggest? _____ .

PROBLEM SET

Refer to Appendix A for detector tube instructions; we recommend you familiarize yourself with these instructions before you begin the problem set.

1. You are a member of a hazardous materials response team called to a butane filling facility. Which tubes available can detect butane? Which tube has the widest concentration range?

2. At a gasoline spill a Dräger Xylene 10/a tube turns reddish brown to 400ppm after 5 ps. What can be determined by these results?

3. A solution of ethanol (ethyl alcohol) and toluene (an aromatic hydrocarbon) has leaked onto the floor of a warehouse. Select at least one tube for each material that will not cross-react with the other contaminate present.

4. A reading of 3,000 ppm is obtained on a Dräger Carbon Monoxide 10/b tube after 5 ps. What is the actual concentration present? What concentration is present if 10 ps were taken?

5. When using the MSA Ammonia/Organic Amine detector tube, what approximate concentration of ethylamine is present when a blue stain of 20 mm is obtained after one-half ps? What is the correct pump stroke interval for a half ps?

6. Above what temperature should the MSA Petroleum Hydrocarbon tube **not** be used? Is this higher or lower than other MSA detector tubes? _____ .

7. A Dräger Ethyl Acetate 200/a tube turns pale green after 10 ps. What does this suggest? What concentration is present? _____ .

8. When using a Dräger Cyanide 2/a tube, a red stain is noted before breaking the ampule. What does this suggest? _____ .

9. Which hydrogen sulfide tube should be used when sulfur dioxide (SO_2) is also present? _____ .

10. What is the actual concentration present of n-amyl alcohol when the MSA Alcohol tube shows a 15 mm green stain after 10 ps? The temperature is 110°F, relative humidity 80%._____ .

4 Combustible Gas Indicators

LEARNING OBJECTIVES

1. Describe the various methods used to measure combustible gas.
2. List the three different measurement ranges available.
3. Recognize the difference between meter readings and actual concentrations.
4. List specific accessories that may be helpful when using a combustible gas indicator or interpreting readings.
5. Describe the limitations of combustible gas indicators.
6. Given a meter reading, calculate the approximate actual concentration present using response curves or conversion factors, and determine the relative risk to responders.

The first portable combustible gas indicators were developed to combat flammable gases encountered in underground mines in Great Britain. These flammable gases were commonly called "fire damp" by miners, but were known to be light hydrocarbon gases similar to marsh gas by scientists of the day. It was also recognized that fire damp, when mixed with seven or eight

Igniting the fire-damp. (A Bettmann Archives photo.)

times its volume of air, became highly explosive, and burned with a pale blue flame. Interestingly, it was also acknowledged that after a fire damp explosion, the air was rendered unfit to breathe, the resulting "choke damp" often producing more fatalities than the actual explosion[1].

The Geordy Lamp, developed around 1815, consisted of a small flame on an oil-fed wick in a glass cylinder; at the top of the cylinder was a perforated metallic chimney. As long as the flame was lit, air entered the lamp only through small holes at the bottom of the cylinder. The Davy Safety Lamp, also introduced about 1815, had a similar design; it had a very fine wire mesh cylinder instead of glass. When flammable gases were present, the flame burned blue; as the concentration of gas increased, the height of the flame gradually increased to fill the entire cylinder. If insufficient oxygen was present, the flame would not remain lit[2].

Another gas indicator took advantage of the principle of osmosis. A very thin natural rubber ball filled with air was placed directly under a lever; any upward pressure on the lever caused a warning bell to sound. When present, flammable gas entered the ball through the rubber, causing it to swell and ring the warning bell[1]. This indicator detected flammable gas at non-explosive concentrations and because it had an audible alarm, it did not have to be constantly watched.

A variety of combustible gas indicators (CGIs) for measuring combustible gases and vapors are available today. CGIs are capable of detecting the presence of flammable gases in the ppm, %LEL, and % gas by volume in air. The most commonly used meters measure the concentration of gas in the %LEL range to determine if there is a significant risk for fire or explosion.

HOW A COMBUSTIBLE GAS SENSOR WORKS

Catalytic Sensors

Nearly all %LEL CGIs are based on catalytic combustion of gases on a filament. The catalytic filament and compensating filament are incorporated into a basic Wheatstone bridge. Most instrument manuals and many texts give an illustration of a Wheatstone bridge similar to the one below. Frankly, we have found most explanations of catalytic sensors somewhat lacking; this is unfortunate, since one cannot really understand the limitations and use of a CGI without first understanding how it detects combustible gases and vapors.

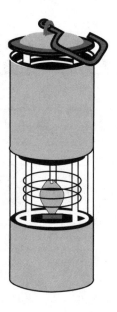

Figure 4-1: Line drawing of a safety lamp used to warn miners of flammable gases and oxygen deficiency.

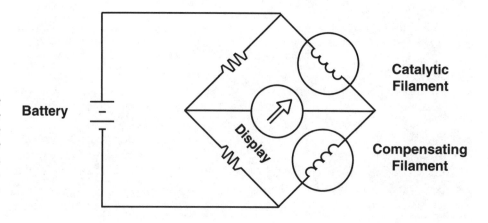

Figure 4-2: A Wheatstone bridge circuit in a catalytic combustible gas sensor. Both filaments are incorporated into the circuit. The circuit uses the principle that a wire's electrical resistance increases with increasing temperature.

Battery

Display

Catalytic Filament

Compensating Filament

The catalytic sensor in a CGI is quite small, with a surface area for gas diffusion usually no larger than the size of a quarter. The sensor is compact and sturdily built, with a strong exterior housing; a coarse metal filter and flame arrestor covers the top of the sensor. The filter allows the sample air to diffuse into the sensor. Although sample air is never pushed into the sensor, a battery-operated pump or aspirator bulb may be used to draw air into the proximity of the sensor.

There are two filaments inside the sensor. The filaments are separated from each other by an interior wall or are positioned in individual wells. One filament, often called the sensing filament, is coated with a catalyst, or contains a catalytic bead; the catalyst facilitates oxidation or combustion of very low concentrations of gas. The other filament has no catalyst and is called the compensating element because it compensates for ambient conditions such as temperature and humidity.

When the meter is turned on, the battery supplies current to the Wheatstone bridge circuit and heats the filaments to a very high temperature. Combustible gases diffuse through the filter and come in contact with the super-heated filaments. Both filaments are heated to the same temperature. Gases that can be oxidized by the catalyst will actually burn on the catalytic filament; this causes the temperature of the filament to increase. The gas does not burn on the compensating filament and there is no increase in temperature. The increase in temperature of the catalytic filament produces an increase in resistance and a decrease in current flow relative to the compensating element. This change in current is translated by the Wheatstone bridge circuit into a meter reading.

Figure 4-3: An intact MSA catalytic CGI sensor and another MSA sensor opened to show the two filaments. (Photo by Steve Napolitano.)

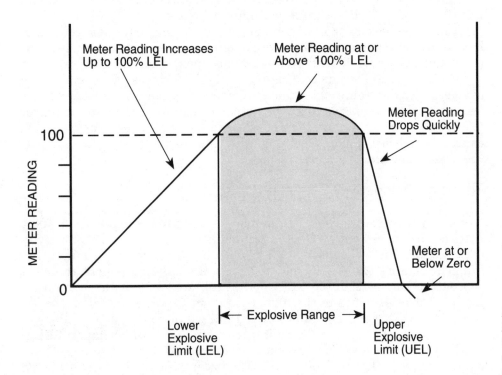

Figure 4-5: % LEL meter response to increasing concentrations of calibrant gas. Above the UEL, the air-gas mixture is too rich and combustion cannot occur on the catalytic filament.

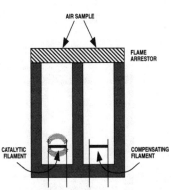

Figure 4-4: Line drawing of a % LEL combustible gas sensor. Combustion occurs only on the catalytic filament. There is no combustion on the compensating filament. The difference in the temperature between the two filaments produces a meter reading.

Figure 4-6: Typical readout display of a methane-calibrated %GAS meter. The top scale displays concentration measured by the catalytic sensor and reads as either 0 to 100% LEL or 0 to 5% gas; the lower scale is expanded to read up to 100% gas. (Photo by Steve Napolitano.)

Semiconductor or Solid State Sensors

Solid state sensors utilize a semiconducting material to detect combustible gases and vapors. A semiconductor has the electrical conductivity that is in between a conductor and an insulator.

A **conductor** is any material that has many free electrons and allows electrons to move through it easily. Electrons are the smallest part of an atom and have a negative charge. **Electricity** is defined as the flow of electrons along a conductor. Copper is a very good conductor and is used extensively to carry **electric current**, which is an organized flow of electrons in **one direction**. For current to flow, an electromotive force or emf must be exerted on electrons to move them in one direction. This force is measured in **volts** and also known as the **potential** or potential difference between two points.

Insulators are used to cover conductors to prevent current from going to the wrong place. An insulator is any material that has very few free electrons and inhibits electron movement. Dry wood, glass, rubber, and some plastics are good insulators.

A common semiconductor is silica. In its pure crystalline form, silica atoms form a lattice of atoms with very few free electrons available; hence, pure silica is a poor conductor and a good insulator. When a tiny amount of an impurity with excess electrons is added, however, the conductivity of silica is dramatically increased.

The impurities added to silica are called doping agents; commonly used impurities are arsenic, antimony, bismuth, and phosphorous. The type and number of impurities, as well as the temperature of the semiconductor and the voltage applied, affect the selectivity and sensitivity of the sensor.

Most solid state sensors consist of a semiconductor coated with a metallic oxide such as zinc oxide or aluminum oxide; these sensors are called metallic oxide sensors or MOS. Heater coils embedded in the semiconductor maintain a constant temperature which controls the number of free electrons and also prevents condensation of water vapor on the semiconductor surface.

The metallic oxide coating has a highly porous surface onto which oxygen molecules from ambient air are trapped. Voltage is applied across the sensor which is connected to a Wheatstone bridge circuit. When the sensor is exposed to contaminants, the gas reacts with the trapped oxygen; this causes the release of more electrons, which decreases the resistance across the bridge, and elicits a meter response.

Thermal Conductivity Sensors

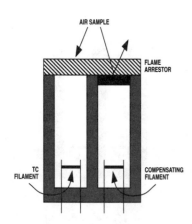

Figure 4-7: Line drawing of a % GAS sensor. Cooling occurs only on the thermal conductivity filament. The compensating filament does not come in contact with the test atmosphere. The relative decrease in the temperature produces a meter reading.

High concentrations of flammable gases, that is greater than 100% LEL, are measured with %GAS instruments using a thermal conductivity (TC) sensor. The TC sensor is often used in conjuction with a catalytic filament sensor. The %GAS meter utilizes a catalytic filament to detect concentrations less than 100% LEL of the calibrant gas; at higher concentrations the user switches to the higher range mode which turns on the TC sensor to measure concentrations up to 100% gas by volume of the calibrant.

The meter in Figure 4-6 is calibrated to methane, which has an LEL of 5% by volume in air. The top scale readings of 1 through 5% GAS are equivalent to 20 through 100% LEL. The lower scale reads between 5 and 100% by volume. The meter can therefore be used to indicate %LEL as well as %GAS concentrations.

The %GAS sensor uses a heated thermal conductivity (TC) filament instead of a catalytic filament. Just as in the catalytic sensor, the %GAS sensor has two filaments; each is connected into a Wheatstone bridge circuit. Both filaments are designed to maintain a high, stable temperature. This time, however, the compensating filament is sealed to prevent sample air from entering. Combustible gases cool the TC filament, decreasing its resistance. As the concentration of gas increases, the TC filament becomes cooler than the compensating filament; this decreases the resistance across the Wheatstone bridge circuit and produces a meter reading.

Sensor Location

The location of the combustible gas sensor varies depending on the type of meter employed. The sensor in sample-draw instruments is located well inside the instrument and is not in direct contact with ambient air. Sample air is drawn through a hose and into the instrument to the sensor using a

Figure 4-8: Examples of meters with sensors located inside the instrument (left to right): Scott D-16 meter with bulb aspirator; MSA Minigard diffusion type meter; Heath combination meter with battery-powered vacuum pump. (Photo by Steve Napolitano.)

bulb aspirator or battery-powered vacuum pump. In diffusion-sensor meters, the sensor is located within the instrument, but the sensor membrane is in direct contact with ambient air. No pump is used to bring the sample into the sensor; the meter itself is carried into the atmosphere to be sampled. Many diffusion instruments are small, compact, and designed as personal monitors to be worn on the belt, slung over the shoulder, or hand-carried. A remote sampling attachment which fits over the sensor is available as an accessory to allow testing of an atmosphere prior to entry.

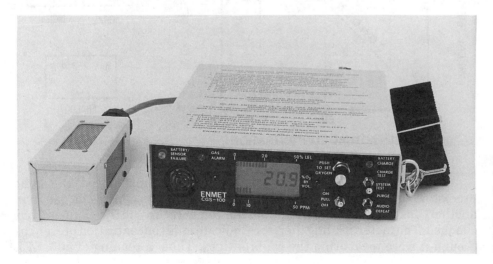

Figure 4-9: The Enmet CGS-100. The CGI sensor is connected to the instrument by a sensor cable allowing the sensor to come into direct contact with the sample atmosphere. (Courtesy Enmet Corp.)

Another type of diffusion sensor instrument is used in situations when it is not practical to use an aspirating pump, or it is preferable that only the sensor come in direct contact with the atmosphere, rather than the entire instrument. In these cases, a diffusion-type meter with an exterior sensor module connected to the instrument by a cable is used. This type of remote sensor design is especially valuable when sampling hot atmospheres.

WHAT THE READINGS REALLY MEAN

All combustible gas indicator readings are relative to the **calibrant gas**. The most commonly employed calibrant gases for %LEL and combination %LEL/ppm meters are methane, pentane, propane, and hexane; methane, propane and natural gas are typically used for %GAS meters. Combustible gas indicators are calibrated to one gas at the factory; the meter will accurately read this calibrant or reference gas throughout the range of the meter. When dealing with the calibrant gas, a properly functioning meter will correctly measure the actual concentration of gas present; thus when methane is known to be present, a 50% LEL reading on a methane-calibrated instrument indicates that the concentration of methane present is actually 50% of the LEL, or 2.5% gas by volume.

The meter reading corresponds to the resistance change produced by the calibrant gas as it reacts with the sensing filament. The greater the change in temperature, the greater the change in resistance. The greater the change in resistance, the greater the concentration present. The easiest way to think about meter response is that the reading really indicates a change in the temperature of the sensing filament.

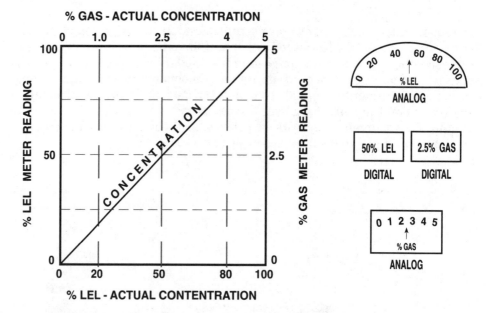

Figure 4-10: A typical calibration curve for a methane-calibrated catalytic filament/%LEL instrument illustrating typical meter response as concentration of methane increases from 0 to 100% LEL which is equivalent to 0 to 5% GAS. The meter response to 50% LEL or 2.5% gas may be shown on an analog or digital display.

When another gas that is not the calibrant is introduced, the same process occurs within the sensor. The meter cannot recognize one gas from another! Instead, it recognizes that the temperature of the filament has changed. The meter reading therefore indicates the concentration present as though it were measuring the calibrant gas. Thus, a reading of 50% LEL indicates that the temperature of the catalytic filament, and the corresponding change in resistance, is the same as when 50% LEL of the calibrant gas is present.

%LEL Meters

Remember that most %LEL meters use the catalytic filament sensor. For these instruments, the meter reading represents how **hot** the filament gets as the combustible gas interacts with the catalyst and burns.

Some gases are more compatible with the catalyst and burn more readily on the filament than the calibrant gas; other gases release proportionately more heat than the calibrant gas. These gases and vapors will produce a hotter filament at a lower concentration, relative to the calibrant gas. When compared to the calibration gas then, these materials can be considered **hot burning gases** and vapors. When these gases and vapors are present, the catalytic filament heats more readily and the meter will give a %LEL reading that is **greater** than the actual concentration present.

Other gases and vapors burn less readily on the catalytic filament, or they release proportionately less heat when burned, relative to the calibrant gas. These materials can be considered **cool burning gases** and vapors. Since these materials produce less heat, a higher concentration is required, relative to the calibration gas, to get a significant increase in catalytic filament temperature. When cool burning gases are present, the meter will give a reading that is **less** than the actual %LEL present.

The concept of hot and cool burning gases and vapors can be helpful in understanding how the instrument responds to different gases and vapors. It is important to remember, however, that instrument response to individual gases and vapors depends not only on which gas the instrument is calibrated to, but also on the type of catalyst employed in the sensor.

Response Curves or Conversion Factors

The concept of hot and cool burning gases and vapors is also helpful in understanding instrument response curves or conversion factors. Many manufacturers of %LEL meters provide response curves or conversion factors which indicate the meter response to individual gases and vapors throughout their LEL range (i.e. 0-100% LEL). Response curves or factors are determined by the manufacturer by introducing a known concentration of gas or vapor into a series of factory-calibrated instruments, and recording the meter reading or the change in resistance.

Response curves indicate the meter response to a particular gas or vapor throughout the entire LEL range. The calibration curve depicted in Figure 4-10 can be considered a response curve for the calibrant gas, which in this case is methane.

If the identity of the gas is known, a response curve can be used to approximate the actual concentration present. For instance, in Figure 4-11, a meter reading of 50% for Gas A corresponds to an actual concentration of approximately 30% LEL; the same meter reading represents an actual concentration of 90% LEL when Gas D is present.

A response or conversion factor is a number which describes the meter response to a particular gas or vapor. While conversion factors are determined in the same way as response curves, the manufacturer provides a number rather than a curve. Conversion factors are usually carried out to one decimal place, that is, 0.8, 1.5, 2.6.

Conversion factors are also used to determine the approximate actual concentration present by multiplying the meter response by the factor. Conversion factors are available for %LEL as well as ppm-equivalent meters. For instance, if the meter reading obtained for a gas is 20% LEL, and the conversion factor is 0.8, then the approximate actual concentration present is calculated as follows:

Meter Reading	x	Conversion Factor	=	Actual Concentration
20% LEL	x	0.8	=	16% LEL

Conversion factors can be derived from response curves by simply dividing the actual concentration present by the meter reading. Look at Curve A and C in Figure 4-12, and follow along:

Curve A:

Meter Reading	Actual Concentration	Conversion Factor
20% LEL	12% LEL	12/20 = 0.60
50% LEL	30% LEL	30/50 = 0.60
75% LEL	50% LEL	50/75 = 0.66
90% LEL	60% LEL	60/90 = 0.66

Curve C:

Meter Reading	Actual Concentration	Conversion Factor
20% LEL	20% LEL	20/20 = 1.0
50% LEL	60% LEL	60/50 = 1.2
75% LEL	100% LEL	100/75 = 1.3

Notice that response Curve A for the MSA Model 100 falls to the left or above the calibration gas curve in Figure 4-11. Gas A for this meter can therefore be considered a hot burning gas; Gas A yields a response factor of less than 1.0. Gas C, on the other hand, falls to the right or below the calibration gas curve; Gas C can therefore be considered a cool-burning gas which yields a response factor which is greater than 1.0. Thus it is possible to recognize hot and cool burning gases by examining their conversion factors. Hot burning gases give a meter reading that is higher than the actual concentration present and therefore have conversion factors less than 1.0. Cool burning gases have conversion factors greater than 1.0 since they give a meter reading that is less than is actually present.

It should be noted that a conversion factor is the number which **best** describes the response curve. In the example given above, a conversion factor of 0.6 can be used to describe Curve A, however, no single number can be used to define Curve C. Newer generation %LEL meters are designed to linearize or straighten out curves; nevertheless, most conversion factors do not accurately represent the **entire** LEL range. Instead, the manufacturer assigns a factor which represents the greatest portion of the curve, and/or which affords the most protection to the user. For example, most manufacturers would assign a conversion factor of at least 1.2 for Curve C in Figure 4-12. This would result in the user **over**estimating the actual concentration of gas for approximately one-half the LEL range.

Response curves and conversion factors are derived using multiple concentrations of gases on representative factory-calibrated instruments. Curves and factors represent the typical response of a particular make and model instrument. Individual instruments may give slightly different responses. The deviation in accuracy of conversion factors and response curves may be as great as 20% of the actual concentration. For example, if Vapor B gives a meter reading of 10% LEL, and the conversion factor is 0.5, then the actual concentration present is 10% x 0.5 or 5% LEL. If the conversion factor has a deviation of up to 20% or 1/5, then the actual approximate concentration range present is 5% ± 1% LEL or between 4% and 6% LEL.

Manufacturer-supplied curves and factors should be used only with the model for which they were generated. Factors and curves vary dramatically when different factory-calibrants are utilized; make sure your model is calibrated to the gas specified by the manufacturer. It is not appropriate, and may be dangerous, to interchange response curves and conversion factors between models or manufacturers. Similarly, curves and factors apply only when the identity of the gas or vapor is known and no other combustible gases or vapors are present.

There may be situations when there are several gases known to be present. It would be prudent to check the conversion factor or response curve for each gas to determine if it is hot or cool burning. If all the gases present have similar response characteristics, i.e. they are all hot burning or cool burning, this information can be used when qualitatively interpreting meter readings. For example, suppose a spilled solvent mixture contains acetone, methanol, and xylene; the response factors for the instrument available are acetone: 1.1, methanol: 1.2, and xylene: 1.5. Since the response factors are all greater than 1.0, the instrument user knows that all three gases are hot burning relative to the calibrant gas. The meter reading will therefore be **less** than than the actual concentration present. On the other hand, if all the response factors are less than 1.0, the operator knows the gases are cool burning and the meter reading will be greater than the actual concentration present.

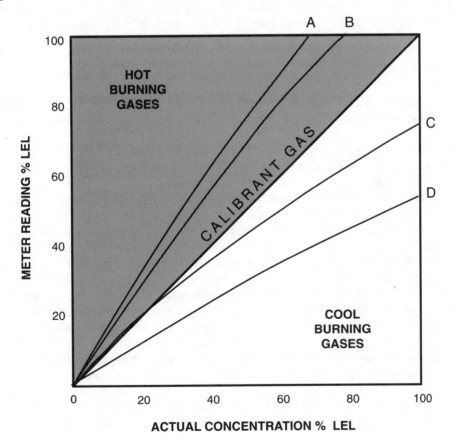

Figure 4-11: Response curves of different gases for a pentane-calibrated MSA Model 100 Combustible Gas Indicator. Note that gases A and B are hot burning and therefore lie above the calibrant curve. Gases C and D are cooler burning and fall below the calibrant curve.

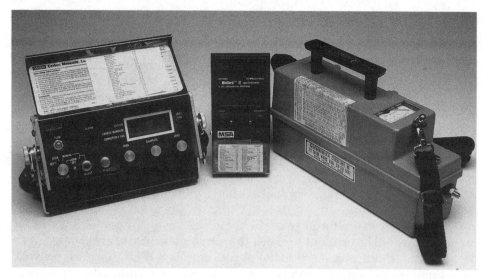

Figure 4-12: Conversion factor charts have been reduced and taped to each meter for easy reference. (Photo by Steve Napolitano.)

Remember that without response curves or conversion factors, your CGI will be unable to give information regarding the actual concentration of a gas known to be present. Curves and factors should always accompany the meter and be available for easy reference.

Most reputable CGI manufacturers supply at least 10-15 conversion factors or curves for their %LEL meters. Typical gases and vapors for which curves or factors are supplied include methane, propane, butane, hexane, benzene or another aromatic hydrocarbon, methanol, ethanol, ethyl acetate, acetone, acetylene, ethylene oxide, and hydrogen. If you do not have factors or curves for your CGI, contact the manufacturer and request them; this information should be supplied at the time of purchase. Response curves and conversion factors from several manufacturers are available in Appendix B and C.

%Gas Meters

Instruments which measure percent by volume concentrations of combustible gases are used whenever concentrations above 100% LEL are encountered or anticipated. Typical situations where a %GAS meter may be employed include gas pipelines, distribution facilities for LPG, butane, or propane, tank farms, and landfills. It is not unusual to find methane concentrations up to 40-60% by volume at landfills.

The %GAS CGI utilizes a thermal conductivity (TC) filament sensor rather than the catalytic filament sensor employed in the %LEL meter. The %GAS sensor is composed of two separate, heated filaments; both filaments are heated to the same temperature. Combustible gases enter only the TC filament side of the sensor, the compensating filament is sealed and remains heated to a fixed, constant temperature. Incoming gases cool the TC filament; as the temperature of the filament decreases, the resistance across the Wheatstone bridge also decreases, resulting in a meter response. In this case, the meter reading represents how **cool** the TC filament gets relative to the hot compensating filament. Since burning or oxidation is not involved, **%GAS instruments do not require oxygen for a valid reading**.

Combustible gases and vapors differ in their ability to cool or absorb the heat of the TC filament. The gases used as %GAS calibrants, such as methane, natural gas, and propane, absorb heat well and thus readily cool the filament. Gases which cool the filament better than the calibrant will indicate a higher %GAS reading than is actually present. Conversely, gases and vapors which are poor heat absorbers will not cool the filament as well, and will therefore indicate a lower %GAS reading than is actually present.

It is important to recognize that gases which cool the TC filament will elicit a meter response even if they are not flammable. The major limiting gas of many %GAS meters is carbon dioxide (CO_2), which is often used as an inerting gas. Carbon dioxide, as well as other gases which absorb heat readily, can therefore produce a false positive reading. Nitrogen, also used as an inerting gas, does not effectively cool the TC filament and therefore does not produce a meter response.

Meter sensitivity to carbon dioxide and other cooling gases vary between manufacturers; this information should be included in the instruction manual or product bulletin. Some meters, such as the MSA Tankscope®, are designed for use in atmospheres inerted with carbon dioxide.

PPM Meters

The sensitivity of a CGI to detect and reliably measure ppm concentrations is dependent on the type of sensor employed and the calibrant. Catalytic sensor instruments with a separate ppm range are usually sensitive to concentrations as low as 0.5% to 1% LEL of the calibrant gas. A hexane calibrated CGI should be capable of reliably measuring concentrations as low as 50 to 110 ppm hexane, which is equivalent to 0.5% to 1% of the LEL of hexane (the LEL of hexane is 1.1% by volume or 11,000 ppm, 1% of 11,000 ppm is 110 ppm). Sensitivity can be enhanced by modifying the catalyst used, but this can affect meter response to non-flammable chlorinated hydrocarbons. Interpretation of meter response is the same as that for catalytic sensor %LEL meters.

It is imperative that the user check the instruction manual and determine the limit of sensitivity of the ppm range. That is, how low a concentration of calibration gas will the meter reliably detect? It is foolhardy to assume the instrument will read extremely low concentrations simply because the readout display starts at zero ppm. If this information is not in the instruction manual, check with the manufacturer.

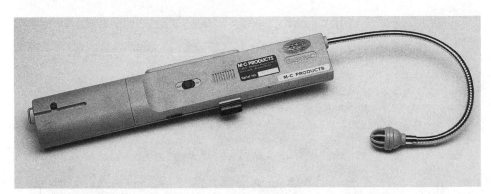

Figure 4-13: The GasTrac is a solid state sensor combustible gas indicator sensitive to methane concentrations as low as 10 ppm. The unit indicates the presence of gas by an audible and visual alarm.

Solid state sensors, on the other hand, are designed to respond to low concentrations of gases and vapors. Solid state sensors can respond to a wide range of combustible gases and vapors. Many solid state sensor instruments do not give a numerical meter reading, but rather use a bar graph, warning lights, variable rate audible tick response, or an audible alarm. Solid state combustible gas sensors can be sensitive to concentrations as low as 10 ppm of the calibrant gas.

When calibrating CGIs with multiple ranges, a separate calibration check should be performed on the ppm range; checking the %LEL range does not ensure the other ranges are functioning properly. Minimum recommended warm-up time prior to ppm range calibration may be as long as 30 minutes. In most cases, ppm-range CGIs are not considered emergency response instruments for initial use because of their long warm-up times. Meter readout of the ppm range will be erratic until warm-up is completed. Some meters give a continuous audible or visible alarm during warm-up which is automatically reset and shut off when warm-up is completed; calibration should not be attempted until the alarm is silenced.

ACCESSORIES

All CGI manufacturers offer a variety of accessories, some of which have already been mentioned. Some accessories are not really optional but are actually required to effectively use and protect the meter. All meters which use a bulb aspirator or battery-powered pump should be used with an **in-line particulate filter** and a **liquid trap**; liquid traps may also be called moisture traps or line traps. Make sure you are familiar with the design features of your meter; some have particulate filters located within the instrument which should be checked and changed periodically. Some manufacturers also offer a **hydrophobic filter** to retain water aerosols. In most cases, a hydrophobic filter will not adequately protect the meter if liquids are drawn into the sample line.

Another valuable accessory item is the **activated charcoal filter**. Activated charcoal adsorbs complex hydrocarbons such as benzene, hexane, and methylene chloride; light hydrocarbon gases such as methane and natural gas are not adsorbed and their progress through the filter is therefore not affected. By using a charcoal filter, the type of combustible gas present can be characterized as a light hydrocarbon gas or a heavier, more complex organic material.

Figure 4-14: Examples of liquid traps available from different manufacturers (left to right) GasTech Tank-Techtor, MSA 260, Bacharach Sniffer 303, MSA Model 60 Gascope. (Photo by Steve Napolitano.)

A few manufacturers offer **dilution tubes** or **diluter fittings** which mix incoming sample air with clean ambient air surrounding the meter. MSA offers several models with dilution ratio options of 1:1, 10:1, or 20:1. Most manufacturers offer only a 1:1 option, which results in mixture of 1 part sample air to 1 part clean air, or a 50% dilution. When using a diluter, the ratio must be known in order to determine the actual meter reading. For example, a 5% LEL meter reading obtained with a 1:1 diluter represents only one half the meter reading that would be obtained without dilution; the corrected meter reading, then, is 10% LEL.

Extension hoses are used to sample remote areas such as tanks, sewers, basements, and other confined spaces prior to entry. Flexible extension hoses come in varying lengths, from 3 feet to 50 feet, and may be made of natural or neoprene rubber, teflon, polyethylene, and other man-made materials; teflon-lined hoses are also available. Do not use excessive hose lengths; maximum length should be specified by the manufacturer. Remember that initial and final response times will be increased when using a long extension hose.

Inflexible **sample probes** are offered in a variety of length, from 10 inches up to 4 feet; probes are valuable for sampling inaccessible areas under doors, in cracks and crevices, within small diameter pipes, or through vent holes in manhole covers. Probe composition may be aluminium, brass, copper, and non-conducting plastic or fiberglass. It is prudent to have several types of probe lengths and composition available, since one probe will not be effective for all types of situations. Many manufacturers suggest the use of a 3 or 4 foot solid probe with inlet holes in the first few inches of the probe end for situations encountered in sumps, sewers, and other locations where the probe may inadvertently come in contact with water.

Inhibitor filters are recommended for atmospheres known or suspected to be contaminated with leaded gasoline. The filter resembles an ammonia inhalant ampule. The inhibitor filter is activated by breaking the ampule just before use to release a material often found in moth flakes, para-dichlorobenzene. The chemical reacts with tetraethyl lead, producing a more

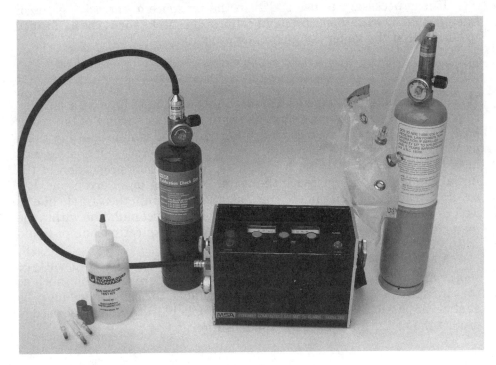

Figure 4-15: Examples of calibration equipment (left to right): Plastic bottle with glass ampule which releases calibrant gas into container when broken by metal weight, sample bag containing calibrant gas, pressurized cylinder of calibration gas with regulator.

Figure 4-16: The GasTech Tank-Techtor has a combination dilution fitting and sample line float. The dilution fitting automatically provides a 1:1 dilution of the incoming sample with ambient air. When the sample line is submerged (bottom), the float is lifted; this shuts off the pump and prevents water from being drawn into the meter. (Photo by Steve Napolitano.)

volatile product which is less susceptible to fume formation. Not all manufacturers supply inhibitor filters; check the manual to determine if they are available as an accessory item.

All reputable manufacturers supply **calibration gas kits**. Kit designs vary and are often dependent on the type of sensor and its design. Meters with diffusion-type sensors usually require an adaptor to isolate the sensor and prevent ambient air from diluting the calibrant gas. Meters which draw sample air with an aspirator bulb or pump typically use a pressurized cylinder with a regulator designed to deliver the gas at the appropriate flow rate. It is imperative that flow control conditions specified by the manufacturer be used during calibration.

GENERAL PRECAUTIONS AND LIMITATIONS

Inadequate Battery Power

As with every other battery-powered instrument, it is important that there be sufficient energy for proper function. It is unsafe and foolhardy to use a meter that has insufficient battery power. Remember that the battery supplies the energy to the sensors that are responsible for indicating a meter response. Inadequate battery power may result in a decreased or negligible meter response which can jeopardize personnel. All CGI users should know how to check battery condition as well as how to replace or recharge batteries, as appropriate. This information is found in the instruction manual.

Corrosive Gases

Corrosive gases can deteriorate filaments or solid state components, causing sensor failure. Thermal decomposition of halogenated hydrocarbons can also corrode the sensor. In some cases, corrosive damage does not immediately result in sensor failure, but rather decreased sensitivity.

If it is necessary to use a CGI in the presence of corrosive gases or halogenated hydrocarbons, frequent calibration checks should be performed to ensure that the meter is functioning properly.

Heated Atmospheres

Hot vapors and gases, when introduced into a sample hose at a location remote from the meter, may condense onto the inner surface of the cooler hose. The risk of gas or vapor condensation within the sample hose increases as ambient temperature decreases, sample gas temperature increases, and hose length increases. In such instances, a significant proportion of the initial sample may never reach the meter, resulting in an erroneous, low reading. Water and other liquids may also collect within the sample line as well as within the instrument itself. Condensation accumulation within the meter can interfere with proper function of the aspirating pump; liquid within the sensor may affect the flame arrestor.

Monitoring of heated samples should be performed with a diffusion type sensor module that can be lowered into the atmosphere on a cable, such as that in Figure 4-9. If such a meter is not available, the sample line must be as short as possible, and a water trap or hydrophobic filter should be used to protect the meter from excessive condensation.

Temperature

Recommended operating temperatures for CGIs can range from –20°C to +50°C (–4°F to +122°F); a more typical operating range, however, is 0 to 40°C (32 to 104°F). Battery life is significantly decreased under extremely cold conditions, and meter response may be sluggish. Diffusion-type instruments which have the sensor exposed to ambient air appear to be more susceptible to cold, windy conditions, which tend to cool the sensor and require more battery power to maintain proper sensor temperature. Sample atmospheres drawn from below ground may be warmer than the ambient temperature, producing condensation within the colder sampling hose.

When operating a meter under temperature extremes, the initial calibration check should be made under the conditions of use, i.e. the meter should be allowed to cool down or heat up to the ambient temperature. Calibration should then be rechecked throughout the use interval to ensure that the meter is still functioning properly. Do **not** use an instrument that cannot be calibrated under ambient conditions.

Operating temperatures vary between manufacturers and are **usually** specified by the manufacturer in the operating manual; incredibly, some manufacturers do not furnish this information. This is another reason to always examine the instruction manual before purchasing an instrument.

Electrical Interferences

Magnetic fields, high voltage wires, static electricity, even radios and cellular telephones, can interfere with meter readout. When such interferants are present, the meter display can fluctuate wildly, often showing a positive and then a negative reading. Magnetic and electromagnetic fields may be encountered in mines, buildings, utility vaults, and sewers. Static electricity can be generated by the user's clothing, walking across certain types of surfaces, or by equipment such as conveyor belts and pneumatic tools.

In some instances, electrical interferences can be eliminated by grounding the user and instrument, moving the meter to a different location and sampling the atmosphere remotely, or shielding the instrument by surrounding the meter with grounded, conductive material, such as chicken wire, which absorbs stray electromagnetic waves.

Liquids and Particulates

Remember that the CGI we are discussing is the type that detects combustible gases and vapors. **They are not designed to detect or to be used in the presence of combustible dusts, fibers, mists, or aerosols**.

Whenever possible, particulates and liquids should not be allowed to enter the instrument. Always use a particulate filter and liquid trap or hydrophobic filter when operating a CGI. When using filters and traps, check them frequently and replace or empty them as appropriate. Never operate a CGI with liquid in the trap; if the meter or trap is tilted, the liquid may enter the instrument and disable the sensor and pump. If liquid is drawn into a trap, remove and empty the trap immediately. If the liquid is a flammable or combustible material, the trap should then be thoroughly cleaned to remove residual material that may continue to off-gas and give false positive readings. A common source of false positive or fluctuating background readings is a contaminated liquid trap.

LIMITATIONS OF %LEL METERS

Oxygen Concentration

Most %LEL meters use a catalytic filament sensor; the gas in the incoming sample air is actually oxidized or burned on the filament as it comes in contact with the catalyst. Oxygen is required for oxidation to occur. All CGI users should know the minimum concentration of oxygen below which oxidation cannot occur on the catalytic filament; at this concentration the meter will give no response regardless of the concentration of combustible gas present. It is also important to know the minimum oxygen concentration that does not affect meter performance; that is, the lowest oxygen concentration at which the meter can be calibrated. Meter response will be decreased, since oxidation will be inhibited, at oxygen concentrations between these two values.

It is dangerous to use a meter without knowing this information! A CGI can give false readings when oxygen levels are outside its operating range. Oxygen requirements vary between manufacturers and are dependent upon the type of sensor and catalyst(s) employed. For instance, the following oxygen concentrations are pertinent for the user of the MSA Model 260, which has a catalytic filament sensor[3]:

- less than 10%: instrument should not be used; there is insufficient oxygen present for meter response.
- greater than 10% and less than 15%: meter response will be decreased; the lower the oxygen concentration, the greater the decrease.
- 15% up to 20.9%: optimal oxygen concentration range; meter response will be normal (instrument can be calibrated at concentrations as low as 15% oxygen).
- 21% to 25%: meter response will be exaggerated due to enhanced oxidation on the catalytic filament; the higher the oxygen concentration the greater the effect on the reading.
- greater than 25%: meter should not be used; there is so much oxygen present that sensor can be damaged.

Catalyst Poisons

Organic materials which contain organic heavy metals, silicones, and silicates form fine fumes when oxidized or burned. A fume is a mixture of vapors or gases and minute solid particles. These particles settle onto the filament and cover the catalyst. Eventually, so much of the catalyst is covered that the sensor no longer functions.

Silicon-containing compounds may be found in hydraulic fluids, surfactants, waxes and polishes, and some firefighting and vapor-suppression foams. A common heavy metal is tetraethyl lead found in leaded gasolines. Fume formation occurs when gases and vapors formed after catalytic combustion rise off the heated filament and begin to cool. As the vapors cool, condensation produces tiny particles which then fall back onto the filament.

Susceptibility to catalyst poisons depends on the temperature of the filament. Sensors which utilize very hot temperatures are less susceptible to poisons since the vapors and gases produced are hotter and therefore less susceptible to condensation. Instrument manufacturers usually supply information regarding sensitivity to catalyst poisons; sensors resistant to a particular poison, such as silicone or organic metals, are available from some manufacturers.

Concentrations Exceeding 100%LEL

Meter response to combustible gas concentrations above 100% LEL may vary between manufacturers and between old and new instruments. Remember that in the catalytic filament sensor, both filaments are heated. One filament has the catalyst, while the compensating filament does not. At concentrations less than the LEL, the catalyst allows the gas to burn. Once the actual concentration reaches the flammable range (i.e. between 100% LEL and the UEL), the gas burns on **both** filaments resulting in a reduced or negligible difference in temperature between the filaments.

The Wheatstone Bridge circuit in some instruments is designed to detect an increase in temperature across both filaments. In this case, at concentrations within the flammable range, the meter response will remain at 100% LEL or indicate a reading greater than 100%.

Other instruments, however, are designed only to indicate the relative difference in temperature between the catalytic and compensating filaments. In this case, when both filaments reach the same temperature the meter response will return to zero, even when both filaments are heated to temperatures hotter than normal. In such a situation, the meter will initially show a reading up to 100% LEL, followed by a rapid decrease to 0%.

Some texts suggest that the meter reading in this situation must go to 100% and then drop back to zero. That is not always correct. Remember that not all gases release sufficient heat to elicit a meter reading of 100%! A gas which is cool-burning relative to the calibrant may produce a meter reading of only 60% or 80% LEL, even though the actual concentration present is greater than 100% LEL.

The danger to personnel in such a situation is obvious. It is imperative that all users of catalytic sensor LEL meters know the response of their instrument when the actual concentration present exceeds 100% LEL. This information can be found in the instruction manual which should always accompany the instrument.

Concentrations Exceeding the UEL

When the concentration of gas exceeds the UEL, the mixture is too rich to burn and oxidation on the catalytic and compensating filaments can no longer occur (see Figure 4-5). The temperature of both filaments will decrease and return to normal. In this case, there is no increase in temperature across either filament and the meter will indicate that there is no gas present. In some situations, the instrument may initially indicate a response up to 100% LEL and then fall back to a lesser number or zero. When very high concentrations are present, the meter reading may start to increase, and then quickly return to zero.

In situations where concentrations of gas are suspected to be within or in excess of the flammable range, a %GAS meter should be used. The %GAS CGI utilizes a thermal conductivity sensor rather than a catalytic sensor, and is designed to detect high concentrations.

Chlorinated Hydrocarbons

Some instruments utilize catalysts which enhance the oxidation of chlorinated hydrocarbon solvents, such as methylene chloride, trichloroethylene, and perchloroethylene. This can result in an exaggerated instrument response which indicates a flammable condition where none exists[4].

High concentrations of chlorinated hydrocarbons may also overwhelm the catalyst, resulting in decreased sensitivity to other gases and vapors for a period of time after exposure. When this occurs, the instrument should be allowed to run for a period of time in clean air; the calibration should then be checked. If there are unknown gases and vapors present, or if the presence of chlorinated hydrocarbons is suspected, frequent calibration checks are warranted to demonstrate continued proper functioning of the meter.

Other gases, such as acetylene, may also produce a response similar to chlorinated hydrocarbons. All catalytic sensors do not respond in the same way; susceptibility to chlorinated hydrocarbons or other gases and vapors should be noted in the limitations section of the instrument manual.

Oxygen-Acetylene Mixtures

Most %LEL meters are not designed to measure mixtures used with oxy-acetylene welding equipment. Such oxygen-rich mixtures produce excessive temperatures on the catalytic filament which can damage the sensor. Some inherent safety warranties forbid meter use in atmospheres containing oxy-acetylene gas.

There are meters designed for use in oxy-acetylene atmospheres such as the MSA Model 4 Explosimeter®. The MSA Model 3 is recommended for use in oxygen-enriched hydrogen atmospheres.

Dealing With Limitations

Recognizing the limitations of a %LEL instrument is important for user safety. In some cases, when the user suspects the meter is giving a false reading, the most prudent action may be to leave the area. In other cases, however, a meter with a different sensor or another type of instrument may be useful to overcome the limitation which is producing the problem.

For instance, a %GAS meter will give a valid response in the presence of combustible gas concentrations which would overwhelm a %LEL meter. If the user suspects that very high concentrations of gas are present, then the use of such a meter is warranted.

If the test atmosphere is being sampled from a remote location, there are %LEL meters available such as the GasTech Tank-Techtor, MSA Explosimeter®, and MSA Model 40 Combustible Gas Indicator that can be equipped with dilution fittings. The diluter mixes the incoming sample with ambient air from around the meter. Sample dilution is very useful in situations where combustible gas concentrations are above the LEL, or when oxygen deficient conditions exist. This option is especially valuable in confined spaces such as sewers, tanks, and pipe galleries.

CALIBRATION

Checking calibration of combustible gas indicators is usually simple and quick. Most manufacturers offer kits which include a known concentration of calibration gas as well as a method to deliver the gas to the instrument. The calibration check is a method to ensure that the meter is still within factory calibration standards. At the factory, the meter is calibrated with several known concentrations of calibrant gas. The calibration check utilizes a single concentration of gas, usually in the middle of the response range. If the meter responds appropriately to one concentration, we can be confident it will respond properly to other concentrations, since the calibration response is by definition a straight line.

It is prudent to check the calibrant gas container for the meter response range which is considered acceptable. For example, a pressured cylinder of calibration gas may read "Methane, 50% LEL ± 2% of value". Two percent of 50% LEL is 1% LEL (50% LEL x 0.02 = 1% LEL). This means that the actual concentration in the cylinder is between 49% and 51% LEL. A slightly higher or lower reading than 50% LEL may therefore reflect a slight discrepancy in the actual calibration gas concentration.

Most %LEL and %GAS meter sensors are very reliable and require little adjustment. Adjustment screws or controls are usually located inside the instrument housing; to reach these controls the meter often must be at least partially disassembled. If frequent adjustment becomes necessary, the sensor is beginning to fail and a new sensor should be ordered.

It is a waste of time to attempt to calibrate or adjust internal controls of a ppm CGI unless the meter is completely warmed up. Inadequate warm-up results in erratic readings and calibration will not be possible.

The calibration check is the **only** method available to verify that the meter is working properly. Watching the meter respond to butane gas from a cigarette lighter or gasoline vapors from a gas tank is **not** evidence that the meter is responding appropriately and such checks are **not** recommended by CGI manufacturers. Indeed, such methods may easily overwhelm the sensor, rendering the meter useless for a considerable period of time.

After the calibration check, we often test our meters against **known** concentrations of other commonly encountered gases to confirm the conversion factor or response curve for those materials. For example, we typically test our pentane- or hexane-calibrated meters against propane and methane calibration gases, and our methane-calibrated meters against pentane or hexane.

OTHER WAYS TO LOOK AT %LEL READINGS

Most CGIs in use today measure combustible gas in %LEL. The %LEL meter is, first and foremost, a **safety** meter, designed to detect hazardous concentrations of gases and vapors up to a meter reading of 100% LEL. When the calibrant gas is present, a reading of 100% LEL indicates that the LEL has been reached, and flammable or explosive concentrations of gas or vapor are present. Suppose that the calibrant gas is pentane, and pentane vapors are known to be present. A reading of 100% LEL indicates that the LEL has been reached, which is 1.5% by volume in air, or 15,000 ppm (1% by volume = 10,000 ppm). If a methane-calibrated instrument is exposed to methane, a reading of 100% LEL indicates that the LEL of methane has been reached, which is equivalent to 5% methane by volume in air, or 50,000 ppm.

Now suppose we are using the methane-calibrated meter and we expose the meter to pentane. The conversion factor for pentane is 0.6. A reading of 100% LEL now indicates that 60% LEL or 9,000 ppm of pentane are present (the LEL of pentane is 1.5% or 15,000 ppm; 0.6 x 15,000 ppm = 9,000 ppm). To perform this type of calculation and to determine the approximate actual concentration present in ppm, the identity and LEL of the material must be known, there must be no other flammable gases or vapors present, and there must be a response curve or conversion factor available for the material known to be present.

The calculations can be easily broken down into each part. In the beginning, always use pencil and paper; do not skip steps.

> Meter reading x conversion factor = actual %LEL present:
>
> 100% LEL x 0.6 = 60% LEL
>
> LEL x 10,000 = 100% LEL in ppm:
>
> 1.5 x 10,000 ppm = 15,000 ppm
>
> Actual %LEL x LEL in ppm = actual concentration present in ppm:
>
> 0.6 x 15,000 ppm = 9,000 ppm

Remember that the %LEL must be converted to a decimal value; it represents a percentage or fraction of 100% and must always be equal to or less than 1.0:

100% LEL	=	100/100	=	1.00
50% LEL	=	50/100	=	0.50
10% LEL	=	10/100	=	0.10
5% LEL	=	5/100	=	0.05

Here is another example; fill in the blanks with the correct numbers, get the answer, and then check your answer at the end of the problem:

At a methanol spill, a %LEL meter gives a reading of 5% LEL. The conversion factor for methanol is 2.0. The LEL of methanol is 6.0%. What is the actual approximate concentration of methanol present in %LEL and ppm?

Meter reading x conversion factor = actual %LEL present

_____ x _____ = _____

LEL x 10,000 = 100% LEL in ppm:

_____ x _____ = _____

Actual %LEL x LEL in ppm = actual concentration in ppm:

_____ x _____ = _____

The actual concentration of methanol present, based on the %LEL meter reading, is 10 %LEL or 6000 ppm. The conversion factor was **greater than 1.0**; this indicates that the actual %LEL present must be **greater than the meter reading** (in this case, twice the meter reading). Notice that a relatively low %LEL reading reflects a very large concentration present in ppm; this is because methyl alcohol or methanol has a fairly high LEL. Even a minimal response on an LEL meter may reflect a dangerous concentration in terms of human health effects to unprotected personnel.

Response curves and factors may be less accurate at low meter readings, i.e. between 1% and 10% LEL. The actual approximate concentration derived from low meter readings may be higher than the calculated value. At low meter readings, the actual approximate concentration obtained from the conversion factor or response curve can be doubled to avoid underestimating the actual concentration present.

Calculate how many ppm methanol are present when the same %LEL meter indicates a reading of 1% LEL (the conversion factor for methanol is 2.0, the LEL of methanol is 6% or 60,000 ppm).

$$Actual\ \%LEL\ x\ LEL\ in\ ppm\ =\ actual\ concentration\ in\ ppm$$

$$0.02\ \ x\ \ 60,000\ ppm\ \ =\ 1,200\ ppm\ methanol\ present$$

Since LEL meters are less accurate when reading low concentrations, the actual approximate concentration may be twice the calculated value, or 2400 ppm. The 8-hour OSHA and ACGIH TWA for methanol is 200 ppm; the 15-minute STEL is 250 ppm. Workers in the methanol spill area where the %LEL meter reads 1% LEL should be using respiratory and skin protection, since the actual approximate concentration present is at least 6 times the TWA and nearly 5 times the STEL. **Never assume that a low LEL reading indicates a safe condition in terms of potential health effects to exposed, unprotected personnel.**

An easier way to determine the approximate ppm concentration present is to first convert the actual %LEL value to ppm. For methanol, with an LEL of 6%, the following quick conversions can be done without the aid of a calculator, or even a pencil and paper:

$$100\%\ actual\ LEL = LEL\ x\ 10,000\ ppm = LEL\ in\ ppm$$

100% actual LEL	=	*60,000 ppm*
50% actual LEL	=	*30,000 ppm*
25% actual LEL	=	*15,000 ppm*
10% actual LEL	=	*6,000 ppm*
5% actual LEL	=	*3,000 ppm*
1% actual LEL	=	*600 ppm*

Thus, a meter response, once converted to the actual %LEL using a response factor or curve, can be quickly translated into ppm. **It is not necessary to calculate the exact number!** This method can be used to get an idea of the approximate concentration present. Remember that even within the optimal response range of 20 - 100% LEL, most response curves and conversion factors have an accuracy of ± 20%. The actual %LEL concentration calculated is not one exact number by rather a concentration **range**. For example, if the actual %LEL is determined to be 10%, then concentration present will be within the range of 8% LEL to 12% LEL (10% LEL ± 20% of the actual %LEL value calculated).

Calculate the approximate concentrations present in %LEL and ppm at a hexane spill where a %LEL meter gives a response of 20% LEL in one area and 4% LEL at another location. The LEL of hexane is 1.1% (11,000 ppm), the meter conversion factor for hexane is 1.3.

A 20% meter response represents an actual concentration of 26% LEL, since 20% x 1.3 = 26% LEL. After the approximate actual concentration is determined, then convert the %LEL value to ppm. The LEL of hexane is 1.1% by volume, or 11,000 ppm.

$$100\% \text{ actual } LEL = LEL \text{ x } 10,000 \text{ ppm} = LEL \text{ in ppm}$$

100% actual LEL	=	11,000 ppm
50% actual LEL	=	5,500 ppm
25% actual LEL	=	2,750 ppm
10% actual LEL	=	1,100 ppm
5% actual LEL	=	550 ppm
1% actual LEL	=	110 ppm

A 20% meter response represents an actual concentration of 26% LEL or approximately 2800 ppm. A reading of 4% LEL is approximately 5% actual LEL, or about 550 ppm. These values represent the middle of the concentration range. For those who wish to calculate more precise (but not necessarily more accurate) answers, the actual concentrations present are approximately 26% LEL or 2860 ppm, and 5.2% LEL or 572 ppm.

OTHER WAYS TO LOOK AT PPM READINGS

CGI meters which detect ppm concentrations of calibrant gas usually have conversion factors for a limited number of other gases and vapors. Calculating the conversion factor for these meters is performed exactly as described for %LEL instruments. There are now a number of meters which offer both the ppm and %LEL ranges. The conversion factor for the ppm scale is different than the factor for the %LEL range.

Meter readings are always relative to the calibration gas; it is incorrect to report a response as so-many ppm. Unless the meter is reading the calibrant gas, the response should be reported as ppm-equivalents. The difference in terminology may seem insignificant but it is not. Too often this distinction is lost, and responders assume that the meter reading is the actual concentration present, regardless of the identity of the material present.

A CGI with **both** the ppm and %LEL range calibrated to the **same** material can be a powerful tool. Such an instrument allows the user to detect concentrations less than 1% LEL meter reading; it also affords the opportunity to compare meter readings on both scales.

For example, suppose we respond to a reported propane gas leak; the local propane company's %GAS/%LEL meter, which is calibrated to propane, gives no response. The lack of response on the %LEL range suggests that the concentration present is less than the limit of sensitivity of the meter, which is usually 1-2% LEL of the calibrant gas. The LEL of propane is 2.1% gas by volume in air, or 21,000 ppm; 1% LEL is therefore equivalent to 210 ppm. A minimum concentration of 200-400 ppm propane are required, then, to produce a meter reading of 1-2% LEL.

$$LEL \text{ of propane} = 2.1\% \text{ by volume x } 10,000 = 21,000 \text{ ppm}$$

Therefore, 100% LEL	=	21,000 ppm
10% LEL	=	2,100 ppm
1% LEL	=	210 ppm

In this case we can use the %LEL range on our propane-calibrated meter to confirm the absence of a %LEL reading (after all, their meter may not be working properly!). We can then switch to the ppm scale on our propane-calibrated meter. The meter response on ppm scale is amplified, the limit of detection is usually around 1% of the LEL, or in this case, approximately 200 ppm.

Since our meter is calibrated to propane, our readings represent the actual approximate concentration present. In this case, if we detect no %LEL meter readings, the concentration of propane present must be less than 1-2% LEL or 200-400 ppm. A meter reading of less than approximately 200-400 on the ppm scale would confirm the results obtained from the %LEL range.

Suppose the propane company's meter was not working properly and our meter gives a reading of 4% LEL; approximately how many ppm should be obtained on the ppm range?

$$Since\ 1\%\ LEL\ =\ 210\ ppm$$
$$4\%\ LEL\ =\ 840\ ppm$$

The ppm range should therefore give a reading of approximately 800 ppm. A ppm reading significantly different than anticipated (i.e. a reading that is more than 2.5-fold different than expected), suggests that a gas other than propane, or mixture of gases, which may include propane, is present.

Remember that when reading close to the limit of detection of a CGI, the meter response may be off by a factor of two on the %LEL or ppm range. Instrument response should always be treated as an approximate concentration, which is often lower than the actual concentration present. The danger of underestimating the actual concentration present is greater when dealing with low meter readings and using conversion factors to measure gases and vapors other than the calibrant gas. When measuring gases and vapors other than the calibrant, a conversion factor must be used. Conversion factors for the LEL and ppm ranges will be different. When using a combination ppm-LEL CGI, a separate calibration check should be performed for each concentration range.

A meter reads 3% LEL and 1000 ppm-equivalents at an acetone spill. Meter conversion factors are 1.3 for the %LEL scale and 3.2 for the ppm range. Are these readings consistent? The LEL of acetone is 2.5% by volume or 25,000 ppm.

$$100\%\ LEL\ =\ 25,000\ ppm$$
$$10\%\ LEL\ =\ 2,500\ ppm$$
$$1\%\ LEL\ =\ 250\ ppm$$

Meter reading x conversion factor = actual concentration present

$$3\%\ LEL\ x\ 1.3\ =\ 3.9\ or\ 4\%\ LEL$$
$$1000\ ppm\ x\ 3.2\ =\ 3200\ ppm$$

Using the conversion factors, the actual approximate concentration of acetone present is within the range of 4-8% of the LEL, or 1000 to 2000 ppm. In this case, the converted ppm and %LEL scale are not consistent. This suggests that acetone alone is not present. One or more other gases or vapors are present. Indeed, there may be no acetone present at all.

When the meter has both ppm and %LEL ranges, and a conversion factor is available for a material known to be present, the instrument can be used to help confirm the identity of the material. This technique cannot be used as the only means of confirmation; rather, it should be utilized as one of several methods to corroborate the identity of the material producing the reading.

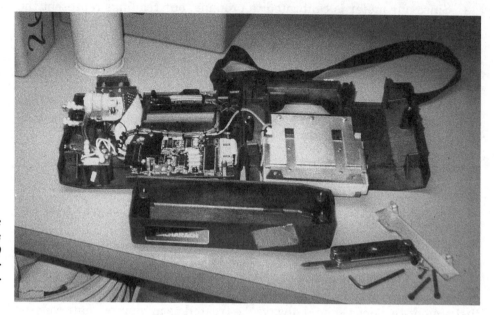

Figure 4-17: Ease of maintenance should also be considered. It takes an experienced user at least 30 minutes to disassemble this meter and replace the sensor. (Author photo.)

USING AND CHOOSING A COMBUSTIBLE GAS INDICATOR

Anticipated Applications and Limitations

The most widely used CGI is the meter that measures %LEL concentrations. This type meter is primarily a safety meter, used to detect and warn the user of the presence of potentially hazardous concentrations of combustible gases. If you can purchase only one CGI meter, it should include a %LEL range. Current regulations require that a %LEL meter be used when dealing with hazardous materials, as well as during excavation activities where combustible gases may be present (i.e. landfills or industrial facilities).

Whenever possible, select an instrument that is calibrated to a gas or vapor you anticipate encountering. Fire departments and gas companies typically select meters calibrated to methane, natural gas, or propane. If you cannot find a meter calibrated to a specific gas or vapor of interest, check to see if there is a response curve or conversion factor for that material. Most manufacturers will calibrate their instrument to a specific gas at additional cost; however the calibration check gas will be expensive and must usually be obtained by special order through a specialty gas supplier.

Consider the anticipated limiting conditions that may be encountered. For instance, if low oxygen concentrations are expected, select a meter which is operational at low oxygen levels, or a meter with a fixed or variable diluter option. If the meter is to be used for emergency response, fast warm-up time is essential. Also check what accessories are available; although not needed now, they may be required later.

Also consider the availability of servicing and sensor replacements. The sensor should be easy to get to and replace; do not purchase an instrument which requires factory servicing for sensor replacement. Look at the manual and read the instructions for sensor removal and replacement. Use your judgement; if it looks and sounds awfully complicated, it probably is. When in doubt, ask the vendor or distributor to demonstrate sensor replacement.

We cannot overemphasize the importance of examining the manual **before** instrument purchase and use. The manual contains more than just operating instructions; it should contain other invaluable information. A copy of the manual should accompany the instrument at all times. Every CGI user should have an opportunity to read and understand the instruction manual and practice using the instrument **before** use in the field.

Finally, decide who or what the meter is going to protect. A belt mounted diffusion-type meter is recommended for use as a personal monitor; such an instrument would not be adequate as a survey instrument at a hazardous materials release. Similarly, a meter with a bulb-type aspirator is adequate for intermittent use but not practical in situations which require continuous monitoring for hours at a time.

Preparing to Use a CGI

The first thing a CGI user should do is look at the meter and determine what concentration range it detects. Analog display meters will indicate either %GAS, %LEL, or ppm. Digital readout instruments may be more difficult to decipher. If necessary, read the side or top panel of the instrument. Do not use an instrument unless you know what it detects! Consider the difference between a 10% GAS reading and a 10% LEL reading. At 10%GAS, combustible materials within the flammable range may be present; such a reading can reflect concentrations in excess of 100,000 ppm.

Examine the Factory Mutual or Underwriters Laboratory intrinsic safety warranty on the meter. Do not use a meter that has no warranty; it may be a source of ignition in the presence of hazardous concentrations of combustible gases.

Check the recommended operating temperature specifications and determine if the ambient temperature, or the temperature at which the meter will be used, is within that range. If the ambient temperature is outside the recommended range, allow the meter cool down or heat up to ambient temperature, then proceed with start-up and calibration; **do not use the instrument if it cannot be calibrated**.

We recommend that the major use limitations be pasted inside the instrument carrying case. These should include operating temperature range, oxygen operating range for ppm and %LEL meters, restrictions regarding hot burning gases such as hydrogen, acetylene, or oxy-acetylene. Also note if the instrument can be used in the presence of tetraethyl lead or if an inhibitor filter is required.

Make sure all necessary accessory items are available; these may include sampling hose and inflexible probe, dust filter, water trap, inhibitor or charcoal filters. If the identity of the material is known, check to see if there is a response curve or conversion factor.

Use Considerations

Combustible gases and vapors can be heavier or lighter than air; when monitoring for gases, move the sample probe slowly up and down. Excavations, trenches, tanks, and other confined spaces should be sampled at multiple levels. Contaminated air is not stationary; it moves in response to wind, subtle air drifts, changes in temperature, even the movement of the instrument user. Response time varies with the meter, sensor type, sample hose length and accessories used. Walk slowly while monitoring or it may be necessary to retrace steps in order to find the source of a reading.

Before entering a building suspected of containing hazardous concentrations of gas, check around ground level and basement windows and under doors prior to entry; assess atmospheres within sumps, catch basins, and manholes if nearby. If vent or exhaust pipes are accessible, check these also. When using a meter outdoors, check low-lying areas or other locations where gases may collect.

ACTION LEVELS

The %LEL CGI is a safety meter; it is designed to answer a basic question: Is it safe to be here? A meter reading of 100% LEL in a workzone is obviously unsafe. It is not unusual, however, to obtain high readings during remote sampling of the interiors of drums and tanks, pipelines, sewers, and other areas where personnel are not present and entry is not required or anticipated. The location of the sampled air, then, must also be considered when responding to CGI readings. Action levels for combustible gases are typically used when the personnel are in the sampled atmosphere, i.e. when the meter readings are obtained in the workzone.

An action level is a designated instrument reading that requires or triggers specific activities by personnel. Action levels are for conditions within the actual or potential workzone, i.e. where an individual can go on purpose or by accident. The action required may be to upgrade levels of protection, to watch instrument readings on a continuous basis, implement engineering controls, or to leave the area. Action levels are often delineated in health and safety plans or SOPs; in the case of combustible gases, OSHA has specified action levels for specific circumstances. The OSHA confined space rule (Permit Required Confined Spaces 29 CFR 1910.146)[5] specifies that an entry permit is revoked when a direct reading instrument indicates a hazardous environment of greater than 10% LEL. A meter reading of 10% LEL is used as an action level for combustible gases, vapors, or mists in confined spaces. Subpart P of the Construction Standards for Excavations (29 CFR 29 CFR 1926.651)[6] requires specific action to prevent worker exposure to atmospheres containing in excess of 20% of the lower flammable limit, or LEL. The actions include ventilating the area and monitoring to ensure safe levels are maintained.

When an action level is reached, all personnel in the workzone should be warned; the supervisor or safety officer should also be notified. Personnel should not enter a potentially hazardous atmosphere unless they are aware of the action level and the action to be taken if the level is reached. Every effort should be made to ensure the absence of potential sources of ignition, and if possible, the source of flammable vapors should be identified. In some cases, it may be possible to ventilate the workzone with intrinsically safe fans or blowers. In the majority of cases, the required action is to withdraw all personnel from the area. If it is necessary for personnel to remain in the area (i.e. emergency responders), all non-essential workers should leave the immediate area. When personnel remain in the workzone after the action level is reached, a new, higher action level should be established. The required action taken when the new level is reached should be well defined and understood.

REFERENCES CITED

1. Chambers, W. and R. Chambers. 1879. Chambers's Encyclopedia, Revised Edition, Vol. VIII. London: W. and R. Chambers.
2. Carrier, E. O. 1965. Humphry Davy and Chemical Discovery. New York: Franklin Watts, Inc.
3. Portable Combustible Gas and Oxygen Alarm, Model 260. Data Sheet 08-00-08. Pittsburgh, PA: Mine Safety Appliances Company.
4. GasTechtor Portable Gas Alarm Model 1314SMPN Instruction Manual. Newark, CA: GasTech Inc.
5. 29 Code of Federal Regulations, Part 1910.146. Federal Register, Volume 58, Number 9, Thursday, January 14, 1993, pages 4462-4563.
6. 29 Code of Federal Regulations. Part 1926, subpart P. Federal Register, Volume 54, Number 209, Tuesday, October 31, 1989, pages 45894-45991.

CHAPTER REVIEW

1. Most %LEL combustible gas indicators employ a _____ sensor incorporated into a Wheatstone bridge circuit.

2. True or false? Catalytic sensors require oxygen for proper function.

3. The _____ in temperature of the catalytic filament relative to the compensating filament produces the meter reading.

4. %GAS meters employ a thermal conductivity sensor. The _____ in temperature of the TC filament relative to the compensating filament produces the meter reading.

5. True or false? Thermal conductivity sensors require oxygen for proper function.

6. True or false? %GAS meters usually have both a catalytic sensor and a thermal conductivity sensor.

7   CGI meters with a ppm range may employ a catalytic sensor or a _____ sensor.

8. True or false? It is acceptable to interchange response curves between different models or manufacturers.

9. A CGI will respond accurately throughout the entire detection range to only one gas or vapor. This gas is the_____ gas.

10. Meter responses to individual gases and vapors other than the calibrant gas are provided by the manufacturer in the form of response curves or _____.

11. True or false? A liquid trap should be emptied immediately if water or another liquid is drawn into it.

12. When using conversion factors, the instrument response is _____ by the conversion factor to determine the actual concentration present.

13. What three criteria must be satisfied before an actual %LEL present can be determined with a response curve or conversion factor?_____.

14. A _____ filter can be helpful in determining if the gas present is a light hydrocarbon such as methane, or a heavier, more complex hydrocarbon such as benzene.

15. List several accessories which would be helpful to have when using a CGI. _____
    _____.

16. An _____ filter is recommended for atmospheres known or suspected to contain leaded gasoline.

17. Name two catalyst poisons other than tetraethyl lead._____.

18. A major limitation of many ppm CGIs is _____ time.

19. The _____ check is the only method available to verify that the meter is working properly.

20. True or false? It is acceptable to use a meter with a low battery as long as the meter readout is stable.

21. True or false? All combustible gas indicators carry FM or UL intrinsic safety warranties.

22. Start-up and calibration of a meter should be performed in a clean, non-contaminated area.

23. Combustible gases and vapors may be _____ or _____ than air.

24. Popular calibrant gases for CGIs include_____ _____.

PROBLEM SET

Refer to Appendix B and C for meter response curves or conversion factors; we recommend that you review the Appendices before beginning the problem set.

1. A hazardous materials response team is called to a butane filling facility. An MSA Model 260 CGI gives a reading of 20% LEL. What is the actual %LEL present?

2. Ethyl alcohol has leaked onto the floor of a warehouse. The GasTechtor (methane calibration) shows a reading of 550 ppm and 1% LEL. What are the actual concentrations present in ppm and %LEL?

3. At a solvent storage depot , a 10 foot sample line is lowered into a 10,000-gallon capacity tank which contains approximately 500 gallons of acetone. A %LEL meter gives no response. What can we say about the presence or absence of flammable vapors inside the tank?

4. The LEL of ethyl alcohol is 3.3%. Convert the actual %LEL reading obtained in question #2 to ppm. Is the actual ppm reading of the instrument consistent with the ppm reading derived from the actual %LEL?

5. A Scott-Alert meter gives readings between 22% and 24% LEL at hexane spill. What is the actual %LEL present? What meter reading would you expect on an MSA 360 that is also sampling at the time and location?

6. A propane-calibrated %GAS meter is used at a release at a propane filling facility. The LEL of propane is 2.2%. The %GAS meter reads 2% GAS. What %LEL is present? What meter reading would you anticipate on the MSA Model 260?

7. A GasTechtor CGI is used to screen for formaldehyde vapors. There is no response on the ppm or %LEL range. What can we say about the presence or absence of formaldehyde vapors?.

8. Local residents adjacent to a gasoline station are complaining of the odor of gasoline. A CGI is used to test for combustible vapors in the sanitary sewer. A reading of 2% LEL and 900 ppm is obtained on a methane-calibrated CGI. How can you determine if the reading is from gasoline vapors and not from naturally-occurring sewer gas (methane)?

5 Oxygen Deficiency, Toxic Gas, and Combination Meters

LEARNING OBJECTIVES

1. Describe how electrochemical sensors work.
2. List the limitations of electrochemical sensors.
3. Discuss the difference between electrochemical and solid state sensors.
4. Describe paper tape gas detectors.
5. Describe the two types of mercury vapor detectors available.
6. List the advantages of combination meters.

The presence of lethal or toxic atmospheres in mines and other confined spaces has been recognized for millennia. The earliest detection devices used were rodents or birds confined in small cages which were carried by workers; alternatively, animals were permanently caged within the suspect area. As long as the sentinel animal was unaffected, the area was considered safe.

Although unrecognized at the time, the actual hazard present was often not only toxic gas but also oxygen deficiency. Interestingly, the use of sentinel animals was not always successful; workers performing heavy labor require more oxygen and are more susceptible to toxic gases than a sedentary animal or bird sitting in a cage.

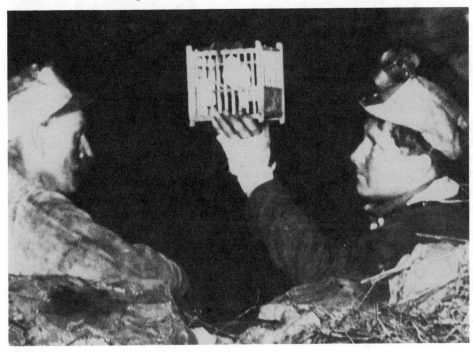

Figure 5-1: John T. Ryan, co-founder of MSA, uses a canary in a cage to detect hazardous atmospheres, ca.1911. (Courtesy MSA.)

83

BASIC ELECTROCHEMISTRY

Electrochemical sensors are often employed to measure a variety of specific toxic gases as well as oxygen. These sensors employ an electrolyte and electrodes to detect the specific gas of interest.

Some sensors rely on a direct chemical reaction between the gas and the electrolyte solution. The ensuing reaction changes the potential of the sensing electrode, that is, its ability to produce ions and electrons. This change in potential causes a change in current and induces a meter reading. The presence of the specific gas of interest is measured by the change in current across the electrolyte to the counting electrode. Hydrogen sulfide, hydrogen cyanide, carbon dioxide, and sulfur dioxide are detected with this type of sensor.

An **electrolyte** is a chemical substance which dissolves in water to form a solution that will conduct electric current. Current is the movement of electrons, which have a negative charge. The usual apparatus to test for such conductivity is a light bulb placed in series with two prongs that are immersed in the test solution. This is also the basic design of a simple battery.

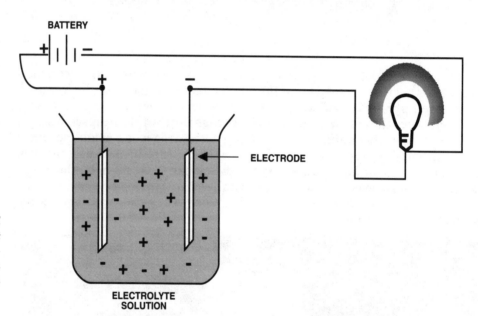

Figure 5-2: The light bulb will glow only if sufficient current from the battery is able to complete the circuit by crossing from one electrode to the other through the electrolyte solution.

In general, acids are good conductors, while bases or alkaline solutions are poorer conductors (your car battery contains a concentrated acid electrolyte). Substances vary in their conducting ability because of the number of ions in solution. An ion is an atom or group of atoms that carries an electrical charge. Table salt is composed of two atoms, sodium (Na) and chloride (Cl) and has no charge. When sodium chloride is dissolved in water, however, the positive sodium ions (Na^+) and negative chloride ions (Cl^-) are dispersed throughout the solution.

Atoms become charged ions when they gain or loose **electrons**. When an atom looses an electron it becomes a positively charged ion; a negatively charged ion is an atom which has gained one or more electrons. In our table salt example, when sodium chloride dissolves in water, the chlorine ion can be thought of as an electron thief, having robbed an electron from sodium. Loss of an electron is called **oxidation**; electrons are gained through a process called **reduction**.

The composition of the electrolyte solution is very important in terms of its conductivity. Electrolytes composed of metallic salts or acids readily form ions and are good conductors. Metals which readily lose electrons include potassium, calcium, sodium, and magnesium.

The composition of the **electrodes** is also very important. Electricity is produced by the movement of charges; in this case the movement of electrons through the electrolyte solution. One electrode interacts with the electrolyte solution and facilitates the production of ions and electrons; the other electrode accepts the electrons and ions and completes the electrical circuit. Electrodes made of different metals will vary in their capacity to lose or gain electrons. Platinum, gold, mercury, silver, and copper, for example, are good electron acceptors while zinc, lead, aluminum and magnesium are good electron donors.

The external force which causes electrons to flow from one electrode to another is known as electromotive force (emf) and is measured in **volts**. This force is also called voltage drop or potential difference.

ELECTROCHEMICAL SENSORS

Electrochemical sensors are usually employed to detect a variety of specific gases such as oxygen, carbon monoxide, hydrogen sulfide, hydrogen cyanide, sulfur dioxide, and chlorine. A typical electrochemical sensor consists of a coarse particulate filter, a semi-permeable membrane of Teflon™ or polypropylene, an electrolyte, and electrodes. The electrolyte may be a liquid, gel, or adsorbed onto a matrix to form a semi-solid or paste.

In most cases, the gas of interest in a sample atmosphere diffuses across the membrane, dissolves in the electrolyte and contacts the sensing electrode. The gas reacts with the sensing electrode producing ions and electrons. These charged particles diffuse across the electrolyte solution to the electron-accepting or **counting** electrode and completes the circuit. A **compensating** electrode is usually also present to ensure that the sensing electrode is maintained at a constant potential.

Limitations

The design of the sensor, composition of the electrodes, and concentration and composition of the electrolyte solution are modified to achieve sensor specificity and detection range. **Interfering gases** of similar molecular size and chemical reactivity may, however, produce a false-positive response. Lack of specificity is a significant limitation for confined space work or other situations where unknown toxic gases, or multiple toxic gases may be present. Major interferants are usually gases which can also be detected by electrochemical sensors; that is, gases which can be oxidized or reduced. Table 5-1 lists some commonly encountered gases and vapors which may interfere with CO or H_2S sensors.

The electrolyte solution, which is either acidic or basic, can be poisoned or **neutralized**. For instance, a sensor containing an alkaline solution will be poisoned by acid gases. This effect may occur rapidly or gradually, depending on the electrolyte, permeability of the membrane, and the pH of the neutralizing gas.

The sensor **membrane can become clogged** with particulates as well as condensation from hot gases, water vapor, aerosols and mists. This inhibits the diffusion of gases into the sensor, resulting in erroneous, low

Acetylene
Dimethyl sulfide
Ethyl alcohol
Ethylene
Hydrogen cyanide
Hydrogen sulfide
Isopropyl alcohol
Mercaptans
Methyl alcohol
Propane
Nitrogen dioxide
Sulfur dioxide

Table 5-1: Common interferants to CO or H_2S electrochemical sensors.

Ammonia
Arsine
Bromine
Carbon dioxide
Chlorine
Fluorine
Freon
Germane
Hydrazine
Hydrogen chloride
Hydrogen cyanide
Hydrogen fluoride
Nitrogen dioxide
Nitrogen oxide
Nitrous oxide
Oxygen
Ozone
Phosgene
Silane
Silicon tetrafluoride
Sulfur dioxide
Tetrachloroethylene
Trichloroethylene
Tungsten hexafluoride

Table 5-2: Some gases which can be detected by electrochemical sensors.

readings. Moving the sensor from low to high humidity levels can inhibit diffusion of oxygen molecules into the sensor; recalibrating the sensor can prevent erroneous readings.

Sensors have a **limited longevity**, usually 6 months to one year. Most sensors, once removed from their protective wrapper, are constantly exposed to the air, even when not in use. This results in a water evaporation and gradual dehydration and concentration of the electrolyte solution. The electrochemical reactions within the electrolyte or at the sensing electrode are also not completely reversible. Eventual sensor failure occurs when there is insufficient reactivity to produce a detectable change in current.

Sensors must be replaced or reactivated periodically. Some sensors can be regenerated by supplementing or replacing the electrolyte; this may be done by the user; in some cases the sensor must be returned to the manufacturer. In most cases, however, the exhausted sensor is simply discarded and replaced.

Most electrochemical sensors are designed to operate within a recommended **temperature range**. The minimum temperature is usually 0°C (32°F); maximum temperatures range from 40 to 60°C (104 to 140°F). Operating temperature limitations are usually due to the constraints imposed by the instrument battery or the sensor electrolyte. Dramatic temperature changes can produce erroneous readings unless the sensor is given time to reach the new ambient temperature. At low temperatures, ion and electron diffusion through the electrolyte will be slowed and instrument response will be delayed. At very low temperatures the electrolyte may actually freeze. Many sensors can be used outside the recommended operating range when the instrument is recalibrated for the ambient temperature of use.

Most electrochemical sensors contain **corrosive** electrolyte solutions which can cause chemical burns upon contact with skin or eyes. Care should be exercised when handling a leaking or ruptured sensor, especially if it contains a liquid electrolyte.

The thin semi-permeable sensor membrane is designed to keep liquids out but allow gases in. The membrane, although usually protected by an exterior housing, is **susceptible to damage**. Always handle the sensor gently and never push down on the membrane.

Electrochemical sensors are sensitive to changes in barometric pressure or **altitude**. At higher altitudes, the partial pressure of gases is decreased; a lesser amount of gas diffuses across the membrane and the meter reading is correspondingly reduced. This effect is best observed with the oxygen sensor (see Table 5-3) but can also affect readings of other sensors as well.

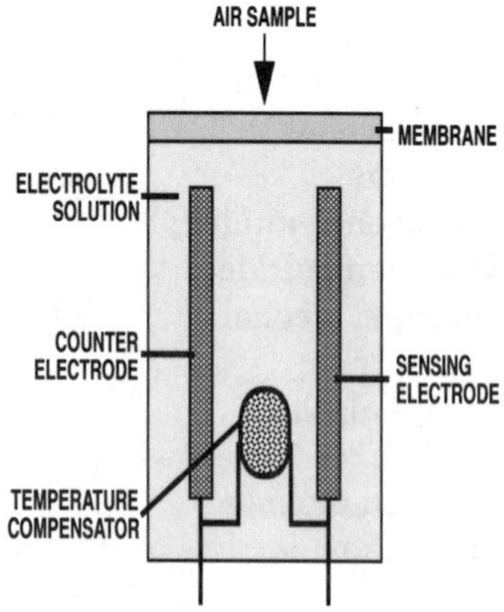

Oxygen Sensors

A typical oxygen sensor utilizes a lead or zinc sensing electrode and a gold or platinum counter electrode immersed in an alkaline electrolyte solution of potassium hydroxide and water; a thin Teflon™ membrane separates the electrolyte from the sample atmosphere. The current generated, which is directly proportional to the partial pressure of oxygen, is passed through a temperature-compensating circuit, amplified, and displayed as percent by volume oxygen within a range of 0 to 25%.

Oxygen deficiency meters are calibrated to measure oxygen concentrations between 0% and 25% volume in air. Normal air contains 20.9% oxygen; significant changes in meter readings are obtained, however, when a meter calibrated at sea level is moved to increasing elevations **above** sea level[1]

Figure 5-3: Line drawing of a typical oxygen sensor.

(Table 5-3). A meter calibrated in New York City, for instance, which is at sea level, would indicate oxygen deficiency in mile-high Denver, Colorado.

Similarly, a meter calibrated at a sea level and then moved to a **lower** elevation will give readings **greater** than 20.9%; a reading of 21.6% oxygen can be anticipated at 1000 feet **below** sea level. An oxygen meter calibrated in Denver and used in New York City would therefore indicate an oxygen concentration above normal. These problems can be avoided by calibrating the meter with clean air at the elevation at which the instrument will be used.

Oxygen deficiency meters with visual or audible alarms are usually factory set to the OSHA limit of 19.5% oxygen; some meters have an oxygen excess alarm set at a 22, 23 or 25% oxygen. Oxygen sensors are routinely calibrated before use with clean ambient air which contains 20.9% oxygen. The validity of oxygen deficient readings can be assessed by using a calibration check gas containing a known oxygen concentration; manufacturers supply check gases containing 10% or 15% oxygen with the balance being an inert gas, usually nitrogen.

Never use exhaled air to check the sensor! Exhaled air contains approximately 16% oxygen and 5% carbon dioxide. Carbon dioxide is an acid gas which will neutralize the alkaline electrolyte in the sensor, causing a loss of sensitivity and erroneously low readings. Always use a check gas to verify that the sensor is working properly.

High concentrations of acid gases such as hydrogen sulfide and carbon dioxide can quickly neutralize the electrolyte and cause very rapid sensor failure. False positive readings may be obtained in the presence of strong oxidants, such as fluorine, bromine, or chlorine. Many strong oxidants are also acid gases; the false positive effect is often counter-balanced by a loss of electrolyte sensitivity which delays recognition of sensor failure.

Other Ways to Look at Oxygen Readings

Recall that normal air contains 20.9% oxygen; the remainder consists of nitrogen and trace gases. Oxygen, then, represents about **one fifth** of the air envelope. A decreased oxygen reading of 0.1%, from 20.9% to 20.8%, actually represents a change in the total air envelope of approximately 0.5% or 5000 ppm. This represents no hazard if the displacing gas is inert; if the displacing material is toxic, however, such a concentration could represent a very real health hazard.

Oxygen deficiency readings should therefore be viewed in two ways; the immediate hazard as a result of decreased oxygen availability, and the potential health hazard if the displacing gas is something other than a non-hazardous inert gas such as nitrogen, argon, neon, or helium. The concentration of displacing gas in air can be obtained by multiplying the decrease in oxygen concentration by five.

Carbon Monoxide Sensors

Carbon monoxide exposure is commonly associated with motor vehicle exhaust and smoke from fuel combustion. The most common detection method for carbon monoxide (CO) is a three-electrode electrochemical sensor and an acid electrolyte, usually sulfuric acid. A separate circuit maintains a constant voltage at the sensing electrode which controls its capacity to oxidize CO to carbon dioxide (see Figure 5-5). The current generated is directly proportional to the concentration of CO present.

Elevation (feet)	Oxygen Reading (%)
–1,000	21.6
Sea Level	20.9
500	20.4
1,000	20.1
2,000	19.3
4,000	18.0
6,000	17.3
8,000	15.4
10,000	14.3

Table 5-3: Changes in oxygen sensor readings when calibrated at sea level and then moved to increasing elevations above sea level.

Figure 5-4: An MSA Oxygen Indicator; the oxygen sensor is equipped with a cable for remote monitoring. (Photo by Steve Napolitano.)

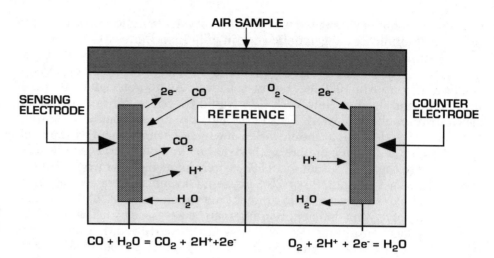

AIR SAMPLE

SENSING ELECTRODE

COUNTER ELECTRODE

REFERENCE

2e⁻   CO O_2 2e⁻

CO_2

H^+ H^+

H_2O H_2O

$CO + H_2O = CO_2 + 2H^+ + 2e^-$ $O_2 + 2H^+ + 2e^- = H_2O$

Figure 5-5: Reaction between CO and sensing electrode in a carbon monoxide sensor (after Ness[2]).

Hydrogen Sulfide Sensors

The electrochemical sensor used to detect hydrogen sulfide (H_2S) is essentially identical to that used for CO. Oxidation of H_2S, however, produces more electrons and ions, and therefore more current at the counting electrode. In some cases, a special bias, or resistance to current flow, is placed on the counter electrode; this bias prevents small amounts of current produced by oxidation of CO or other interferants from eliciting a meter response. In other sensors the composition of the sensing electrode is changed to facilitate oxidation of H_2S over other materials.

Other Toxic Gas Sensors

Electrochemical sensors are now available to detect up to 50 different gases and vapors. Detection diversity is based upon modification of the sensor electrode, electrolyte, membrane permeability, and application of a bias on the counter electrode. A pyrolyzer unit is used to thermally degrade stable gases to facilitate electrochemical detection. Table 5-3 is a partial list of gases which can be detected by electrochemical sensors.

PAPER TAPE GAS DETECTORS

Figure 5-6: The 900-series GMD AUTOSTEP® offers three different modes of operation: fast response for detecting leaks, survey, and continuous area monitoring. (Courtesy GMD Systems.)

Tape monitors employ a paper tape impregnated with a chemically reactive mixture designed to change color in the presence of a specific gas. Sample air reaches the tape by diffusion or a pump pulls the air through the tape. When the tape-specific gas is present, the tape changes color. The tape is then scanned by a photodiode detector; the intensity of color change is proportional to the concentration present.

Although paper tape systems are best known as stationary monitors for toxic gases, they are also used as passive dosimeters or badges as well as portable survey meters and personal monitors[3-6]. In most systems, a continuous ribbon of tape moves across a sampling port at a fixed rate. Although reel-to-reel systems are still in use, most manufacturers have switched to cassettes which minimize handling. As it moves across the sample port, the tape is exposed to the sample air for a specified interval. Sampling intervals vary from less than 30 seconds to up to 4 or 5 minutes and depend on the colorimetric reaction employed for the specific gas detected as well as the monitoring strategy.

Some instruments have different detection modes; sampling rates can be modified depending on whether the instrument is used in a search, survey, or monitor mode. Even in tape systems which offer continuous sampling, however, there must be a lag time between sampling and instrument response. During this lag time the color develops on the tape which is then scanned by the detector. Paper tape systems cannot offer the near instant-readout response that other detection systems offer. Nevertheless, these systems are widely used to detect low concentrations of toxic gases such as isocyanates, phosgene, hydrazines, and inorganic acids.

Limitations

The paper tape contains a chemically reactive mixture which is susceptible to **environmental factors** including humidity and temperature. Many tape systems take a reading of the background color of the tape prior to sampling, which is then subtracted from the color reading. When the instrument is not in use, unused tape or the tape cassette should be protected in a sealed container. Some manufacturers recommend that a tape should not be used for more than two weeks after initial opening of the sealed pouch.

Chemical reactants impregnated into the tape will eventually degrade over time. Each paper tape reel or cassette should have an **expiration date**. Do not use a tape beyond its expiration date. Potential sources of error also include variability in sample flow rates, paper sampling intervals, and photodiode accuracy; these should be assessed on a regular basis.

Paper tape colorimetric reactions are also susceptible to **interfering gases**. Some interferants produce a color change indistinguishable from that of the gas of interest. Other interfering gases such as oxidants, may produce a bleaching effect, while still others will produce an abnormal color reaction. In some cases, a suspected interferant can be identified by the atypical tape color it produces.

SOLID STATE SENSORS

Semiconductor sensors can be used to detect low concentrations of toxic gases. Recall that a response from a semi-conductor sensor is elicited as a result of a change in the concentration of free electrons at the surface of the sensor. This decreases the resistance, resulting in an increase in current.

Slight modifications, in the proprietary mixture of metallic oxides, semiconductor temperature, or voltage applied across the semiconductor, affects the selectivity and sensitivity of the sensor for different gases and vapors. Although designed to respond to a particular gas, solid state sensors will also respond to a variety of other materials, particularly hydrocarbons. The sensor cannot discriminate between or identify a particular material which elicits a meter response.

Solid state instruments which detect carbon monoxide and hydrogen sulfide are typically used as general survey instruments because of their lack of specificity. These meters may also be adapted to monitor for chlorofluorocarbons (Freons).

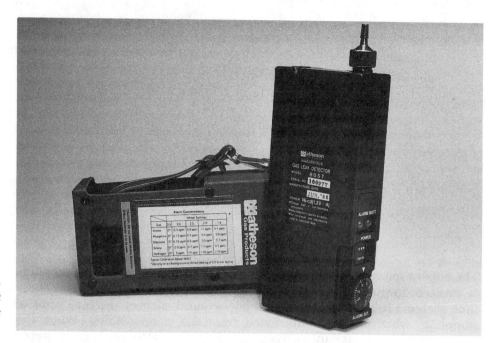

Figure 5-7: A Matheson leak detector. The wheel is used to adjust alarm intensity. (Photo by Steve Napolitano.)

Matheson Model 8057

The Matheson Model 8057 Hazardous Gas Leak Detector was originally developed to detect minute leaks in closed systems containing toxic process or doping gases used in the semiconductor industry. The so-called doping gases are inorganic hydrides, i.e. materials containing hydrogen such as arsine (ArH_3), phosphine (PH_3), diborane (B_2H_6), and silane (SiH_4). The instrument is factory-calibrated to hydrogen[7].

The Matheson Model 8057 employs a solid state sensor designed to detect extremely low concentrations of gases; the meter does not, however, actually **measure** the concentration of gases present. Rather, it produces an audible and visual alarm when detectable concentrations are present.

The leak detector uses an alarm wheel to adjust the sensitivity to detectable gases. A wheel setting of approximately 3.5 should produce a very slow alarm in clean air. After adjusting the wheel until the alarm is very slow, the meter is introduced into the area where a leak is suspected. An increased alarm rate indicates the presence of contamination. If the source of contamination is not evident, the wheel setting is adjusted counter-clockwise until the alarm is very slow or just goes off. The meter is then used to sweep the area until it points in the direction which causes the greatest increase in the alarm. This procedure is repeated until the source of the leak is found or the air becomes too heavily contaminated.

The Matheson leak detector is exquisitely sensitive to inorganic hydrides and will respond to concentrations of less than 0.5 ppm. It will, however, also respond to other organic and inorganic materials, although significantly higher concentrations are required. High concentrations of some of these interferants can overwhelm the sensor; when this occurs, a constant alarm, or no alarm, will be obtained.

When the wheel setting is adjusted until the alarm is very slow, the sensor will also respond to chemicals in exhaled air, producing a more rapid alarm. The sampling probe should always be pointing away from the user when the wheel setting is adjusted.

Acetone
Ammonia
Benzene
Carbon Monoxide
Chlorine
Dichloroethane
Ethanol
Ethylene
Ethylene Oxide
Formaldehyde
Freons
Hexane
Hydrogen Chloride
Hydrogen Selenide
Hydrogen Sulfide
Isopropyl Alcohol
Methane
Methanol
Methyl Bromide
Methyl Chloride
Methyl Ethyl Ketone
Methyl Isobutyl Ketone
Propane
Sulfur Dioxide
Toluene
Trichloroethylene
Xylene

Table 5-4: Common interfering gases of the Matheson 8057 leak detector.

The Matheson is not calibrated but rather checked before use with a Check Gas Vial which contains camphor (naphthalene). The alarm should sound when the instrument probe is placed in the vial. The meter must then be purged for several minutes in clean air before it is ready for use. If the unit is urgently needed, an alternative check is to exhale gently toward the probe. A slow alarm should sound and the unit will not require extensive cleansing.

The presence of interfering gases can limit the sensitivity of the sensor to low concentrations of hydrides. Most often, high concentrations of interferants produce a prolonged, rapid alarm even after the unit has been removed from the contaminated area. When this occurs, an extended purging interval may be needed before the instrument can be used again.

Limitations of Solid State Sensors

The most important limitation of solid state sensors is their lack of specificity. The presence of **interfering gases** may cause a false positive meter response and prevent accurate measurement of the specific gas of interest. **High humidity**, and **condensation** of gases and vapors onto the surface of the sensor can inhibit diffusion and reduce sensitivity. **Dust and particulates** can also clog the porous sensor surface and limit gas diffusion. As with all instruments, solid state sensors have a recommended operating temperature range.

Oxygen is necessary for the sensor function; the minimum concentration required varies with the sensor design and composition. Manufacturers often do not include this information in the instructions; if oxygen deficient atmospheres are anticipated, determine the sensor oxygen requirements before purchase.

MERCURY VAPOR DETECTORS

Mercury vapor detectors are utilized at mercury spills as well as at fixed facilities where mercury is used, to detect contamination on shoes and clothing. Spilled mercury is often spread by foot traffic or collected inappropriately using vacuums or sweepers. A relatively small amount of mercury can contaminate a large area. The OSHA TWA is 0.05 mg/m^3; the NIOSH IDLH is 28 mg/m^3.

One detection method used is ultraviolet light (UV) absorption[8]. A 257 millimicron UV light passes through a chamber and onto a photoresistor which measures the intensity of UV radiation. Mercury vapor absorbs strongly in the UV spectrum. When sample air drawn into the absorption chamber contains mercury, the intensity of UV light reaching the photosensor is decreased in proportion to the concentration present. The meter displays the concentration of mercury vapor in milligrams per cubic meter.

Most UV-absorption mercury analyzers have a range between 0 and 1.0 mg/m^3; these meters are designed to detect mercury around the TWA. Higher concentrations must be measured using colorimetric detector tubes.

Any material which absorbs UV light as well or better than mercury vapor will interfere with the measurement of mercury vapor contamination. Commonly encountered interfering gases and vapors include acetone, benzene, ethanol, ethyl acetate, gasoline, hexane, trichloroethylene, and carbon disulfide. Water vapor will affect the transmission of UV light to the photosensor. Humidity effects can be corrected by zeroing the instrument at a humidity comparable to that of the sample atmosphere.

Figure 5-8: A Bacharach MV-2 Mercury Vapor Sniffer®.(Photo courtesy Bacharach.)

Figure 5-9: The Bacharach Sniffer® 503A; all CGI responses are indicated on a single readout display; a separate display gives continuous oxygen readings. (Photo courtesy Bacharach.)

A gold foil sensor can also be used to detect mercury vapor[9]. An exact volume of sample air is drawn over the sensor at a constant flow rate. Mercury in the sample is adsorbed onto the sensor which in turn changes the resistance. The change in resistance is proportional to the concentration of mercury present.

Eventually, the sensor becomes saturated with mercury and other contaminants; these are burned off during a short heating cycle. The gold foil sensor is not affected by commonly encountered materials such as hydrocarbons, carbon monoxide or water vapor. Internal filters or scrubbers remove potential sensor interferants such as particulates and acid gases.

Although mercury analyzers can be calibrated by users with access to a mercury vapor generator, we recommend that the unit be returned to the factory on a yearly basis for factory calibration. Instrument responsiveness to mercury vapor can be assessed using a small glass vial containing a drop of mercury. The condition of the charcoal-iodine absorbent, used to zero the instrument in an atmosphere containing mercury, can also be checked at this time. If the unit cannot be zeroed, the absorbent is saturated and must be replaced.

COMBINATION METERS

Meters which can detect multiple types of hazards are considered combination instruments. This type of instrument has multiple sensors, each of which detects one particular type of contaminant, such as combustible gas, oxygen deficiency, carbon monoxide, or hydrogen sulfide. They are available as personal monitors, area monitors, or portable survey instruments. Such instruments are valuable in situations which require monitoring for multiple hazards.

Combustible Gas-Oxygen Deficiency Meters

The most popular combination instrument is the %LEL CGI-oxygen deficiency indicator. Recall that catalytic filament and solid state sensor CGIs require a minimum concentration of oxygen; the oxygen sensor therefore not only serves to warn of oxygen deficiency for personnel but also allows the user to determine if sufficient oxygen is present for CGI sensor function. In most cases, the oxygen alarm will be activated at a pre-set alarm point of 19.5% even if the oxygen sensor reading is not continuously displayed.

CGI-oxygen deficiency meters are also available with multiple combustible gas ranges. For example, the Bacharach Sniffer® 503A is a sample draw survey meter which offers CGI calibration gas (methane or hexane) detection ranges of 0-2000 ppm, 0-10,000 ppm, and 0-100% LEL; it also measures 0-25% oxygen. Combustible gas readings, regardless of range, are indicated on 0 to 100 analog display. The operator must remember which range is in use and multiply by the appropriate factor for the ppm ranges (20 or 100), or read directly for the %LEL range. Continuous oxygen readings are displayed separately. Other analog meters, for example the GasTech Tank-Techtor, use only one display for all readings.

GasTech offers a variety of combination combustible gas/oxygen deficiency instruments; an especially useful model is the GasTech Landfill Monitor (also known as the Model 1939OX), which has three combustible gas ranges (%GAS, %LEL, and ppm) as well as an oxygen sensor. The oxygen sensor operates continuously, however, the user must select the desired

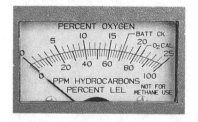

Figure 5-10: Readout display of the GasTech Tank-Techtor. Two ppm ranges, %LEL, % oxygen, and battery condition are all indicated on a single display. (A Steve Napolitano photo.)

combustible gas range. All instrument responses are read on single analog display. Another GasTech meter, the Tank Techtor™ was originally designed for underground storage tank removal; it is also popular at landfill sites because it has a **methane elimination switch**. The user initially assesses full meter response to the sample atmosphere; when the elimination switch is activated, the meter will not respond to methane or natural gas. Workers find this instrument useful in discriminating between naturally-occurring landfill gases and other gases and vapors that may be encountered.

Combustible Gas-Oxygen Deficiency-Toxic Gas Meters

A variety of electrochemical and solid state sensor options are now available for survey instruments as well as personal monitors. The most popular meter used for confined space entry and survey monitoring is the so-called tri-detector which measures %LEL combustible gas, oxygen, and either hydrogen sulfide or carbon monoxide. Such meters are available with bargraph, analog, or digital readout displays. Alarm levels are usually 10 ppm H_2S and 25 or 50 ppm CO.

The Enmet CGS-100 Tritector® utilizes independent solid state sensors to continuously monitor for oxygen, combustible gas, and toxic gas. The %LEL sensor in the Below Ground/Public Works Model is calibrated to methane while the toxic sensor detects CO or H_2S.

A Petrochemical Calibration option is available which utilizes propane as the combustible gas calibrant; the toxic sensor detects **methyl chloride** (not methylene chloride). The toxic sensor alarm point is 100 ppm methyl chloride, which is twice the TWA. The sensor at this setting is sensitive to low concentrations of hydrocarbons typically found at petrochemical facilities. The toxic sensor is designed as a leak detector for total hydrocarbons and should be used only as a yes-no indicator for the presence or relative absence of such contaminants.

Manufacturers are offering more instruments with multiple toxic sensor capabilities. Enmet for example, offers the Quad-400 with independent sensors for O_2, %LEL, CO, and H_2S. The H_2S sensor is a solid state sensor; the CO electrochemical sensor has a constant bias to prevent interference from low concentrations of H_2S. Readings from all sensors are

Figure 5-11: The MSA Model 360 and Scott Alert Model S109 measure %LEL, oxygen, and carbon monoxide; similar models (not shown) have the same design but detect hydrogen sulfide instead of carbon monoxide. (Photo by Steve Napolitano.)

Figure 5-12: The Enmet Quad-400 (left) with remote sensor module and the MSA Passport™ (right) have multiple toxic gas sensors as well as oxygen and %LEL detectors. (Photos courtesy of Enmet Corp. and MSA.)

displayed simultaneously. Audio and visual alarms are activated if concentrations exceed pre-set alarm set points. The Quad-400 has a sensor cable and detachable sensor module which can be utilized for remote sampling; alternate models with hand aspirator or sample pump are also available.

MSA has recently introduced the Passport™ Portable Alarm which is designed as a diffusion-type personal monitor. The Passport monitors for O_2, %LEL, and can also accommodate up to three toxic gas sensors. Besides CO and H_2S, other toxic gases which can be monitored include chlorine gas (Cl_2), hydrogen chloride (HCl), hydrogen cyanide (HCN), sulfur dioxide (SO_2), and nitrogen dioxide (NO_2). The standard readout displays all five concentrations simultaneously; the user may also check the peak reading obtained for each sensor, average exposure over a 15-minute interval, or average exposure since the unit was turned on. A pump module and sample lines can be used for remote monitoring. The CGM Tox-Alert and Neotronics Exotox 60 have similar capabilities.

Datalogging Options

Many meters are equipped with a microprocessor which records exposure peaks, as well as 15-minute STEL, and 8-hour TWA information. Measurements from each sensor are taken at fixed sampling intervals (i.e. 60 or 120 per minute), averaged over a specific interval (usually one minute), and stored in the memory. Memory capacity varies with the average interval as well as the number of sensors monitored. Storage capacity should be assessed by both the number of values stored and number of hours of continuous function.

Data can then be displayed for immediate retrieval or transferred via an RS-232 interface to a serial printer or personal computer. When transmitted to a printer, the data is usually printed in tabular form, showing fixed interval readings. In many cases the raw data is dumped into the printer and the user must perform all other calculations.

Formatted print-outs, on the other hand, usually provide additional information such as the TWA, STEL, peak readings, and readings above exposure limits. The same data information is also available when transferred to a PC. Specialized software is often required to transfer, store, and manipulate data. Some software includes data management features which enable the user to create a library data base as well as output files which may be imported and manipulated by a separate spread sheet program. Graphics packages to plot exposure data over the recorded time are also offered.

AIM Gas Detectors

The AIM Safety Company offers the AIM gas detectors which resemble a large flashlight or microphone. The basic AIM Gas Detector Model 815Z has a metal oxide solid state sensor which detects toxic and combustible gases. The unit is equipped with a microprocessor and pre-calibrated to 33 toxic/combustible gases; the meter response to a particular gas is indicated by an appropriate concentration range (i.e. 0-9%, 10-19%, 20-29%, 30-59%, 60-89%, 90+% LEL). When the identity of the contaminant is unknown, a general sensing mode is utilized, which represents a composite response. The microprocessor will also display concentrations as a percentage range of the IDLH. Colored warning lights, warning messages, and a pulsing audible alarm are also used to indicate the relative hazard present. Alarm frequency increases and warning lights change from green to yellow to red as the concentration range increases.

The general sensing mode should be used unless the identity of the contaminant is known and there are no other materials present which may interfere with meter function. When the identity of the material is known, the user may select from the menu of up to 33 toxic/combustible gases listed in Table 5-5. The user must then select either the toxic mode which displays concentrations in %IDLH, or the explosive mode which gives readings as %LEL ranges. The unit automatically defaults to the explosive calibration mode (%LEL) if a selection is not made within 5 seconds.

Most AIMs in use are combination meters which also monitor for oxygen concentrations (AIM 2000); the 3000 model monitors for toxic/combustible gases, oxygen, and CO or H_2S. The AIM 3000 will display the actual concentration of CO or H2S present or provide information on the **percentage** of the TWA and STEL accumulated since the instrument was turned on. Percentage values are based on TWA and STEL values in use at the time of original manufacture, or 10 and 50 ppm for H_2S and 50 and 400 ppm for CO. Thus a reading of 80% TWA for the CO sensor indicates the average concentration present over the elapsed time is 80% of the TWA or 40 ppm. Exposure values for CO, however, have been changed to 25 and 200 ppm for the TWA and STEL, respectively. Thus a reading for a TWA or STEL percentage of 50% on older units, actually indicates that the current maximum allowable concentration has been reached.

The CO and H_2S sensor are cross-specific. That is, the CO sensor will respond in the presence of H_2S and the H_2S sensor responds to CO. Unfortunately, each sensor reacts to the cross-specific gas in a unique way. The only way to determine how each sensor responds is to test it with a

1. Butane
2. Methane
3. Ethane
4. Propane
5. Ethylene
6. Ethyl ether
7. Acetone
8. Methanol
9. Pentane
10. Hexane
11. Octane
12. Acetylene
13. Propylene
14. Butylene
15. Toluene
16. Xylene
17. Diethylamine
18. Dimethylamine
19. Ethanol
20. Propanol
21. Benzene
22. Gasoline
23. Petroleum naphtha
24. Methyl ethyl ketone
25. Ethyl acetate
26. Methylene chloride
27. Ethyl chloride
28. Ethylene chloride
29. Hydrogen
30. Hydrogen sulfide
31. Aviation fuel
32. Sulphur dioxide
33. Ammonia

Table 5-5: To select a precalibrated-gas, the AIM 3000 user enters the appropriate number from the list of 33 toxic/combustible gases.

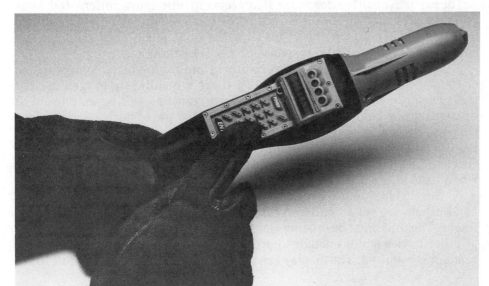

Figure 5-13: The AIM 3000 has an oxygen, combustible/toxic gas, and a CO or H_2S sensor. The keypad is difficult to use while wearing firefighting gloves; use a pencil to change functions or settings. (A Steve Napolitano photo.)

Carbon monoxide
Hydrogen sulfide
Sulfur dioxide
Nitric oxide
Nitrogen dioxide
Hydrogen cyanide
Hydrogen chloride
Chlorine
Hydrogen
Oxygen

Table 5-6: Electrochemical sensor options available in AIM Logic Series Detectors.

known concentration. Another limitation of these sensors is that they lose approximately 2% sensitivity every 2 months. The sensor must be electronically "rescaled" every two months.

A new series of AIM Logic Gas Detectors have recently been introduced. The "ZeroSeries" has a a basic keypad with up-down arrow, select key, and on-off switches. The 100, 200, and 300 Logic Series models have the numerical keypad as well as other function keys. When purchasing an AIM, a combination of %LEL, oxygen, and toxic sensor options can be selected. The number of sensors is combined with the Logic Series number to arrive at a specific model number, e.g. a 200 Logic Series instrument with three sensors is designated as a Model 3200, a 300 Logic Series instrument with 2 sensors is a Model 2300; the second number indicates the Logic Series Model. The 200 Logic series offers a multi-gas %LEL sensor that has been pre-calibrated to 28 gases and vapors. The 300 offers a multigas %LEL/ppm combustible gas sensor that must be calibrated by the user.

The LED lights on Logic Detectors indicate instrument operation and hazard level. The light farthest to the left is green when the unit is in operation and sensing. The lights to the right turn green when 50% of the alarm value is reached, and red when 100% of the alarm value is reached. In this case, a green light does not indicate a no hazard condition. The 200 Logic Series detectors are factory-set to alarm at 19.5% and 23% oxygen, and 20% LEL; toxic sensors alarm at the TLV concentration.

The Logic Series Detectors give combustible gas readings in %LEL rather than a range. There is a larger selection of toxic gas sensors; a ppm-range sensor is also offered. The "X" series of sensors for carbon monoxide (CX), hydrogen sulfide (HX), and sulfur dioxide SX) are designed to give higher resolution; the CX sensor is less susceptible to interference from hydrogen sulfide and sulfur dioxide.

AIM combustible sensors **cannot** determine or confirm the identity of the gas present. The pre-calibrated channel which gives the highest reading cannot be used as an indicator of the identity of a material. The unit will respond to hundreds of gases, not just those gases for which it is pre-calibrated. It is imperative that the correct gas is selected, and that there are no other interfering gases that will affect the reading. If more than one material is present, the AIM should be used on the general sensing mode.

AIM detectors are not sensitive to concentrations less than 10% LEL on the explosive calibration scale. If low concentrations are anticipated, or no response is elicited on the %LEL scale, the more sensitive toxic mode should be employed. The user should monitor the toxic mode until readings are sufficiently high to elicit a response on the %LEL range.

AIM combustible gas sensors require a minimum oxygen concentration of 10%; at concentrations less than 10% there is insufficient oxygen to elicit a response. In some AIM detectors, sensor response will be inaccurate at concentrations below 15% oxygen. Halogenated gases such as methylene chloride, carbon tetrachloride, and ethyl chloride can desensitize the sensor, resulting in erroneously low readings. The combustible gas sensor should be checked frequently using the calibration check kit to ensure proper function when used in the presence of halogenated materials or unknowns.

Several frequently-overlooked limitations warrant mention. The keypad buttons are very small and cannot be individually pressed by a gloved finger. The unit must be set-up prior to entry, or the user must carry a pencil or other device small enough to press buttons when a function change or readout is required. The display screen and colored LEDs are difficult to see in bright sunlight. Erroneous sensor readings may be caused by radio-frequency interferences produced by two-way radios or cellular phones. The effect of communication equipment on AIM sensor readings should be established before actual field use.

USING METER READINGS

Oxygen Indicators

Oxygen levels should always be monitored if there is any potential for oxygen deficiency or excess. OSHA standards[10-12] stipulate that oxygen concentrations of 19.5% or lower are indicative of oxygen deficiency; personnel must use supplied air respirators when the concentration falls below the action level of 19.5%.

Oxygen deficiency may occur as the result of displacement by another gas or consumption of oxygen by a chemical reaction. Confined spaces and low-lying areas outdoors are especially susceptible to oxygen deficiency. An action level of 19.5% oxygen should be used only when dealing with known, non-toxic displacing gas such as nitrogen or argon; these materials exert no adverse health effects at non-asphyxiating concentrations and do not pose a flammability hazard.

Oxygen excess is also a cause for concern; the flammability of materials is greatly enhanced at concentrations above 20.9%. Oxygen concentrations greater than 23% should be considered a cause for concern. The OSHA action level for oxygen excess in permit-required confined spaces is 23.5%[12].

Combination Meters

When potentially toxic or flammable materials are known or suspected to be present, it is inappropriate to use oxygen concentration as an action level. For instance, suppose oxygen levels decrease from 20.9% to 19.9% as a result of a methane gas release; the concentration of methane in the oxygen portion of the atmosphere is 1% by volume or 10,000 ppm. Since oxygen represents only one-fifth of the atmosphere, the actual concentration of methane gas present is actually 5% by volume, or 50,000 ppm. The area could still be considered "safe", since the oxygen action level of 19.5% had not been reached, despite the presence of LEL concentrations of methane (the LEL of methane is 5% by volume). In this case, a combination CGI/oxygen deficiency indicator should be used; the action level employed should be based on either %LEL readings or oxygen concentration, whichever is reached first.

Users of combination meters frequently view each sensor reading as separate and do not compare them with other readings. The presence of a minimum concentration of oxygen for proper function of a catalytic filament CGI sensor can be verified using an oxygen sensor. Similarly, a relative decrease in oxygen concentrations can be compared to %LEL readings. For example:

A combination instrument gives readings of 20.4% O_2 and 90% LEL; the combustible gas present is known to be methane. The CGI conversion factor for methane is 0.6. Are the readings compatible?

Yes. The oxygen reading indicates that the concentration of displacing gas is actually 0.5% x 5=2.5% or 25,000 ppm. The LEL of methane is 5% or 50,000 ppm. The actual LEL present is 90% meter reading x 0.6 conversion factor=54% LEL, or about 27,000 ppm.

If readings from multiple sensors are not compatible, then several possibilities should be considered: one or more sensors have malfunctioned, one or more interfering gases are present, radio frequency interferences are affecting sensor resistance and readout, or other factors such as temperature, humidity, or condensation are affecting sensor function or delivery of sample air to the sensor.

Toxic Gas Monitors

The 8-hour OSHA or ACGIH TWA is often utilized as an action level to determine when respiratory protection should be used; these exposure values are also often used to define areas where by-standers and unprotected personnel should be excluded. When the OSHA and ACGIH values differ, the lower value is usually selected. TWAs are defined for commonly encountered toxic materials, including carbon monoxide (25 ppm), hydrogen sulfide (10 ppm), chlorine gas (0.5 ppm), sulfur dioxide (2 ppm), phosgene (0.1 ppm), and mercury vapor (0.05 mg/m^3 as mercury). For instruments with an audible or visible alarm, the alarm level is set at the TWA concentration at the time of manufacture. When using a toxic gas monitor, the alarm and action level for a material that has a TWA should be **one-half** the TWA; this allows for individual variability in response to the toxic material present, as well as errors in sensor measurement. Most toxic sensor manufacturers will adjust the alarm setting if the request is made at the time of purchase.

Solid State Leak Detectors

Solid state or semiconductor sensors are often employed to alert the user of the presence of small leaks of toxic and flammable metal hydride doping gases in cylinders or pressurized systems, or to locate the source of a known leak. Detectors are sensitive to extremely low concentrations (less than 1ppm). Leak detectors are non-specific, however, and **cannot** indicate the relative concentration of doping gas present; the detector can only give a yes or no response regarding the presence or absence of detectable concentrations of these gases. Since leak detectors are non-specific and do not give an indication of the amount of material present, the user should not use an alarm condition as an action level. Because of the extreme toxicity of most doping gases, leak detection activities should always be conducted using adequate respiratory and skin protection.

While leak detectors can respond to a variety of organic and inorganic materials, the concentrations required to elicit an alarm response are extremely variable, and many organic solvents can overwhelm the sensor. For this reason, we cannot recommend the use of a leak detector as a generic toxic and flammable gas indicator.

REFERENCES

1.   Oxygen Indicator Model 246RA Instructions. Mine Safety Appliances Company, Pittsburgh, PA.
2.   Ness, S. A. 1991. Air Monitoring for Toxic Exposures. New York: Van Nostrand Reinhold.
3.   Remote Intelligent Sensor Area Monitor Operating Manual. GMD Systems, Inc. Hendersonville, PA.
4.   Personal Continuous Monitor Operating Manual. GMD Systems, Inc., Hendersonville, PA.
5.   User Instructions for 550 Series Sure-Spot TDI Badge System. GMD Systems, Inc. Hendersonville, PA.
6.   Autostep Portable Continuous Monitor Operating Manual. GMD Systems, Inc. Hendersonville, PA.
7.   Model 8057 Instruction Manual. Matheson Gas Products. East Rutherford, NJ.
8.   Instruction 23-9510 Model MV-2 Mercury Vapor Sniffer®. Bacharach, Inc. Pittsburgh, PA.
9.   Jerome 431-X Mercury Vapor Analyzer Operation Manual. Arizona Instrument Corp. Tempe, AZ
10. 29 Code of Federal Rgulations, part 1926, subpart P. Construction Standards for Excavations.
11. 29 Code of Federal Regulations, part 1910. 120. Hazardous Waste Operations and Emergency Response.
12. 29 Code of Federal Regulations, part 1910. 146. Permit Required Confined Spaces for General Industry.

CHAPTER REVIEW

1.   In an electrochemical sensor, a semi-permeable membrane separates the sample air from the
     _____.

2.   Name three gases, other than oxygen, that can be measured using an electrochemical sensor
     _____.

3.   List three limitations of electrochemical sensors: _____.

4.   True or false? Exhaling into an oxygen sensor is a good way to determine if it is working properly.

5.   Clean, ambient air contains _____% oxygen.

6.   True or false? Paper tape cartridges have an expiration date.

7.   True or false? The Matheson leak detector is designed to detect less than a ppm of toxic inorganic hydride gases such as arsine and phosphine.

8.   True or false? The pre-calibrated AIM channel which gives an optimal response can be used to identify the material present.

9.   Name some commonly encountered interfering gas which can affect the response of the Matheson leak detector: _____.

10. The detection principle of a mercury vapor detector can be based on a _____ sensor or _____ light absorption.

PROBLEM SET

1.   An oxygen sensor is lowered into an underground storage vault prior to entry. The meter reads 20.1% oxygen OSHA requires that a minimum of 19.5% oxygen be present or supplied air must be used. Should workers enter the vault? Why or why not?

2.   What approximate concentration of displacing gases are in the under ground vault to produce an oxygen reading of 20.1%?

3.   The interior of an above-ground storage tank has been cleaned with acetone, then pressure rinsed with hot water and steam. A residue of hot water remains in the tank, and steam can be seen escaping from the top of the tank. A worker lowers an oxygen sensor into the tank and gets a reading of 16.6%; is this oxygen reading valid? Why or why not?

4.   A truck carrying compressed gas cylinders has a flat tire on a well-traveled road adjacent to a residential area. Some of the cylinders contain arsine, phosphine, and diborane, which are poisonous gases. Neighborhood residents become concerned and call the fire and health departments. What can responding agencies do to convince the residents that there is little risk?

5.   A mercury-contaminated shipment of hazardous waste was received at a solvent waste recovery facility. The facility health and safety officer must assess the extent of mercury contamination. Which type of mercury vapor detector would be best suited for this use?

6.   An isopropyl alcohol (IPA) supply line is leaking adjacent to an underground electrical vault. ScottAlert combination meter readings are 20.6% O_2, 40 ppm CO, and 100% LEL. Using the Scott-Alert conversion factors in Appendix C, and an LEL of 2.0% for IPA, determine if the %O_2 and %LEL readings are or are not compatible.

6 Photoionization Detectors

LEARNING OBJECTIVES

1. Describe how photoionization detectors work.
2. List the major limitations of photoionization detectors.
3. Describe the effect of humidity on meter function.
4. Discuss the importance of ionization potential and UV lamp eV capacity.
5. List several important gases that cannot be detected by PIDs.

Chemical analysis based on ionization has been used since the 1960s; most analyses were based on flame ionization although photoionization was also used in stationary instruments. The first portable analyzer based on the principle of photoionization was developed in the early 1970s[1].

Photoionization detectors (PIDs) are low concentration instruments which typically have a detection range of 0.1 to 2000 ppm for the factory calibrant gas. They are used, when relatively small concentrations are anticipated, as leak detectors, for plant surveys to identify potential problem areas, to monitor chamber concentrations in toxicology studies, to monitor the effectiveness of ventilation controls or changes in workplace practices, to detect residual contamination in soil or water, to check for the presence of accelerants in arson investigations, and to determine the need for personal protective equipment for hazardous materials workers.

Figure 6-1: PIDs from different manufacturers (left to right, top): HNU Systems HW-101, Thermo Environmental 508, Mine Safety Appliances Photon™. The Photovac TIP™ (bottom) is still in use but no longer manufactured. (A Dietrick Lawrence photograph.)

HOW A PHOTOIONIZATION DETECTOR WORKS

To understand photoionization we must revisit our discussion of basic electrochemistry from Chapter 5. Recall that atoms can gain or lose electrons and become charged ions; materials vary in their ability to become ions. The movement of ions or electrons from one electrode to another can generate an ion current which is measured in volts.

Our discussions in Chapter 5 revolved around ions and electrolyte solutions; ions were formed as a result of a chemical reaction between contaminants in the sample air and constituents in the electrolyte solution. The electrolyte solution served as the conductor of electric current; materials which readily lost electrons and formed ions were good conductors.

Now think what would happen if ions were formed in another way and were freely available within a very small air space instead of an electrolyte solution. Under such conditions, ions could also be collected at an electrode and generate current. This is essentially what happens within the ionization chamber of a PID. Ions formed in air are driven in one direction by a bias electrode and accumulated at a collecting electrode; the ion current generated is amplified, then translated to a meter reading. Current is directly proportional to the number of ions formed and collected.

Chemical contaminants in sample air form ions or are **ionized** as a result of being bombarded by high energy ultraviolet (UV) light. These compounds actually absorb the energy of the light, which excites the molecule and results in the temporary loss of an electron and the formation of a positively charged ion. This process is called **photoionization** and is frequently represented by the formula:

$$RH + h\mu \longrightarrow RH^+ + e^-$$

Where RH represents the chemical molecule, $h\mu$ the photon of UV light, and RH^+ the positively charged ion.

Ionization Potentials

Although all elements and chemical compounds can be ionized, they differ in the amount of energy required. Some materials lose electrons, or are ionized, relatively easily while others are not. The amount of energy required to displace an electron is called the **ionization potential (IP)** and is measured in **electron volts (eV)**. Each element and chemical compound has its own IP; the lower the IP, the lower the amount of energy required to ionize the material.

Ionization potentials vary widely between different materials. The IPs of atmospheric gases—nitrogen, oxygen, carbon dioxide—are greater than 12.00 eV, while organic chemicals have ionization potentials less than 12.00. A very important exception is **methane**, which has an IP of 12.98 eV.

Ultraviolet Lamp Capacities

The light energy emitted by a UV lamp is also measured in electron volts (eV). The energy of the UV light determines if a particular chemical will be ionized. When the UV light energy is **greater than** the IP of the material, ionization will occur. Electrons will not be displaced, and there will be no ionization, if the UV light energy is **less than** the IP of the material.

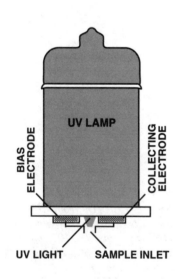

BIAS ELECTRODE

UV LAMP

COLLECTING ELECTRODE

UV LIGHT   SAMPLE INLET

Figure 6-2: UV lamp and ionization chamber of a PID.

Lamp eV energy **greater than** IP of material = ionization
Lamp eV energy **less than** IP of material = no ionization

This is a very important concept. It is imperative that PID users know the eV capacity of the lamp used, and have an idea of the IP of materials known or suspected to be present. There are a number of UV lamp capacities available; the most commonly used are 9.5, 10.2, 10.6, and 11.7 eV. One manufacturers also offers a 10.0 and 11.8 eV lamp.

UV lamps contain an elemental gas such as oxygen, nitrogen, hydrogen, or argon under low pressure. When the gas is excited by radio waves or electric current, UV light is emitted. As it leaves the lamp, the light is then focused into a powerful, narrow beam. UV energy levels depend on the gas utilized; hydrogen is used to emit 10.2 eV, nitrogen for 10.6 eV, and argon for 11.7 eV.

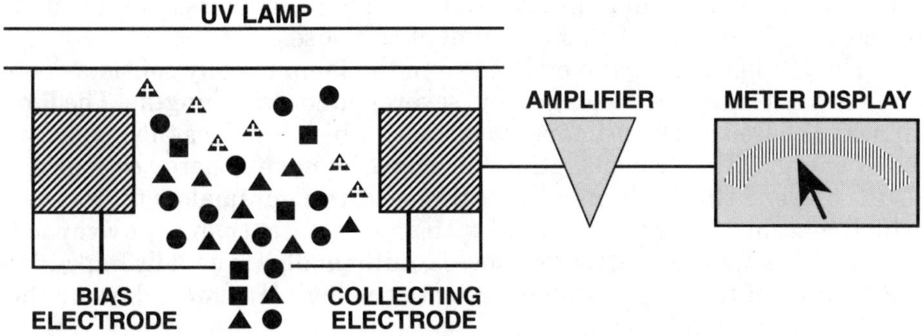

UV LAMP

AMPLIFIER **METER DISPLAY**

BIAS ELECTRODE **COLLECTING ELECTRODE**

Figure 6-3: Positively charged ions formed within the ionization chamber are collected at an electrode, producing an ion current. Chemicals that cannot be ionized (● and ■) will block the UV light from reaching ionizable materials (▲).

The 11.7 and 11.8 eV lamps are the highest-energy lamps available today; they are capable of ionizing the largest number of organic chemical species and also respond to inorganics. These high eV lamps appear, at first glance, to be the best to use. Unfortunately, this may not always be the case. While the body of the lamp is usually made of glass, the UV light emitted is focused into a beam which is directed out through a small opening or **window**. The 11.7 and 11.8 eV lamp windows are made of a special material, lithium fluoride, which can transmit high energy, short wavelength UV light. Lithium fluoride is very hygroscopic and readily **absorbs water vapor** in the air, even when not in use. Water absorption causes the window to swell and decreases the amount of light transmitted through the window which in turn reduces the amount of light available for ionization.

The lithium fluoride window is also **degraded by UV light**; the more the instrument is used, the greater the damage. Eventually, the combination of water vapor and UV damage is so great that the lamp must be replaced. A service life of approximately 3-6 months should be anticipated when using an 11.7 or 11.8 eV lamp. Service life can be somewhat extended by storing the lamp in desiccant whenever it is not in use.

Finally, although these high energy lamps are capable of ionizing a wider array of chemicals, the **intensity** of light emitted is less than that emitted from lower energy lamps[2]. The weaker ion current generated therefore requires significantly greater amplification which can lead to problems such as meter reading drift and increased sensitivity to temperature changes.

The windows of lamps emitting 9.5, 10.2, and 10.6 eV are made of magnesium fluoride; the 10.0 lamp window is usually made of calcium fluoride[3]. These lamp windows are not hygroscopic and are not degraded by

Figure 6-4: A UV lamp; note the lens (arrow) which focuses the light into a narrow beam. (Photo by Steve Napolitano.)

Elements	eV
Nitrogen	15.58
Oxygen	12.08
Chlorine	11.48
Iodine	9.28

Simple Inorganic Molecules

Carbon monoxide	14.01
Carbon dioxide	13.79
Nitrogen dioxide	9.70
Water	12.59
Hydrogen sulfide	10.46
Hydrogen cyanide	13.91

Saturated Hydrocarbons

Methane	12.98
Ethane	11.65
n-Butane	10.63
Hexane	10.17
Cyclohexane	9.88
Octane	9.82

Unsaturated Hydrocarbons

Acetylene	11.41
Ethylene	10.52
Propylene	9.73
Butadiene	9.07
Cyclohexene	8.95

Chlorinated Hydrocarbons

Methylene chloride	11.35
Chloroform	11.42
Carbon tetrachloride	11.47
Vinyl chloride	10.00
Trichloroethylene	9.45
Tetrachloroethylene	9.32

Aromatic Hydrocarbons

Benzene	9.25
Xylene	8.45
Toluene	8.82
Phenol	8.50
Naphthalene	8.12

Aldehydes and Ketones

Acetone	9.69
Methyl ethyl ketone	9.53
Methyl isobutyl ketone	9.30
Formaldehyde	10.87
Acetaldehyde	10.21

Oxygenated Compounds

Methyl alcohol	10.85
Ethyl alcohol	10.48
Isopropyl alcohol	10.16
Diethyl ether	9.53
Ethylene oxide	10.56
Propylene oxide	10.22
Ethyl acetate	10.11
Acetic acid	10.37

Sulfur and Nitrogen Compounds

Carbon disulfide	10.08
Methyl mercaptan	9.44
Dimethyl sulfide	8.69
Ammonia	10.15
Acrylonitrile	10.91
Methyl amine	8.97

Table 6-1: Ionization potentials (IPs) of elements, and organic and inorganic compounds, in electron volts (eV).

UV radiation and therefore have a much longer service life. The 9.5 eV lamp is typically used at fixed facilities or at locations where the user wishes to selectively discriminate between low IP and higher IP materials. It is rarely used in other situations since it detects relatively few materials. The 10.2 and 10.6 eV lamps are most commonly found at hazardous materials sites. The 10.2 lamp has a good service life; both have relatively broad sensitivities to organic materials.

Ultraviolet Lamp Sensitivities

There is much confusion over the sensitivity of PIDs in general, as well as lamps with differing eV capacities. When we say a PID is sensitive to a particular material, we really mean that the instrument will respond to it. Recall our discussion regarding IP of a material versus the eV capacity of the lamp; a material with an IP greater than the lamp eV will essentially not be detected, and there will be no instrument response.

The UV light spectrum emitted from the lamp usually consists of one dominant or primary wavelength and several minor wavelengths. The lamp capacity is based on the primary wavelength which provides the majority of the eV energy. Minor wavelengths which have higher eV energy ratings may also be present; this explains why a PID may give a minimal meter response to high concentrations of a material with an IP greater than the eV capacity of the lamp. Since the relative response is quite small, it is usually impossible to determine if the meter is responding to the high IP material or another contaminant with a lower IP.

Manufacturers usually offer tables listing the sensitivities of their PID with a particular eV capacity lamp to various chemical compounds. Sensitivities are usually indicated as meter response to a known concentration, usually 10 ppm. Sensitivities range from greater than 10 to 0.02 (the limit of detection).

Several manufacturers of digital display PIDs offer "response factors". Each factor represents the number, which when multiplied by the meter response, gives the actual concentration present (usually 10 ppm). For the most part, these factors are correctly represented in the instruction manual; the manufacturer warns that the instrument will respond to any ionizable materials present, not just the chemical of interest.

One manufacturer which offers a 10.0 eV lamp provides an extensive list of "response factors" as separate sheets with the product brochure; these factors do not, however, appear anywhere in the instruction manual. Factors range from 0.5 to 100; a factor of 100 indicates that the meter reading obtained was 0.1 which must be multiplied by 100 to get 10 ppm. Compounds with very large response factors have IPs much greater than the lamp 10.0 eV capacity. For instance, response factors for carbon tetrachloride (IP 11.47), 1,2-dichloroethane (IP 11.12), and ethylene oxide (IP 10.57) are 100, 50, and 33, respectively. Such "response factors" are absurdly high; providing this information is, in our opinion, irresponsible and misleading. This instrument should **not** be used to detect compounds with IPs greater than lamp capacity.

Relative photoionization sensitivities for organic compounds depend on the number of carbon atoms present, type of bond (saturated or unsaturated), presence of functional groups (alcohol, halide, amine, thiol), and structure (cyclic, branched, linear). These sensitivities are **not** response factors; they should be used as a guide to determine if the PID in question will give an adequate response and is appropriate for the anticipated use.

It is important to carefully review the criteria used to determine the relative photoionization sensitivities. For instance, meter responses recorded after calibration with the factory calibrant without additional adjustment will be very different than when the meter reading is optimized by amplifying the response or altering calibration settings.

What the Readings Really Mean

All PID readings are always relative to the factory calibrant gas specified by the manufacturer. The typical range of PIDs is 0.1 to 2000 ppm for the calibration gas. When calibrant gas is introduced into the ionization chamber, ions are formed and an ion current is produced. The current is amplified and then displayed on the analog or digital display. Instrument readings in response to other materials correspond to the ion current relative to that produced in response to a known concentration of calibrant gas. A reading of 10 units indicates that the ion current produced is the same as when 10 ppm of the calibrant gas was present.

Benzene and isobutylene are the most often used factory calibrants. A factory calibrant is the gas used when an instrument is initially calibrated at the time of manufacture; factory calibrations use multiple concentrations throughout the response range. Once calibrated, the meter responds accurately, under optimal conditions, **only** to the calibrant gas. The meter cannot discriminate between the calibrant and other gases and vapors; it simply detects an increase in the ion current and displays a reading. Since all readings are relative to the calibrant, meter responses should never be recorded in ppm, but rather as ppm-calibration gas equivalents, or PID units, or simply units. Unfortunately, most PID readout displays indicate readings in ppm, even though this is inaccurate except when only the calibration gas is present.

Chemical	Sensitivity
Xylene	11.4
Benzene	10.0
Toluene	10.0
Diethylamine	9.9
Trichloroethylene	8.9
Acetone	6.3
Methyl ethyl ketone	5.7
Vinyl chloride	5.0
Propylene	4.0
Hydrogen sulfide	2.8
Hexane	2.2
Ammonia	0.3
Nitrogen dioxide	0.02
Methane	0.0
Acetylene	0.0

Table 6-2: Optimized photoionization sensitivities for the HNU PI-101 equipped with a 10.2 eV lamp[4].

LIMITATIONS AND PRECAUTIONS

Most of the limitations of PIDs are simple to understand when one realizes that the entire detection principle is based upon the transmission of UV light. Think for a moment about light from a flashlight or your car headlights and what can affect it.

Water Vapor (Humidity)

The most important limitation which afflicts all PIDs is water vapor or humidity. Water vapor has an IP of 12.59; it is not ionized by UV light, but it does deflect, scatter, and absorb the light. Imagine driving your car at night in the fog; light from the headlights is scattered. In some cases the light is actually reflected back toward you, making it difficult to see. The same phenomenon occurs within the ionization chamber. The UV light beam is usually focused into a narrow beam which bombards contaminants in the sample air as it enters the ionization chamber. Water vapor or ambient humidity scatters the light, allowing less UV light to reach contaminants that are ionizable, and results in a lower meter reading.

In the presence of a significant concentration of contaminants, a meter reading will still probably be obtained. However, when there is little or no contamination present, and a significant concentration of water vapor is present, ambient (clean air) background readings may be significantly reduced or even eliminated. This concept is revisited in a later section.

A frequently overlooked effect of water vapor is that it often contains mineral salts and carries a charge. Atmospheric water vapor in areas around the sea or brackish water may produce a positive meter response in the absence of chemical contaminants. The water vapor essentially is a very dilute electrolyte solution which becomes ionized when it enters the electric field inside the ionization chamber. When this occurs, the ions move toward the collector electrode and produce an ion current, which results in a meter response. The same effect may be seen when conducting ground water investigations in areas where the water is "hard" because it contains a significant concentration of minerals.

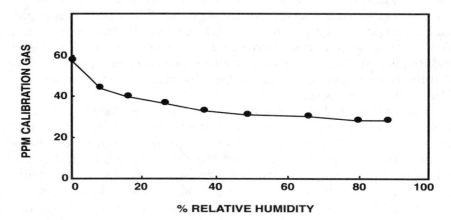

Figure 6-5: Effect of relative humidity on photoionizer response[5].

Non-Ionizable Gases and Vapors

Chemical compounds with IPs greater than lamp eV capacity act in the same way as water vapor. They scatter UV light and in large concentrations act as UV blockers. UV blocking occurs when there is significant excess of non-ionizable materials present. Imagine a football team with an awesome front line to protect the quarterback and running backs, and only the quarterback and running backs can be ionized. The UV light must get past the front line blockers before it reaches the ionizable team members. Chemical UV blockers act in the same way, essentially acting as shields which protect ionizable materials. Meter response can be significantly decreased when significant concentrations of non-ionizables are present. For example, it has been demonstrated that Photovac TIP meter response to toluene is decreased by 30% in the presence of 0.5% methane[6].

Lamp Condition

A dirty UV lamp will also affect meter readings by decreasing the amount of UV light transmitted. Dirt can enter instruments that do not have an inlet filter, and gases and vapors leave a residue on the lamp window. Imagine how well you would see out your car window if it were never cleaned, or how easy it would be to see if you never cleaned your eye glasses! Lamps should be cleaned on a regular basis; some manufacturers recommend cleaning at the end of each use, or at the end of a specified number of instrument hours.

Dust and Particulates

Although most PIDs have an inlet filter, dust and particulates smaller than the filter size can still enter the ionization chamber. The importance of dust

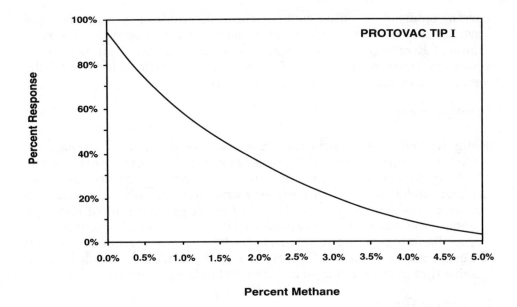

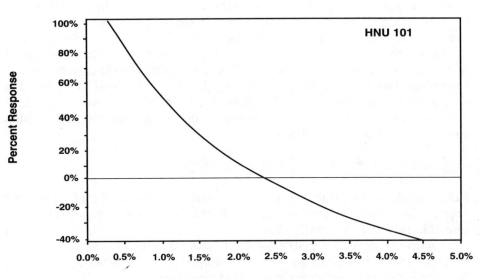

Figure 6-6: Percent decrease in instrument response to 102 ppm toluene diluted with increasing concentrations of methane compared to meter response to 102 ppm toluene diluted with hydrocarbon free air[6].

and particulates should not be overlooked. Think about how we see dust; in most cases, dust particles are so small that they cannot be seen individually. Only when they aggregate on a surface can they be seen; but we can see each particle as it dances in a beam of sunlight streaming into a room. Look away from the beam of sunlight at the rest of the room and we see no dust; it is still there, but the light is not reflecting off of each dust particle, therefore we cannot see it. The same light-scattering effect occurs within the ionization chamber; the dust not only scatters the light but also acts as UV blockers and protects ionizable materials from coming in contact with the UV light.

Dust and particulates may also be charged and affect the ion current within the ionization chamber. Your television or computer screen is like a dust magnet because the screen becomes statically charged while it is on; dust particles with an opposite charge are attracted and settle on the screen. The same phenomenon can occur within the ionization chamber; when charged particles enter the ionization chamber they can augment the ion current, producing a positive meter response. Alternatively, strongly charged particles may actually stick to the collector electrode, producing erratic readings. Erratic readings in response to charged particles are most often seen in PIDs which lack an inlet filter and allow large particles to enter the ionization chamber which can then not escape.

Heated Atmospheres

Hot gases and vapors are difficult to monitor because they often condense within the sample hose, or on the cooler lamp window. Condensation on the UV lamp window results in a significant decrease in UV transmission. Suspect condensation when there is a dramatic increase, then a sudden decrease in the meter reading. The final meter reading is often less than original background, since the amount of UV light escaping from the window and available to generate an ion current is dramatically decreased. Although the condensate will eventually evaporate off the window, it usually leaves a residue that must be removed by cleaning the lamp window.

Corrosive Gases

Corrosive gas and vapors can affect the electrodes within the ionization chamber as well as the lamp window. Prolonged exposure to corrosive materials may result in permanent fogging or etching of the window. If a PID is exposed to a corrosive material, the lamp and ionization chamber should be thoroughly cleaned immediately after use. During use in the presence of corrosives, the instrument should be regularly re-calibrated to ensure it is still working properly. Some PID manufacturers offer corrosion-resistant models, nevertheless, these instruments should not be exposed to high concentrations of corrosive gases and vapors.

High Concentrations of Gases and Vapors

PIDs are designed to detect relatively **low** concentrations of gases and vapors. The response curve for a typical PID is linear only to about 400 or 500 ppm. Manufacturers usually camouflage this information by indicating that the **usable** range of the instrument is 0 to 2000 ppm. This means that the instrument will indeed give a response to 2000 ppm of the factory calibrant, but it will not be a particularly **accurate** response.

Inadvertent exposure to excessively high concentrations of ionizable vapors or gases often results in a very high or maximum meter response, followed by a gradual decrease until the meter stabilizes at a lower reading. This response is often seen when a PID user samples the headspace of a drum, and is a result of the overwhelming concentration (i.e. greater than 10,000 ppm) entering the ionization chamber. Only a small proportion of molecules can be ionized, the remainder escapes ionization because the UV light cannot reach all the molecules at once. Remember that the ionization chamber is exceedingly small; a limited number of molecules can be ionized as they pass through the UV light.

Response Factors?

The combustible gas indicator catalytic sensor described in Chapter 4 has a compensating filament which is exposed to **identical conditions** as the catalytic filament. This allows the catalytic sensor to compensate for ambient conditions such as temperature and humidity. Such compensation allows the CGI to remain accurate under a variety of conditions encountered in the field.

PIDs have only **one** lamp and **one** ionization chamber; there is **no compensation**. Accordingly, instrument ionization efficiency varies widely depending on environmental conditions and the presence of interfering gases and vapors. Some manufacturers provide so-called response factors, but it should be evident that these are inappropriate to use in field situations, since the presence of other contaminants can invalidate meter readings. Except under stringently controlled laboratory conditions, response factors are of little practical use. They should only be used as a guide to assist the PID user in determining what to expect in terms of optimal instrument response.

CALIBRATION

Since all meter readings are relative to the calibrant gas, it is obvious that meter calibration is very important. Calibration is performed initially at the time of manufacture using multiple concentrations within the response range. In most cases, the manufacturer provides calibration gas or span gas in air at one or more concentrations. The one-point calibration procedure ensures that the UV lamp and ionization detector are operating properly, and the appropriate ion current is produced when a known concentration of gas is introduced into the ionization chamber. Put simply, it is the only way to ensure that the instrument is working properly.

Calibration of PIDs was originally performed by the user employing a known concentration of the factory calibrant gas. The gas was introduced into the instrument and the meter readout compared to the actual concentration present. If necessary, an external span potentiometer or span "pot" was adjusted until the meter reading corresponded directly to the actual concentration present. This procedure was also known as "spanning the instrument" and the calibration gas was often called "span gas".

The function of the span potentiometer is similar to the fine adjustment knob on your television set. Imagine having a set which receives only one channel; the fine tuning knob allows you to get a better signal which gives a sharper and clearer picture. The span potentiometer amplifies the signal derived from the ion current, allowing you to get a more accurate reading in response to the factory calibrant. It will not, however, allow you to detect materials with IPs beyond its lamp eV capacity.

Manual calibration also allowed the user to assess the status of the UV lamp. Frequent span potentiometer adjustments suggested that the lamp was dirty and required cleaning. Erratic or drifting meter readings indicated that the ionization chamber was dusty or the collector electrode was dirty. If significant improvements were not obtained after cleaning the lamp and ionization chamber, the user was alerted that the lamp was beginning to age and would soon need replacement.

Digital display instruments such as the Thermo Environmental OVM, MSA Photon, Photovac MicroTIP and HNU DL-101 are now available with microprocessors which essentially conduct calibration procedures independently. The user participates only by punching into the keypad the actual concentration of gas used. Span potentiometer adjustments during calibration are performed by the instrument rather than the user. In most cases, it is not possible to determine the extent or frequency of span adjustments required for the instrument to accurately read the calibration gas.

HNU instruments are factory calibrated to benzene, which is toxic. Another gas, **isobutylene** is used as a span gas for calibration. Isobutylene has a calibration response curve similar to benzene; however, it is ionized less efficiently than benzene. HNU provides cylinders containing 100 ppm isobutylene in air. Representative cylinders of each lot of span gas are tested on HNU instruments, and the ion current is measured. **The concentration of benzene which produces the same ion current is then assigned to that lot of gas.** Concentrations assigned to HNU span gas cylinders are in factory-calibrant (benzene) equivalents; the concentration for instruments with 10.2 eV lamps is usually around 55 or 56 ppm, for 11.7 eV lamps it is around 65 or 66 ppm. HNU instruments can be calibrated directly from the low pressure cylinder, using an HNU regulator which supplies the appropriate flow of gas. Alternatively, the span gas may be placed in a sample bag and then drawn into the instrument.

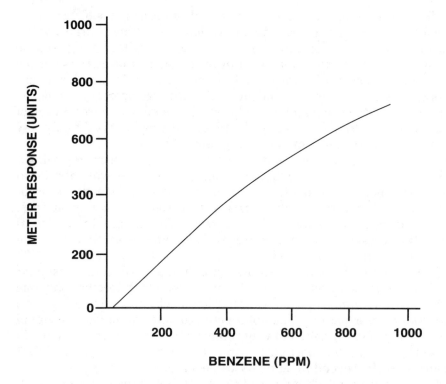

Figure 6-7: Typical calibration curve for an HNU meter in response to benzene, the factory calibrant[7].

Many HNU users purchase 100 ppm isobutylene span gas from specialty gas suppliers for approximately half the price than when purchased from HNU Systems. We have seen significant variability in the actual concentration of span gases received from specialty gas suppliers. In more than one case, instruments have been actually returned for service because they could not be calibrated with an incorrectly prepared or labeled lot of span gas. If another calibrant gas supplier must be used, a few safeguards should be taken. When buying multiple cylinders of span gas, make sure they are all from the same lot; the lot numbers will be on the label or stamped on the side of the cylinder. When a shipment of span gas cylinders arrive, several cylinders should be randomly selected and checked against HNU span gas. This may be done by first calibrating the meter with HNU span gas, then repeating the calibration with the new lot of span gas. The span potentiometer settings should be very close or identical. This procedure requires the purchase of at least a few cylinders of HNU span gas for quality control purposes, but it is worth the investment.

Most manufacturers supply only one concentration of span gas, usually in disposable cylinders. Only HNU supplies the user with a calibration curve which clearly indicates the linear portion of the instrument response. Other manufacturers offer similar information, but it must be gleaned from the specification section of the instruction manual. For instance, the accuracy of the MSA Photon after calibrating with 100 ppm benzene is ± 10% for benzene concentrations between 0 to 100 ppm, ±20% between 100 to 1000, and ± 25% for concentrations between 1000 to 2000. Thus, a meter reading of 1500 units would be considered accurate, under optimal conditions, in response to 2000 ppm benzene.

Figure 6-8: Calibration of an HNU PID using the method recommended by the manufacturer. (Photo by Steve Napolitano.)

Zeroing the Instrument

HNU analog display instruments offer the user the advantage of electronically zeroing the instrument on standby before turning on the UV lamp. While on standby, all systems are operational except for the UV lamp. This allows the user to assess the stability of the meter readout, which reflects the status of the collector electrode and electric field in the ionization chamber. When the UV lamp is turned on, the needle display gives a reading, which depends on the location.

Clean air **always** has some ionizable materials present and a reading of at least 0.2 units should be obtained. Naturally occurring materials, including fruit (especially citrus), trees (especially conifers), grasses, and flowers emit ionizable vapors and gases which will give readings. Vehicles, roadbeds, clothing, deodorants, perfumes, soaps, bug sprays, and equipment can also be sources of background meter readings. Background readings are not static but will vary with location, humidity, temperature, and proximity to vehicles or heavy equipment. It is important to be aware of background readings; such readings should never be "zeroed out". For example:

An instrument is electronically zeroed in an old field containing grasses and wildflowers adjacent to a hazardous waste site that is devoid of vegetation. The UV lamp is turned on; initial background reading in the field is 1.8 units. At the edge of the waste site where there is no vegetation, meter readings drop to 0.4 units; readings increase to 2.0 units over an area where the soil is discolored. The increase above the initial background is only 0.2 units; however, the increase over the site background is 1.6 units. If the 1.8 unit background had been zeroed out, a barely perceptible meter reading of 0.2 units would have been obtained, and possibly missed or ignored.

If the background meter reading seems inordinately high (or low), the operator can use non-contaminated air (also called zero air or hydrocarbon-free air) from a specialty gas supplier. After zeroing with non-contaminated air, the ambient air background reading should be again assessed.

Instruments with microprocessors often tempt the operator to use **clean** air as **zero** air; zero air has no contamination or ionizable materials present. When the user instructs the instrument to use background air as zero air, the instrument subtracts the background ion current value present from all subsequent readings. Since most microprocessor controlled instruments are programmed to assign a value of 0.0 to the background, the user is forced to zero the instrument with zero air, and then obtain the actual background reading present.

Assigning an Arbitrary Background

Instruments such as the HNU analog display meters, and the Photovac TIP allow the user to increase the background reading. Remember that humidity and non-ionizable gases affect meter response and may actually reduce or eliminate background readings. If the background present is very low, such as 0.2 units, a reduction of the background reading may not be obvious on the HNU, since the needle display does not fall appreciably below zero. The Photovac TIP, on the other hand, will give a negative reading if the ion current falls below zero. However, many users, (not to mention clients) are uncomfortable with a negative reading. In some cases, the minus sign may be missed, and the reading mistaken for a positive response.

When the background value is very low, it is appropriate and recommended to increase the background. A background of 1.0 unit for HNU analog meters allows the user to observe upward as well as downward needle deflection. The Photovac TIP background tends to drift downward; a background value of 2.0 to 5.0 units is usually sufficient to maintain positive background readings during use.

Assigning a background of at least 1.0 unit allows the user to detect the presence of high non-ionizable gases and vapors, and water vapor. For example:

An instrument is turned on and zeroed with hydrocarbon free, zero air; the unit is then calibrated with 100 ppm isobutylene. A background reading of 1.2 units is obtained when the sample bag containing calibration gas is disconnected and ambient air is sampled. The user begins to take background readings from a variety of areas adjacent to the worksite. The meter probe is placed through a hole of a sewer manhole cover. The meter reading immediately drops to 0.0. The sewer contains water vapor and probably methane gas. Both materials will act as UV blockers and decrease meter ionization efficiency; large concentrations of such materials can reduce the meter reading to zero.

Humidifying Calibration Gas

Calibrating an instrument does not compensate for the presence of interfering gases and humidity. HNU Systems offers an accessory which can humidify calibration gas which normally contains no water vapor. The calibration gas humidifier consists of a permeable membrane tubing placed in line between the calibration gas cylinder and the meter. This device introduces ambient water vapor concentrations into the calibration gas; after the span is adjusted, the effect of water vapor concentrations present at the time of calibration is eliminated.

The humidifier should be used only after the meter has been calibrated with dry gas and the span setting recorded. The humidifier is then connected and the meter recalibrated. A significant difference in the span setting will be obtained in the presence of high humidity. The effect of humidity will be markedly reduced or eliminated for humidity conditions present **at the time of calibration**. Humidity conditions change quickly, however, with temperature, weather conditions, vegetation, and presence of standing or running water. It is not unusual to encounter a variety of environmentally different areas such as a parking lot, open field, and shaded woodland; the amount of water vapor present can vary significantly between these locations. It is incorrect to assume that a one-time calibration with a humidifier will compensate for humidity effects for an entire day and for all locations sampled.

USING AND CHOOSING PHOTOIONIZATION DETECTORS

Inherent Safety

A PID utilized at hazardous materials sites should have at least a Class I, Division 2, Groups A,B,C,D rating from a recognized organization such as FM, UL. Not all PIDs are certified for use in hazardous locations; an instrument **designed** for such use is not necessarily **certified**.

HNU Systems, Inc.

HNU pioneered the development of portable photoionization detectors and was the first manufacturer to offer a PID as a monitoring instrument for fixed facilities as well as hazardous materials site work. The HNU analog display meters are still the most widely used PIDs today; HNU PIDs can be equipped with 9.5, 10.2, or 11.7 eV lamps. HNU offers a general purpose meter which has no intrinsic safety approvals (GP-101), as well as meters with Division 2 (PI-101, HW-101, DL-101) and Division 1 (IS-101) ratings. The IS-101 is also corrosion resistant; corrosion protection is available as an option on other HNU PIDs. Original-design HNU meters have two basic components, a **probe** which houses the UV lamp and a **box, readout assembly, or control module** which contains the meter display, battery, and controls. The probe cable is attached to the assembly by a 12-pin plug which fits into a connector. A short, hollow tube probe extension can be also be used.

  The probe and readout assembly are provided as a single unit for all HNU instruments; probe and readout assembly have the same serial number. The UV lamp in the probe and readout assembly are factory calibrated as a unit; it is important that they remain together. If probes and readout are interchanged, the instrument is no longer precisely calibrated; significant changes on the external span potentiometer will be necessary.

  In the GP, PI, and IS models, a small fan inside the probe draws sample air into the ionization chamber; the resulting current is translated into a

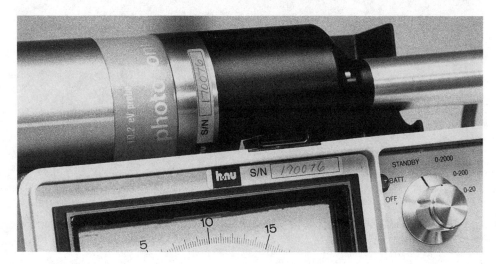

Figure 6-9: Close-up of an HNU HW-101; note the serial numbers on the probe and box; lamp capacity (10.2 eV) is also indicated on the probe. (Photo by Steve Napolitano.)

meter response which is read on the analog display. There are three ranges, 0-20, 0-200, and 0-2000 units, which are selected on the control knob. A battery check as well as standby mode are also available.

The fan, UV lamp, ionization chamber, detector, and amplifier are located in the probe, which is attached by a cable to the readout assembly. Accessing and cleaning the lamp is simple and painless.

The GP-101, PI-101, and IS-101 do not have an inlet filter; without a filter, dust and particulates in sample air can enter the ionization chamber. This is a serious limitation at dusty sites, since particulates can affect lamp and detector function. Fortunately, electrical interference caused by charged particles inside the ionization chamber can be detected by erratic readings while the instrument is on standby.

Steel wool packing inside tygon tubing has been used to prevent large particles from entering the probe and minimize gross contamination of the probe extension[8]. However, HNU Systems recommends that **no** filtering material be used in conjuction with their fan-equipped photoanalyzers. The drawing power of the fan (between 125-180 cc/min) is insufficient to accommodate any filtering material, no matter how coarse.

The HW-101, designed for hazardous materials applications and soil gas monitoring, has a positive displacement pump and inlet filter. A short probe extension, which contains a 20 micron inlet filter, must be used whenever the instrument is on. The pump allows sampling from remote locations up to 50 feet, which is not possible with the fan-equipped models.

The HW-101 also has an LED bar display at the back of the probe which provides an indication of the meter reading. Each bar represents 10% increments of the scale setting, that is, when the range selected is 0-200, each bar represents an increase of 20 units. Increases of less than 10% scale will not be indicated on the LED display. This feature frees the user from continually checking the meter display for possible readings; however, small shifts in the meter reading will not change the LED display and can be missed. An added feature is a probe holder attached to the readout assembly case, which can be used to cradle the probe when the unit is not in use. The holder should not be used to carry the probe while the instrument is carried.

One problem frequently encountered by HW-101 users is a clogged inlet filter. This often occurs during soil gas monitoring, when water or dirt (or

Figure 6-10: The HNU Systems DL-101 photoionization detector is equipped with a datalogger. Readings can be stored and recalled on the LED display or transferred to a PC or printer. (Author photo.)

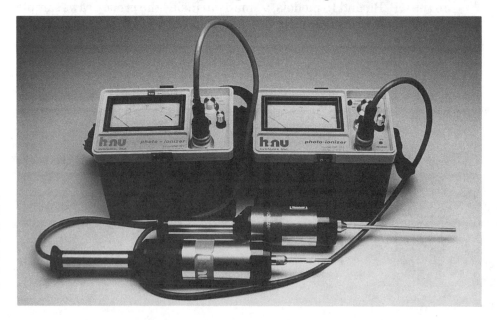

Figure 6-11: HNU Systems GP-101 (right) and HW-101 (left) photoionization detectors. The HW has a shorter probe extension with a 20 micron filter; the longer GP probe extension is a hollow tube without a filter. (Photo by Steve Napolitano.)

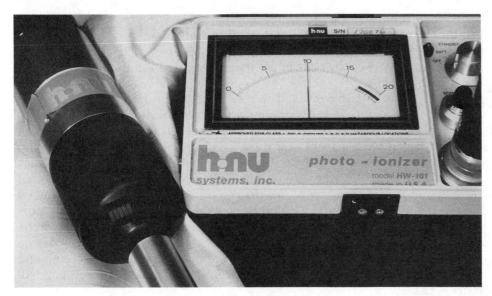

Figure 6-12: Five bars are evident on the LED display just above the probe handle of this HW-101; this indicates that the meter reading is 10 units. Each bar represents 10% of the full scale for the 0-20 unit range selected. (A Steve Napolitano photo.)

both) are drawn into the probe extension. In most cases, the user had previously abused a GP-101 or PI-101 by using the hollow probe extension to make holes in loose soil before taking a reading. This usually resulted in a dirty lamp and clogged probe extension. It must be stressed that the presence of an inline filter does not make the HW-101 immune to such blatant mistreatment. Care is still required to avoid drawing water or soil into the probe. If this occurs, the instrument should be immediately shut off, and the inlet filter replaced. An HW-101 should never be operated without the filter; the ionization chamber is designed to operate under a slightly negative vacuum which the filter maintains.

One criticism of HNU meters is that they require two-handed operation; one hand holds the probe while the other must be ready to change the range setting as necessary. This is true when one is conducting soil gas surveys or sampling in areas where close attention and control as to the location of the probe extension is required. The advantage of a probe separate from the rest of the meter is that it can be placed in relatively inaccessible places.

Figure 6-13: The filter and probe extension of an HW-101; the meter should always be operated with a CLEAN filter in place. (Photo by Steve Napolitano.)

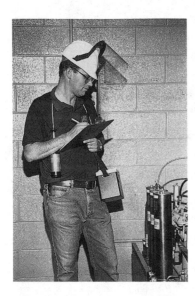

Figure 6-14: Relatively hands-free use of a GP-101; note that the hollow probe extension has been removed. (Photo by Steve Napolitano.)

When conducting general site survey work, the probe can be placed around the user's neck and where it is left to dangle; unless it is an HW-101, the probe extension is not needed and can be removed. This allows free use of both hands most of the time while the meter takes readings directly from the breathing zone.

We are always amazed at the resilience of the HNU probes, having observed their being abused in the most unbelievable ways. Probes used as hammers and rubber ball bats; they have been thrown into the back of pickups; stored with augers and pipe wrenches; left out in the rain; and dropped into ponds and lagoons. Probes have also been used to punch holes through drums and containers, and to make bore holes in soil. We do not recommend any of these uses. We **do** recommend that probes be handled gently and stored properly. The probe represents more than half the purchase price of the instrument; it is expensive to replace or repair. The probe should never be stored in the top of the instrument case which provides no padding or protection and will eventually damage the cable. We recommend transporting the readout assembly, probe, and accessories in a well-padded, sturdy, case.

The recently introduced DL-101 is a microprocessor based PID with features similar to the HW-101. Readings may be automatically stored in the field and then later recalled and displayed on the LED display or transferred to a PC or printer. The DL-101 has a control module with controls and digital display and a separate probe with an LED concentration bar display and pistol grip. The unit has up to four operational modes: survey, hazardous waste, industrial hygiene, and leak detection. In the default or survey mode, the meter is factory calibrated to benzene and uses isobutylene as the span gas. The unit stores up to 12 calibrations in the hazardous waste mode. The industrial hygiene mode is similar to the hazardous waste mode except that it also offers the option to determine concentration averages or TWAs from readings logged at a particular site. In the leak detection mode, a reading is recorded each time a log button is pressed.

Photovac Inc.

Photovac instruments do not have a separate probe but rather incorporate all components into a single unit. The Photovac TIP™ (which stands for Total Ionizables Present) is a lightweight hand-held instrument equipped with a pump, inline filter, and digital LCD readout. TIP models are no longer produced but many are in use and Photovac still provides technical support and repair services for these units. The TIP was usually provided with a 10.6 eV lamp; 9.5, 10.2, and 11.7 eV lamps were also available. The first TIP model made, now known as the TIP 0, did not seem especially designed for field use; there was no shoulder strap and the on-off switch was a flimsy toggle which could be (and frequently was) easily sheared off. The rounded shape allowed the instrument to roll when placed on a surface. A battery life of no more than 4 hours was obtained after a 16 hour battery charge.

Later TIPs were improved and had an anti-roll bar, shoulder strap, and a push-button on-off control. One feature that was not changed was the span and zero control knobs. Locking hubs for controls lock to the **left** and release to the right. This is the reverse of what is normally encountered (i.e. right to tighten, left to loosen) and has resulted in much confusion as well as damaged controls knobs caused by operators forcing the hubs in the wrong direction.

The TIP I has a yellow body color and has no intrinsic safety rating; the TIP II is orange and has a Class I, Division 2 rating. The TIP is equipped with an inlet filter; a hollow stainless steel or Teflon tube extension can also be used. The pump draws approximately 600 cc per minute and can be used for soil gas analysis or other applications where samples must be drawn from a location distant from the instrument.

One drawback of the TIP is its short battery life; a battery pack which plugs into the unit is available, however it is awkward and voids the intrinsic safety rating. There is no battery check option since the unit is designed to be permanently attached to a trickle charger. Another limitation is the LCD display, which is difficult to see in bright sunlight and must be shaded before the numbers can be read. The LCD display will "melt" if subjected to excessive heat or used in the sun for an extended interval. When this occurs, the unit should be left on and placed in a cooler, shaded area; the numerals will eventually coalesce and no permanent damage will be done.

Many users are disconcerted by the rapid changes in the meter readout; meter updates are limited to once per second. For most situations, we recommend recording the range of meter readings obtained at a particular sampling location; if readings cluster around a smaller sub-range, that too can be noted. It is not necessary to record decimal units! Readings of 5.3 or 5.2 can be logged as simply 5 units; it is a waste of time and notebook space to record meaningless fluctuations of 0.2 or 0.3.

The TIP is factory-calibrated with isobutylene; Photovac recommends using 100 ppm isobutylene for a one point calibration; the calibrant gas is placed in a sample bag and then drawn into the instrument. Many TIP users dread calibrating their instrument; this is because the meter reading can fluctuate significantly and it is difficult to adjust the span potentiometer to achieve an exact reading of 100 units. As a further insult, the reading is often changed when the span hub is rotated counter-clockwise and locked.

We recommend using zero air to initially zero the instrument, then calibrating with 100 ppm isobutylene and overshooting to approximately 105 to 110. Carefully rotate the span locking hub until snug; as the hub locks, the knob is usually displaced slightly to the left and reading will be reduced to around 100 to 105 units. Disconnect the sample bag and check the background reading; if necessary, adjust the background to between 2 and 5 units.

Cleaning the TIP UV lamp requires steady hands and a stout heart; once the cover is removed, the circuit board, detector, and lamp holder are exposed. The user must disconnect two small wires from the circuit board and then carefully unscrew the detector cell while keeping the exterior lamp holder from rotating. The instructions for this procedure are somewhat confusing, and a novice user can damage the meter if it is performed incorrectly. The instruction manual should include pictures or diagrams to assist the user in performing this procedure correctly. If there is any doubt about how to remove the lamp, stop and call Photovac for technical assistance.

The Photovac MicroTIP™

The MicroTIP™ is a reliable, user-friendly instrument that has overcome the problems associated with the Photvac TIP. The MicroTIP has an advanced microprocessor which allows the user to store and retrieve data from up to 12 hours of sampling. A two-line digital LCD provides numeric or a bar graph readout of meter readings and date. The meter can also record and display maximum, minimum, and average meter readings at specified intervals; an alarm message can be activated at a pre-programmed meter reading. The keypad is designed for easy access and programming, even when wearing

Figure 6-15: The Photovac TIP II. Many TIPs are still in use. TIP operators must remember that control knob hubs lock to the left, not to the right. (Photo by Steve Napolitano.)

Figure 6-16: Photovac TIP and MicroTIP users must be very careful when removing the UV lamp from its holder. The wires, resistors, and other electrical components are exposed and susceptible to inadvertent damage. (Photo by Steve Napolitano.)

Figure 6-17: The Photovac MicroTIP. (Author photo.)

gloves; each key represents a single function. Data can be downloaded to a PC or printer. The meter can accommodate a variety of lamps, from 8.4 to 11.7 eV.

A high-range option can also be accessed, which provides the highest sensitivity possible. This mode provides a shaded bar display instead of unit values and is recommended for use as a yes-no tool for the presence of low concentrations of contaminants as when attempting to detect or locate leaks.

The Mine Safety Appliances Company (MSA) offers the Photon™ Gas Detector which is similar to the Photovac MicroTIP. The Photon offers a slightly different keypad design and is available only with a 10.6 eV lamp. The Photon offers no datalogging options and cannot download data.

The MicroTIP and Photon are factory-calibrated with isobutylene. Field calibration is essentially controlled by the instrument; the MicroTIP and Photon display messages to the user and stores the results for up to five different calibration gases or lamp capacities in its memory. The instrument must first be calibrated to zero air, then to isobutylene span gas. The ion current generated is determined; the microprocessor then subtracts the stored zero air reading and divides the difference by the user-entered span gas concentration to obtain a PID sensitivity value. The sensitivity value compensates for changes for meter variability[9].

$$\frac{\text{span gas reading - zero air reading}}{\text{actual span gas concentration}} = \text{sensitivity factor}$$

In most cases, when the instrument, lamp, and ionization chamber are functioning optimally, the sensitivity value should be around 1.0 when the factory calibrant is used as the span gas. During use, the meter subtracts the zero signal from the sample air reading and then divides the difference by the sensitivity value.

$$\frac{\text{sample air reading - zero air reading}}{\text{sensitivity factor}} = \text{meter reading}$$

This procedure sounds fool-proof, however the operator must be very careful to enter the correct span gas concentration. We have observed calibration problems with microproccessed-controlled instruments when the user has inadvertently supplied the wrong gas or the wrong concentration information. For instance, if a user mistakenly informs the instrument that the calibration gas concentration is 50 ppm instead of 100 ppm, all subsequent readings will be divided by a sensitivity factor of approximately 2, resulting in erroneously low readings.

A limitation of the original MicroTIP is the relatively short battery life of 6 hours continuous use when the backlight is turned off; newer versions have a battery life of 7 hours. A battery check is available which displays normal battery voltage. The battery pack is easily changeable; Photovac and MSA recommend that two battery packs be available for use in the field. After use, battery packs must be recharged for at least 8 hours, an 8 hour interval, however, may not be sufficient to obtain a complete re-charge. Many users have opted to purchase three battery packs and two battery chargers to ensure the availability of adequately charged packs for an entire field-day.

The MicroTIP battery pack must have a minimal charge in order to maintain all stored set-up parameters and previous calibration information. When not in use, the instrument should be turned on periodically and battery

Figure 6-18: The battery pack of the MicroTIP and Photon snaps out and can be easily replaced. (Photo by Steve Napolitano.)

status checked. When the battery becomes depleted, the pack should be replaced or recharged.

Many Photon and MicroTIP users find it difficult to see the meter readout, especially when bright sunlight reflects off the display window. The meter reading is displayed in the upper right-hand corner of the display; numerals and letters are approximately 3mm high. Reading such small characters is made even more difficult when wearing a fullface respirator. In many cases, the user must bring the meter display closer or continually reorient the instrument to see the reading.

Thermo Environmental Instruments Inc.

Users of the AID 580 (made by the now-defunct Advanced Instrument Design, Inc.), may recognize the Thermo Environmental OVM (Organic Vapor Meter) Model 580. This instrument can be equipped with a 9.6, 10.0, 10.6, or 11.8 eV lamp. The OVM Model 580 also offers a variety of datalogging and data retrieval options similar to the Photovac MicroTIP. Meter readings are displayed as a bar graph as well as a digital readout. There is a 10 to 1 dilution probe available which allows sampling of concentrations up to 20,000 ppm calibration gas equivalents. The OVM Model 580 has a field battery life of 8 hours; there is no replaceable battery pack.

Like the MicroTIP and Photon, the OVM is also difficult to read in bright sunlight. The small screen displays two lines of up to 16 characters each. An audible clicking signal gives a relative indication of changes in meter readings, however, it is difficult to hear in noisy situations.

One feature of the OVM that we consider a limitation is the plug or key that must be inserted into the power receptacle at the rear of the instrument. The meter cannot be started if the power plug is not in place. When not in use, the plug is supposed to be inserted backwards to serve as a dust cap. In reality, the plug, which is connected to the meter by a small chain, can be easily lost; without the power plug, the instrument cannot be used.

Figure 6-19: The MicroTIP meter reading is displayed in the upper right corner; also displayed is the sample number or instrument status, time, and date. (Author photo.)

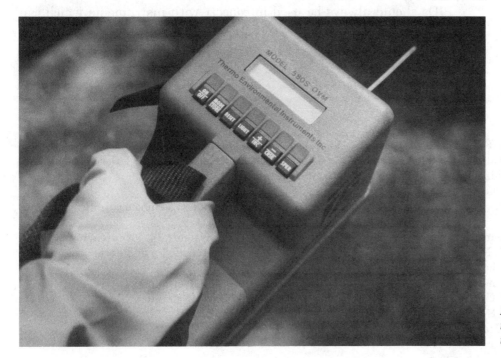

Figure 6-20: The Thermo Environmental OVM Model 580. (Author photo.)

Figure 6-21: The back of a OVM Model 580 with power plug and receptacle; the plug is connected to the meter by a small chain. If the plug is lost, the meter cannot be turned on. (Author photo.)

A significant limitation of the OVM is that the user must press multiple control switches to access different functional modes; some modes have submodes which also must be accessed. The parameters mode is the most important and controls 8 functions, including run/measurement mode, autologging, setting the alarm, lamp selection, and calibration. The user must scroll to the desired function and use a variety of keys to change readings or enter information.

Using PIDs

PID users must constantly remind themselves of two very important facts. First, PIDs **cannot** detect all potential contaminants present; therefore, the absence of a meter reading cannot be interpreted that there is no contaminant present. Second, a meter reading on a PID is **not** the actual concentration of contaminants present in ppm; the actual concentration present is usually significantly **greater** than the reading.

PIDs used for soil gas analysis should always have a positive displacement pump and an inline replaceable filter. Fan-equipped HNU meters have insufficient draw and do not have a filter. When using PIDs for soil gas analysis and headspace monitoring, the effect of humidity and water vapor must be considered; the meter response obtained will be significantly less than the actual concentration present. The lack of a meter response should never be interpreted as a lack of contamination.

Investigators at landfills use PIDs because PIDs do not respond to methane. It must be noted, however, that methane will significantly **decrease** meter response to ionizable contamination. When a Photovac TIP was exposed to 102 ppm toluene and 0.5% methane, meter response to toluene was reduced to about 70 units; when 5% methane was present, a negligible meter response was obtained. The HNU response to methane was to give diminished meter readings up to a concentration of approximately 2.5%; at greater methane concentrations a **negative** meter response was obtained[6].

PIDs should always be used in concert with instruments which can detect non-ionizable gases, such as a %LEL or ppm CGI, and a flame ionization detector (FID).

REFERENCES

1. Driscoll, J.N. and Spaziani, F.F. 1975. Trace Gas Analysis by photoionization. Anal. Instrum. 13: 111-114.
2. Discoll, J.N., Ford, J. Jaramillo, L. Becker, J.H., Hewitt, G., Marshall, J.K., and Onishuk, F. 1978. Developments and applications of the photoionization detector in gas chromatography. Amer. Lab. 10: 137-147.
3. Personal communication. Photovac International Inc., Deer Park, NY. March 1992.
4. Discoll, J.N. and Becker, J.H. 1979. Industrial hygiene monitoring with a variable selectivity photoionization analyzer. Amer. Lab. 11:6976.
5. Application Note-Calibration Gas Humidifier. Product Fact Sheet 101-22. HNU Systems, Inc., Newton, MA.
6. Nyquist, J.E., Wilson, D.L., Norman, L.A., and Gammage, R.B. 1990. Decreased sensitivity of photoionization total organic vapor detectors in the presence of methane. Am. Indus. Hyg. Assoc. J. 51(6): 326-330.
7. Instruction Manual Trace Gas Analyzer HNU Model ISPI-101. HNU Systems, Inc., Newton, MA.
8. Ressl, R.A. and Ponder, T.C., Jr. 1985. Field Experience with Four Portable VOC Monitors (EPA/600/4-85/012). Environmental Monitoring Systems Laboratory, USEPA, Research Triangle Park, NC.
9. MicroTIP MP-1000 User's Manual. Photovac International Inc., Deer Park, NY.

CHAPTER REVIEW

1. Photoionization detectors use _____ light to form ions and create an ion current which is proportional to the number of ions formed.

2. The most commonly encountered lamps used for hazardous materials site work are the _____eV and _____eV.

3. True or false? The 11.7 eV lamp has a much shorter service life than other eV capacity lamps.

4. List two reasons why 11.7 eV lamps have a short service life._____
   _____.

5. For a material to be ionized, the eV capacity of the lamp must be greater than the _____ of the material.

6. APID is equipped with a 10.6 eV lamp. Which of the following materials will elicit a negligible reading? Carbon monoxide, carbon tetrachloride, oxygen, toluene, benzene, acetone. (Hint: check Table 6-1).

7. The PID detection principle is based on UV light. List some limitations of PIDs: _____
   _____.

8. List two commonly encountered gases or vapors that can act as UV blockers: _____
   _____.

9. True or false? PID readings can be considered ppm readings, regardless of the contaminants present.

10. True or false? Response factors can be used in the field to determine the exact concentration of contamination present.

11. A PID is turned on in a clean area, zeroed with hydrocarbon-free zero air, and then calibrated with 100 ppm isobutylene. The background in the clean area is 2.2 units; should the PID be adjusted so it reads 0.0? Why or why not?

12. True or false? Methane and other non-ionizable gases can significantly reduce meter response to ionizable materials.

13. True or false? Condensation or dirt on the UV lamp window has no effect on the efficiency of the meter to ionize materials.

14. The detection range for PIDs is _____ppm for the calibration gas.

15. PIDs are most commonly factory calibrated to _____ or _____. The most common span gas for one-point calibration by the user is _____.

PROBLEM SET
Refer to Appendix D for ionization potentials.

1. You are investigating an industrial dry cleaning facility where a drum of waste solvents has been spilled. The spilled solvent is a mixture of tetrachloroethylene and carbon tetrachloride. Which material will a PID with a 10.2 eV lamp respond to? Will the other material affect meter response?

2. At the incident described in the first problem, the PID with a 10.2 eV lamp gives readings between 720 and 800 units. Can the meter response be converted directly to ppm? Why or why not?

3. At a methyl ethyl ketone spill, absorbent pads used to mop up the spilled material are placed inside an empty 55-gallon drum. PID meter readings become erratic when placed right over the sorbents in the drum. What is happening?

4. A PID with a 10.6 eV lamp is used to screen soil samples saturated with water. Clean glass jars containing 10 cc soil samples are covered with aluminum foil and then heated in a water bath set at a temperature of 30 °C. After 5 minutes, the headspace of each jar is sampled. What difficulties can you anticipate using this protocol? Some samples give no meter response; is it appropriate to conclude that there is no contamination present in these samples? Why or why not?

5. A perimeter survey of a petroleum product tank farm is about to be performed. Investigators park on the side of a busy highway adjacent to the tank farm. A PID shows a reading of 6.2 units when turned on. The user calibrates the meter, then decreases the background meter reading from 6.2 units to zero. Is this procedure correct? Why or why not?

7

Flame Ionization Detectors

LEARNING OBJECTIVES

1. Describe how flame ionization detectors work.
2. List the major limitations of flame ionization detectors.
3. Describe how flame ionization differs from photoionization.
4. List several important gases that cannot be detected by flame ionization.
5. Discuss the difference between response factors and ionization sensitivity.

Flame ionization detectors (FIDs) are low concentration instruments which have a detection range of 0.2 to 1000 ppm or 1.0 to 10,000 ppm for the factory calibrant gas. Like PIDs, these meters are used when relatively low concentrations are anticipated or known to be present.

HOW A FLAME IONIZATION DETECTOR WORKS

PIDs and FIDs are ionization detectors. The real difference between the two is the way ions are produced. Recall that PIDs used high-energy UV light to displace electrons and form ions. FIDs do not use UV light, instead, the organic materials in air are **burned** in a **hydrogen-fed flame**. The flame has sufficient energy to ionize any organic material with an ionization potential or IP of 15.4 or less.

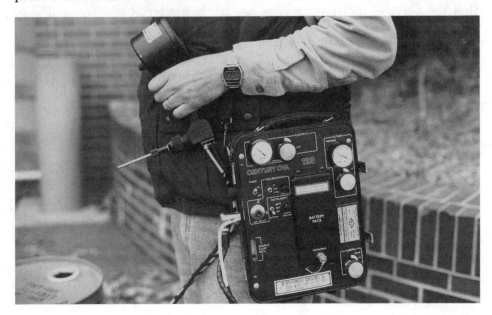

Figure 7-1: The Foxboro OVA-128 is the most commonly used FID; Foxboro is the only manufacturer of intrinsically safe FIDs. (Author photo.)

123

Type of Compound	Efficiency
Saturated Hydrocarbons	
Methane (calibrant)	high
Ethane	high
n-Butane	moderate
Hexane	moderate
Cyclohexane	high
Octane	high
Unsaturated Hydrocarbons	
Acetylene	very high
Ethylene	high
Propylene	moderate
Butadiene	low
Cyclohexene	high
Chlorinated Hydrocarbons	
Methylene chloride	high
Chloroform	moderate
Carbon tetrachloride	low
Vinyl chloride	moderate
Trichloroethylene	moderate
Tetrachloroethylene	moderate
Aromatic Hydrocarbons	
Benzene	very high
Xylene	high
Toluene	high
Phenol	moderate
Naphthalene	moderate
Aldehydes and Ketones	
Acetone	moderate
Methyl ethyl ketone	high
Formaldehyde	not detected
Acetaldehyde	low
Other Oxygenated Compounds	
Methyl alcohol	low
Ethyl alcohol	low
Isopropyl alcohol	moderate
Diethyl ether	low
Ethylene oxide	moderate
Propylene oxide	moderate
Ethyl acetate	moderate
Acetic acid	high
Sulfur and Nitrogen Compounds	
Carbon disulfide	not detected
Dimethyl sulfide	low
Acrylonitrile	moderate

Table 7-1: Flame ionization efficiencies for an FID calibrated to methane. Methane efficiency is rated as high[1].

Organic materials are chemical compounds which contain the element **carbon;** common organic compounds are methane, acetone, methanol, chloroform, acetylene, and benzene. When sample air is drawn into the instrument, it is carried into a combustion chamber where organic materials are burned and released as charged particles or ions. These ions are attracted to a collecting electrode, which produces a small ion current. The ion current is amplified and translated into a meter display. Current is directly proportional to the number of ions formed and collected.

FIDs are truly organic detectors; they do not respond to inorganic materials such as hydrogen sulfide, nitrogen dioxide, carbon dioxide, or carbon monoxide. To be detected, the organic material must have multiple carbon-hydrogen bonds, or multiple carbon-carbon bonds.

FIDs are factory calibrated to methane gas (CH_4) which has four carbon-hydrogen bonds and burns readily in the hydrogen flame. Not all organic materials, however, burn as efficiently as methane; these materials will give a proportionally lower reading than methane. Other materials, such as aromatic compounds, burn more readily and release a larger number of ions. The presence of chemical side groups, such as amines, alcohols, or halogens affect how efficiently the material is burned and therefore how well it is detected. Some very small compounds, such as formaldehyde and monomethanolamine do not have the requisite bond structure and do not release a sufficient number of ions to be detected.

Relative **flame ionization efficiencies** can be crudely classified as indicated in Table 7-1. These qualitative efficiency values are not response factors; they simply indicate how well the instrument will respond to a particular material under optimal conditions. Flame ionization efficiencies can differ markedly between similar compounds. Contact the manufacturer if you are unsure if a material will be detected or determine instrument response using a known concentration of the material.

Relative FID Sensitivities

Although organic materials burn and release ions with differing efficiencies, FIDs are sensitive to, or can detect, nearly all organic materials. Since they are not dependent on a UV lamp, the IP of materials present is not a factor in determining instrument response. FIDs are similar to PIDs, however, in that the relative instrument response to the material present will vary depending on the compound present.

Manufacturers often provide sensitivity factors, often called response values, for a particular FID model, to various chemical compounds. Sensitivities are usually listed as the meter response to a known concentration. These values are useful only under rigidly controlled conditions and should not be used as response factors. Recall that response factors are valid only when it is possible to compensate for instrument and environmental conditions and when the user is positive that there are no interfering gases or vapors present. These criteria cannot be satisfied under field conditions.

Relative response values or sensitivities are useful when the user wishes to anticipate the relative meter response to materials known or suspected to be present. If the material in question is known to have a low sensitivity, the user is alerted that the meter response will be considerably less than the actual concentration present. This information assists the user in determining if it is appropriate to use the meter for the concentrations and compounds known or suspected to be present.

What the Readings Really Mean

As with other instruments, FID readings are always relative to the factory calibrant gas. Methane is the calibrant of choice for all portable FIDs. After factory calibration, the meter will respond accurately only to the calibrant gas. Nearly all organic materials will be detected; the meter cannot distinguish between the calibrant and other organic materials. Instrument response corresponds to the ion current produced when the calibrant is present. In other words, in a properly functioning and calibrated FID, a reading of 50 units indicates that the ion current measured is equivalent to the current produced when 50 ppm methane is present. Since readings are always given in methane equivalents, they should be recorded as ppm methane-equivalents, ppm-equivalents, methane units, FID units, or simply units.

Meter readings do not represent the actual concentration present but rather give an indication of the presence of detectable materials and their relative concentration. This is acceptable, since ionization detectors are used primarily to determine the presence of detectable materials, and are not used to measure or quantify the actual concentration present.

Chemical Compound	Relative Response
Xylene	116
Benzene	150
Toluene	110
Diethylamine	75
Trichloroethylene	70
Acetone	60
Methyl ethyl ketone	80
Vinyl chloride	35
Propylene	45
Hexane	75
Methane	100
Acetylene	225

Table 7-2: Relative response values to selected chemicals for a Foxboro OVA-128 expressed as a percentage relative to methane, the calibrant gas (relative response of 100)[2].

LIMITATIONS AND PRECAUTIONS

Oxygen Concentration

Incoming sample air supplies oxygen to the hydrogen flame. The flame requires oxygen in order to burn. Oxygen deficiency reduces the height of the flame or will actually cause the flame to be extinguished. Most FIDs have flame-out alarms to alert the user if the flame goes out. However, the user will usually not be aware that flame height has been affected; when this occurs, the efficiency of combustion and ionization is reduced, and the meter reading obtained will be erroneously low.

Figure 7-2: Fuel cylinders are refilled using regulators which fit the pressurized storage tank. The cylinder may be disconnected (left, HeathTech PortaFID-II™), or fixed within the instrument (right, Foxboro OVA). (A Steve Napolitano photo.)

Figure 7-3: Foxboro OVA models 108 and 128 are intrinsically safe for Class I materials, and can be used in Division 1 and Division 2 locations. (Photo by Steve Napolitano.)

The Foxboro Company
600 N. Bedford Street
East Bridgewater, MA 02333
Tel: 1-800-321-0322

Hydrogen. Compressed Flammable
Gas UN 1049
Weight or Volume 75 ml
DOT Exemption No. DOT-E 7607
INSIDE CONTAINER COMPLIES WITH
PRESCRIBED REGULATIONS

Figure 7-4: All packages shipped under the Foxboro DOT exemption must carry this label[4]. The exemption is required only if the OVA is filled with hydrogen and taken on board passenger carrying aircraft.

Foxboro has conducted tests with their FIDs; the hydrogen flame is extinguished at oxygen concentrations less than 14%[3]. Manufacturers of other FIDs, however, have not determined the oxygen limitations of their FIDs.

Hydrogen Fuel

The necessity for fuel to feed the flame is probably the biggest limitation of FIDs. Most FIDs use pure hydrogen to maintain a flame in the combustion/detection chamber. HeathTech FIDs manufactured by Heath Consultants Inc. utilize a 40% hydrogen/60% nitrogen mixture. The fuel cylinder may be fixed inside the meter or it may be easily removable for recharging. Regardless of location, the cylinder must be refilled.

Frequency of recharging varies with the instrument and use; in most cases, refilling is necessary every day or every other day. Foxboro and HeathTech supply CGA-compatible refilling hoses or fittings which attach directly onto a 2400 psi pressurized cylinder. Sensidyne recommends that a specialty gas supplier refill the hydrogen fuel tank as necessary.

Fuel gas should be certified by the distributor to contain less than 1 ppm hydrocarbon contamination. Organic contaminants in the fuel will be burned with the hydrogen, producing ions and a **fuel contamination background**. A background of several units is acceptable, severe hydrogen fuel contamination, however, can result in very high backgrounds which render the instrument incapable of detecting low levels of contamination in sample air.

In some cases, hydrocarbon contamination occurs **after** the fuel is placed into the pressurized storage cylinder. The gas supplier certifies the gas that goes into the cylinder, not what comes out. If the interior of the steel cylinder is in poor condition, the carbon in the steel will react with hydrogen to form methane; this will occur most readily when the fuel has significant water vapor contamination. For this reason, we recommend the use of certified UHP (ultra high pure) fuel; for hydrogen this represents 99.999% hydrogen (also called 5-nine hydrogen by gas suppliers) and less than 1 ppm THC (total hydrocarbon contamination). This leaves no more than 9 ppm for other contaminants, such as helium, nitrogen, or water vapor. We also recommend that fuel storage cylinders be maintained for no more than six months before they are emptied and refilled with fresh gas.

Dust and Particulates

FIDs usually have one or more inlet filters which prevent large particles from entering the combustion/detection chamber. These filters should be changed or cleaned regularly. Fine particulates may be formed however, during combustion of organic materials. A fine soot-like residue may be found within the combustion chamber of FIDs which are exposed to very high concentrations of organic materials. Charged particulates can be attracted to the collecting electrode. When this occurs, the meter will give an erratic reading before the flame is ignited.

Particulates lodged in filters can adsorb vapors and gases entering the sample line and then slowly desorb or release these materials over a period of time. When this occurs, recovery time will be prolonged as the particulates slowly release the adsorbed vapors and gases. In some cases, background readings will be variable, even when the meter is stationary in a non-contaminated area.

Flammable Gases

FID users are often concerned about the safety of FIDs. Only intrinsically safe FIDs should be used at hazardous materials sites or incidents. Currently, only FIDs manufactured by Foxboro (Models OVA-108 and OVA-128) are certified as intrinsically safe. Other FIDs are **not** intrinsically safe and cannot be made intrinsically safe.Both Sensidyne and Heath anticipate having an instrinsically safe version of their FIDs by 1993.

A non-intrinsically safe FID is a source of ignition and should be used **only** in situations where there is **absolutely no possibility** of a sudden or unexpected release of flammable gases or vapors. Non-intrinsically safe FIDs should not be used in Division 1 or Division 2 locations.

Large concentrations of flammable gases (i.e. within their flammable range) can act as an additional fuel source for the hydrogen flame. When this occurs with Foxboro FIDs, the flame becomes larger and the meter display shows a rapid increase, usually followed by the reading falling back to zero. This response is caused by an increase in flame size and ion production, followed by an automatic, momentary shut-off of the hydrogen supply as flame size exceeds the confines of the combustion chamber. The abrupt interruption of the hydrogen supply extinguishes the flame and causes the meter response to drop back to zero.

Flame-out may also occur when the concentration of gases is so high that it produces oxygen deficiency. This condition is often encountered when sampling enclosed areas where vapors and gases cannot escape, such as tanks and drums. When flame-out occurs, a flame-out alarm is activated; the sample line should be immediately removed from the sample atmosphere. Wait 5 to 10 seconds, then attempt to reignite the flame. If the flame cannot be reignited, shut off the hydrogen supply and allow the pump to continue flushing the combustion chamber with clean air. Repeat the start-up procedure in a clean, non-contaminated area. Excessively large concentrations of gases and vapors can saturate the combustion chamber and collecting electrode, which overwhelms the detector. The meter should be recalibrated **after** background readings stabilize.

Temperature

FIDs burn hydrogen in air, producing water vapor within the combustion chamber. Since water vapor is formed during use, ambient humidity has little effect on the meter efficiency. The water vapor formed during combustion is evacuated from the unit through the exit port. At low temperatures, water vapor condensation will decrease flow through the exit port, which can inhibit pump function and sample air flow. At temperatures approaching 0°C or 32°F, water vapor will begin to freeze, obstructing the exit port and causing the pump to stall and flame-out.

Recommended minimum operating temperatures for FIDs range between 5 to 10°C, or approximately 40 to 50°F. At minimum operating temperatures, response time and recovery time is increased, and battery life is decreased. The maximum operating temperature for FIDs is 40°C or 104°F; that is **meter** temperature, not ambient temperature. A meter sitting in the sun when ambient temperature is 90°F can easily overheat; keep the instrument shaded whenever possible. At temperatures over 104°F, a gradual erroneous increase in meter readings will be obtained as the unit overheats. If an FID must be subjected to extremely high temperatures, it should be kept in a cooler area and used for only limited intervals at the higher temperature.

CALIBRATION

Warm-up Time

FIDs require a relatively long warm-up interval; a minimum of 5 to 15 minutes is required before igniting the flame. Calibration should not be attempted until the needle display is stable and there is no perceptible upward or downward drift; this usually requires an additional 10 to 15 minutes after ignition.

FIDs are factory-calibrated to methane, which is readily available from specialty gas distributors if it is not offered by the manufacturer. Methane span gas must be supplied in air with normal oxygen concentrations. If the instrument will be used in situations with oxygen concentrations consistently lower than normal (i.e. less than 20.9%), span gas should be prepared with the anticipated, lower oxygen concentration. Span potentiometer controls or set screws may be inside the instrument housing or more readily accessible on the outside.

The one-point calibration procedure is usually sufficient for FIDs with a linear calibration response curve; these units have limited detection ranges, such as 0.2 to 1000 ppm or 1.0 to 5000 ppm. A one-point calibration should be performed within the range of contamination anticipated. For instance, if very low concentrations are expected, a calibrant gas between 2 to 10 ppm or 10 to 50 ppm should be utilized. On the other hand, if the unit will be used in areas known to contain large concentrations of contaminants, a methane concentration of 500 to 1000 ppm, or more should be used.

FIDs with larger detection ranges of 1.0 to 10,000 ppm have logarithmic response curves. These meters should be calibrated with methane concentrations at the high and low end of the detection range, say 100 ppm and 10,000 ppm. The Foxboro OVA Model 88 measures 1.0 to 100,000 ppm; this meter should be calibrated at 100, 10,000, and 100,000 ppm methane. Care should be used when performing calibrations with 100,000 ppm; 10% methane is within the flammable range.

Zeroing the Instrument

All FIDs can be electronically zeroed before the flame is ignited. Prior to flame ignition, all electrical components of the unit are operational; stability of the meter display indicates the status of the collecting electrode in the combustion/ionization chamber. Prior to igniting the flame, the user should carefully zero the instrument on the most sensitive setting. The needle should be stable and not drift up or down. When the instrument holds the zero setting for at least one full minute, the hydrogen can be turned on and the flame ignited; there will be a momentary upward deflection of the needle which will then quickly return to a lower reading.

FID readings should stabilize within 15 minutes after the flame is lit; during warm-up, readings will drift. After the readings stabilize, the initial background reading should be recorded; typical background readings vary between 1.0 and 5.0 units. Fuel for the flame will always contain low concentrations of hydrocarbon contamination; the background reading obtained is therefore a combination of contamination in ambient air as well as in the fuel.

If fuel contamination is significant, i.e. greater than 5 units, the user may wish to zero out the fuel background in order to obtain a workable ambient background reading. While sampling zero air, the needle display should be observed until the reading is stable; the zero adjust knob can then be adjusted until the meter reads zero. When the zero reading appears stable, disconnect the zero air and assess the new background reading, which now represents ambient background only.

If zero-air is not available, an activated charcoal filter can be attached to the sample inlet to remove hydrocarbons from clean ambient air. Activated charcoal will filter out most hydrocarbon contaminants but not methane and ethane which are too small to be trapped. Record the meter response while sampling ambient air in a clean area, then attach the charcoal filter. The meter reading in response to charcoal-filtered air represents the presence of hydrocarbon contamination in the fuel and methane or ethane gas in sample air. If there is little or no change in meter reading with the charcoal filter attached, the presence of fuel contamination should be confirmed by using hydrocarbon-free or zero air.

The fuel background should be zeroed out, or added to the calibrant gas concentration used, in order to achieve an maximum accuracy when calibrating. For instance, a fuel background of 4 units should be added to the meter response to a calibration gas containing 10 ppm methane for a total response of 14 units.

Figure 7-5: A charcoal filter adaptor for a Foxboro OVA (top) and a disposable charcoal filter (bottom). Charcoal filters can also be used to differentiate the contribution of methane and ethane to the total hydrocarbon reading. (Photo by Steve Napolitano.)

Assigning an Arbitrary Background

FIDs typically detect more organic contaminants and usually have higher ambient background readings than PIDs. It is unusual for an FID to display a reading less than 1.0 unit when sampling clean ambient air, however, in rare cases a lower background reading may occur. We recommend adjusting the background reading to at least 1.0 units. **Never zero out ambient background and operate an FID with an initial background reading of 0 units.** For most FIDs a reading of zero on the analog display usually triggers the flame-out alarm, even when flame-out has not occurred. If background readings are close to zero, slight changes in the reading will repeatedly trigger the alarm; when this occurs the user eventually turns off the audible alarm, and will not be alerted when flame-out actually does occur. For example:

An FID is turned on and electronically zeroed in clean air. The flame is then ignited and a reading of 5 units is obtained. The FID is zeroed using a sample bag filled with non-contaminated or zero air. The bag is removed and a reading of 2.8 units is obtained in the clean area. This procedure correctly removed a hydrogen background of 2.2 units; the remaining 2.8 units represents the ambient air background that should not be zeroed out.

USING AND CHOOSING FLAME IONIZATION DETECTORS

Survey vs Gas Chromatography

Portable FIDs are most commonly used as survey meters; in the survey mode, sample air is drawn into the combustion/detection chamber and burned. The meter response reflects the total amount of detectable hydrocarbon contamination present in methane equivalents. In this mode, it is not possible to distinguish between different individual hydrocarbons or classes of hydrocarbons.

Figure 7-6: A Foxboro OVA-128GC; note the circular GC column and cylindrical activated charcoal scrubber assembly. Compare this meter to the Foxboro OVA 128 in Figure 7-9. (A Detrick Lawrence photo.)

Some FIDs can be equipped with a gas chromatography or GC column; the FID can then be operated in either the GC or survey mode. A GC column is a tube tightly packed with a material which attracts hydrocarbon vapors. Vapors are pushed through the column by the **carrier gas**, usually hydrogen, which is not attracted to the packing material. The affinity of the column material varies between different hydrocarbons, each material becomes separated as it passes through the column. The separated contaminants flow through the column and then into the combustion chamber/detector; meter response is indicated on the readout display or a strip chart recorder.

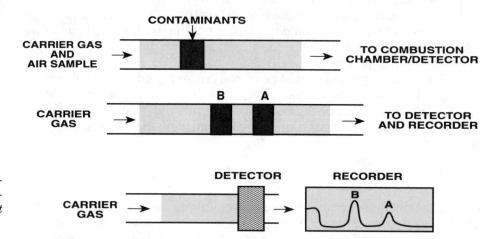

Figure 7-7: Typical column separation of hydrocarbon contaminants and peaks on strip chart recorder.

The time it takes for a material to pass through the column is known as its **retention time**. Retention time varies with the type of packing used and the affinity of packing material for the hydrocarbons passing through the column. If specific compounds are known or suspected to be present, standard concentrations of each can be injected into the column and individual retention times and peak characteristics recorded. Sample air is then introduced into the column and the retention time of each peak compared with the specific compounds of interest. Using this method, an experienced user can **tentatively** determine the number, identity, and approximate concentration of **known** contaminants.

Foxboro FIDs can be alternated between survey and GC mode. In the survey mode, the meter provides a composite reading for all hydrocarbons present. In the GC mode, the number contaminants and relative peak responses can be compared with survey mode meter readings.

After each sample air analysis, the GC column must be cleared of other hydrocarbons with longer retention times that are still in the column. This procedure is called **backflushing** and is effected by reversing the flow of carrier gas in the column. The backflushed material is then sent to the combustion/detection chamber and the relative amount of materials not eluted as peaks can be determined. The height and width of the peak can offer clues regarding the number and relative concentration of uncharacterized contaminants.

FIDs with GC options are designed for applications where there is a limited number of **known** contaminants. The user must first establish retention time and peak characteristics for each compound of interest, using standards of known concentrations. This procedure must be performed at the time of analysis, since retention times vary with temperature, hydrogen pressure, and the type, length, and condition of column used.

After injection, the user waits for the contaminants present to travel through the GC column and elute into the combustion chamber/detector. This may take several minutes or more; during this interval the readout is watched for indications of meter response, and peak readings are recorded. Retention time for Peak A in Figure 7-8, for example, is 2.0 minutes, and 4.0 minutes for Peak B; the last analytical peak (E) has a retention time of approximately 6.5 minutes. Backflushing must be performed after each analysis; the backflush interval must be at least 1.5-2.0 times analysis time. It is not unusual for a single GC analysis to require 5-10 minutes or more.

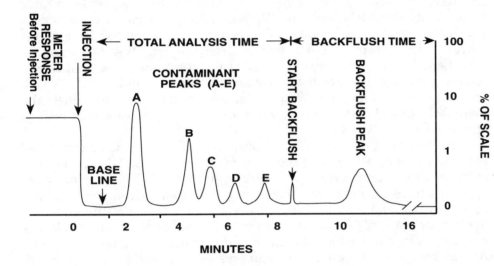

Figure 7-8: A typical chromatogram representing meter response on survey mode, relative peak heights of sample air contaminants separated in the GC column after injection, and backflushed peak.

Some users carry a strip-chart recorder attached to the FID; Foxboro offers an intrinsically safe recorder designed to be used with the sidepack. While this set-up is cumbersome it can be a valuable tool, especially when attempting to discriminate between methane and non-methane contaminants. The strip chart recorder can also be used in the survey mode to provide a continuous record of meter readings.

While it can be valuable to assess the number of contaminants present in an uncharacterized sample, we find the GC analysis to be unwieldy and time consuming on sites containing unknowns. If only a limited number of **known** contaminants are present, and a column is available which affords relatively short retention times, then the GC option may be worthwhile.

A short column can be used to quickly separate a light hydrocarbon such as methane from heavier hydrocarbons while working landfills or other sites known or suspected to contain methane. In this case, good column separation of individual components is not required or desired; the column should be used only to rapidly determine overall meter response from methane versus non-methane contaminants.

The Foxboro Company

Foxboro manufactures three portable organic vapor analyzer (OVA) models, the OVA-88, OVA-108, and OVA-128. The OVA-88 has a detection range of 1.0 to 100,000 ppm methane; this model is not intrinsically safe and is used primarily for walk through or drive through pipeline leak detection surveys.

The Model OVA-108 has a single scale logarithmic readout display and a methane detection range of 0.2 to 10,000 ppm. The Model OVA-128 has a detection range of 0.2 to 1,000 ppm methane. The OVA-128 has three linear

Figure 7-9: Sidepack and readout assembly of a Foxboro OVA-128. The term OVA to describe an organic vapor analyzer should be reserved for Foxboro FIDs. (Photo by Steve Napolitano.)

readout scales, 0-10, 0-100, and 0-1000; this model is more sensitive to low concentrations than the OVA-108. Both models can be equipped with a GC column. The OVA-108 and OVA-128 are the only FIDs currently certified to meet NEC/NFPA and CENELEC standards for intrinsic safety. The Foxboro Rustrak Strip Chart Recorder is intrinsically safe and derives power directly from the OVA battery; this enables the user to obtain a hard copy in the field.

OVAs were originally developed and manufactured by the Century Systems Corporation which was subsequently acquired by Foxboro. Unfortunately, Century did not copyright the name of their portable Organic Vapor Analyzer models; the term OVA is now used to denote any instrument which can detect organic vapors. This is unfortunate and has led to much confusion; for the experienced instrument user, the term OVA is synonymous with Century (now Foxboro) FIDs. Novice instrument users tend to call anything that measures atmospheric contaminants (including PIDs, CGIs, and toxic sensors), an OVA. We have learned by experience to request the manufacturer and model number when discussing or interpreting readings obtained from an "OVA".

Foxboro OVAs consist of a sidepack assembly and a sampling probe/readout assembly attached to the sidepack by a 5-foot umbilical cord. The exterior of the sidepack contains control knobs, sample flow rate indicator, hydrogen pressure indicator gauges, and hydrogen valves; the hydrogen igniter switch and umbilical connectors are on the side of the unit. The battery pack is easily removed or can be recharged in place. The readout is a scaled analog meter which displays readings in ppm-methane equivalents. A small knob on the back of the readout assembly allows the user to program a specific needle position which will trigger an audible alarm.

The interior of the sidepack contains a single stainless steel hydrogen cylinder with a capacity of 75 cc. A separate valve allows the user to refill the cylinder in place using a hydrogen filling hose assembly. The recommended maximum hydrogen fill pressure is 2300 psi; 130-150 psi are used per hour. A full cylinder provides sufficient hydrogen for approximately 12-13 hours use. Always maintain a minimum pressure of 300 psi or more in the cylinder. Never allow the cylinder to reach atmospheric pressure; if this occurs, ambient air and humidity can enter the cylinder, causing eventual degradation of the interior of the cylinder. If ambient air is inadvertently allowed to enter the cylinder, purge the cylinder repeatedly with a dry, inert gas, then refill with hydrogen.

Figure 7-10: A book holder, milk carton, or even a child's chair can be used to keep the OVA upright, minimize contamination, and prevent exhaust port blockage. (Author photo.)

The entire unit weighs approximately 12 lbs. When run in an upright position, the flow meter should read between 1.5 and 2 units. The unit can be operated in any position, however, the exhaust port on the bottom of the sidepack must not be blocked or the hydrogen flame will be extinguished. A leather, sheep-skin padded shoulder strap is provided with each unit. Since the padded strap cannot be decontaminated, we recommend you use it to carry your luggage. Use unpadded, disposable straps to carry the OVA.

Sample air is drawn through sampling probes and into Teflon tubing portion of the umbilical cord and into the combustion chamber located near the base of the sidepack. A short probe is used for close sampling; a longer, adjustable telescoping probe with a tubular sampler is used for area sampling and when monitoring locations that are not readily accessible. Additional sampling accessories include a charcoal filter adapter and sample diluter.

Each sampling probe contains a porous metal particulate filter. To ensure adequate protection against particulates, a probe should always be in place before starting the pump; never use the telescoping probe without the tubular sampler. The primary filter for the OVA is a 10 micron stainless steel filter cup located behind the sample inlet connector on the sidepack. All filters should be regularly examined and cleaned.

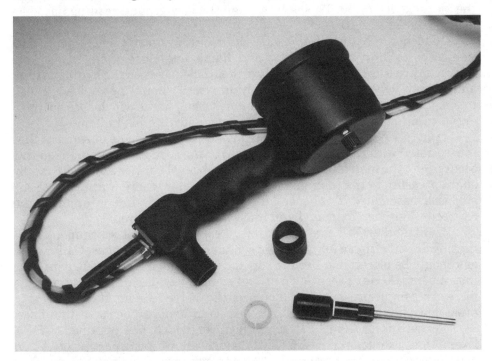

Figure 7-11: Readout assembly and short probe for close area sampler; the probe must be firmly seated against the Teflon ring to prevent leakage. (Photo by Steve Napolitano.)

Most novice OVA users complain that the flame is difficult to ignite, or that flame-out occurs immediately after ignition. In most cases, this is due to inadequate battery power, insufficient warm-up or attempting to ignite the flame too soon after opening the hydrogen valves. Warm-up time prior to igniting the flame should be at least 15 minutes; at temperatures between 10 and 20° C warm up time should be no less than 20-30 minutes. After warm-up, the hydrogen tank and supply valves should be slowly opened. Hydrogen tank pressure should be greater than 300 psi; the low pressure hydrogen supply pressure should be between 8-12 psi.

A minimum of **60 seconds** is required for the fuel to completely fill the supply system prior to igniting the flame. Too often the OVA user presses the igniter button immediately after opening the hydrogen valves. When nothing happens, the button is pressed again and again, accompanied by expletives and ominous mutterings. All this can be avoided! **Wait a full minute**, then depress the igniter button for **no more than 6 seconds**; a small pop may be heard, and needle deflection should be seen on the readout. Waiting only 20 or 30 seconds will only partially fill the fuel supply tubing, when this occurs, the flame will usually momentarily ignite and then go out. Failure to wait the entire 60 seconds will only put the user in a bad mood and eventually result in igniter failure.

Flame-out is another common complaint of OVA users; this is often the result of restricting sample air flow into the combustion/detection chamber. Sample air flow may be cut off completely if the umbilical cord is inadvertently kinked during use. Foxboro recently introduced an umbilical cover on new OVAs which reduces kinking. Flame-out may also occur when the sample line or probe is clogged with liquid or particulates, the exhaust port is blocked, insufficient oxygen is available in sample air to support combus-

Figure 7-12: Kinking of the umbilical line obstructs sample air flow through the Teflon hose and can cause flame-out. (Author photo.)

Figure 7-13: The sidepack / umbilical electrical connector can be damaged if the unit is dropped, or if the meter is lifted by the umbilical cord. (Photo by Steve Napolitano.)

tion, high concentrations of flammable vapors are present, or there is insufficient hydrogen pressure to maintain the flame.

The audible or earphone alarm should always be used to alert the user of flame-out. In some cases, however, flame-out has not occurred although the alarm is activated. This occurs when the background reading has been incorrectly set at zero.

Inadequate hydrogen pressure can result in erratic instrument function, poor calibration reliability, and flame-out. Most often, inadequate hydrogen pressure is caused by failure to sufficiently open the hydrogen tank and supply valves. Hydrogen supply pressure should be at least 8 psi; inadequate hydrogen supply will result in a small flame which does not burn organic contaminants efficiently and is more susceptible to flame-out.

Some OVA users may observe transient, erratic needle movement that is associated with moving the instrument or the readout display. When this occurs, hold the instrument steady on a solid, level surface and gently apply slight pressure to the umbilical readout connector where it attaches to the sidepack assembly. If the needle jumps when pressure is applied, the connector has been damaged and must be factory repaired.

Unlike other FID manufacturers, Foxboro does not recommend calibrating directly from a low pressure cylinder of compressed gas. Calibration gas should be placed in a sample bag and then drawn from the bag into the instrument. Do **not** press down on the sample bag during calibration; this increases sample pressure and increases the meter reading. It is incorrect to press down on the sample bag until the meter displays the correct number.

Foxboro is the only FID manufacturer which has obtained a DOT Exemption Extension (DOT-E 7607) which allows transport of OVAs on passenger carrying aircraft with the hydrogen tank filled to 2100 psi. This allows users to transport and then use their OVA for at least a full day without the need for additional hydrogen. Care is required however, to ensure compliance of all aspects of the exemption. The outside of the package must be marked with the name of the manufacturer and location and DOT exemption number; a copy of the exemption must be supplied to the shipper and the pilot advised of its contents. A copy of the entire exemption should be obtained from Foxboro.

Figure 7-14: A Foxboro Model 128 shows a reading of 25 units when calibrated with 25 ppm methane (left); the same unit shows a reading of 30 units when excess pressure is applied to the sample bag right). (Photos by Steve Napolitano.)

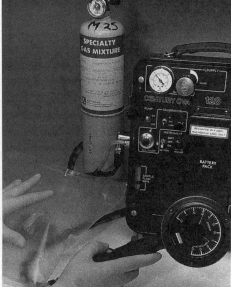

Heath Consultants Incorporated

Heath currently manufactures two FIDs, the Porta-FID II and the Detecto-Pak III™; both models are **not intrinsically safe** and cannot be made intrinsically safe. These instruments are designed for outdoor natural gas pipeline surveys and distribution system inspections, and sidewalk and customer leak complaint/detection investigations. The extendable cone-shaped probe is designed to facilitate detection of small leaks.

Figure 7-15: The HeathTech Porta-FID II is easy to use and weighs less than 5 lbs. (Photo by Steve Napolitano.)

The Porta-FID II has a linear detection range of 1-5,000 ppm methane; the meter has three scales, 0-50, 0-500, and 0-5000. The Detecto-Pak III has a detection range of 1-10,000 ppm methane and five scales, 0-10, 0-50, 0-100, 0-1000, and 0-10,000.

Heath FIDs utilize a fuel mixture of 40% hydrogen/60% nitrogen. Fuel cylinders have an approximate capacity of 10 cc; the maximum recommended fill pressure is 1750 psi[5,6]. Both units have separate, removable fuel cylinders which must be replaced at least once during an 8-hour use. The cylinder, regulator, and fuel supply hose are relatively exposed in the Porta-FID II. These components are more protected and recessed inside the Detecto-Pak III.

Heath user manuals do not indicate the recommended operating temperatures or oxygen concentration limits for either of their FIDs. No specific information could be obtained concerning oxygen limitations, however, a Heath representative indicated that the flame would probably go out at oxygen concentrations less than 16%[7].

Sensidyne, Inc.

Sensidyne currently offers a Standard FID which has two semi-logarithmic detection ranges, 0-1000 ppm and 0-10,000 ppm. The X10 Sensitivity FID is designed to detect methane concentrations of less than 1 ppm and has a detection range of 0-100 and 0-1000 ppm. Both FIDs are calibrated to methane and are also available with a gas chromatography option. Sensidyne FIDs are designed for use in non-explosive, non-flammable areas only; they are not certified as intrinsically safe.

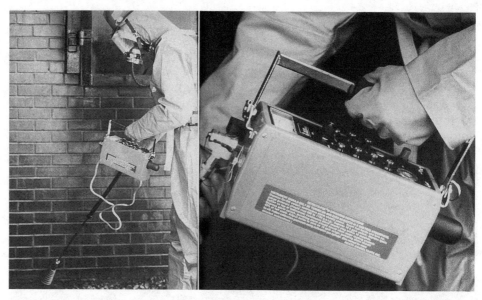

Figure 7-16: The Detecto-Pak III;
note the end of the cylinder pro-
truding from the back of the meter.
(Detrick Lawrence photos.)

Several different sampling probes are available for use. The standard probe consists of a hollow stainless steel tube on a handle with perforations at the sampling end; straight probe, roller probe, and cup probe attachments are also available. No information could be obtained concerning temperature and oxygen limitations of Sensidyne FIDs[8].

Using FIDs

Although FIDs have somewhat different limitations than PIDs, the user should not become overconfident regarding FID detection capabilities. FIDs cannot detect all potential organic contaminants present and inorganic materials will not be detected. Since ionization efficiencies will vary for those organic contaminants that are detectable, FID meter readings do not reflect the actual ppm concentration present. The actual concentration is often significantly greater than the meter reading.

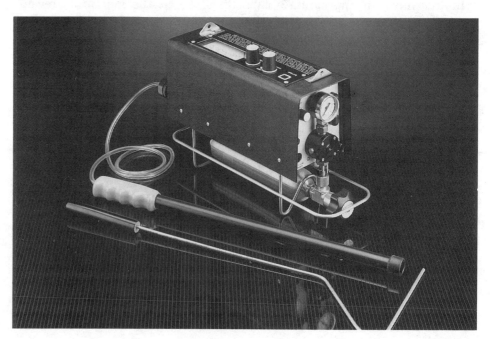

Figure 7-17: Sensidyne's portable flame ionization detector with handle and standard probe attachment. (Courtesy Sensidyne.)

FIDs are typically not used during landfill investigations because of their sensitivity to methane gas. PID ionization efficiency, however, can be adversely affected when significant concentrations of methane are present. For this reason, it is prudent to compare FID and PID responses. When methane gas is suspected, a charcoal filter can be used to determine the relative contribution of methane to the overall FID reading.

REFERENCES

1. Brown, G.E., DuBose, D.A., Phillips, W.R. and Harris, G.E. 1991. Response Factors of VOC Analyzers Calibrated with Methane for Selected Organic Chemicals, Report No. EPA-600/2- 81-002. U.S. Environmental Protection Agency, Office of Research and Development, Research Triangle Park, NC.
2. Instruction Manual OVA 128 Century™ Organic Vapor Analyzer. 1990. The Foxboro Company, East Bridgewater, MA.
3. Personal communication with Foxboro technical staff.
4. Exemption Extension (DOT-E 7607). 1991. U.S. Department of Transportation, Research and Special Programs Administration.
5. Instruction Manual Heath Detecto-Pak III. 1990. Heath Consultants Incorporated, Stoughton, MA.
6. Instruction Manual Heath Porta-FID II. 1987. Heath Consultants Incorporated, Stoughton, MA.
7. Personal communication with HeathTech Manufacturing Division technical personnel, March 1992.
8. Personal communication with Sensidyne technical representative, March 1992.

CHAPTER REVIEW

1. Flame ionization detectors use _____ to produce ions and create an ion current.

2. True or false? FIDs can detect both organic and inorganic compounds.

3. List two major limitations of FIDs: _____

_____.

4. True or false? All currently available FIDs are intrinsically safe.

5. Which of the following materials will elicit no response on an FID? Carbon monoxide, carbon tetrachloride, carbon dioxide, toluene, benzene, methane, acetone, carbon disulfide, nitrogen dioxide.

6. True or false? FID readings can be considered ppm readings, regardless of the type or number of contaminants present.

7. FIDs are usually factory and field calibrated with

8. True or false? Response factors can be used in the field to determine the exact concentration of contaminants present.

9. List some common reasons for flame-out:

10. What type of contaminants can be retained or adsorbed onto a charcoal filter?

11. True or false? FID response is affected by ambient humidity.

12. An FID is turned on in a clean area and shows a high background reading. What are the two potential sources of the reading? _____.

13. When using Foxboro FIDs, wait at least _____ seconds after turning on the hydrogen before attempting to ignite the hydrogen flame.

14. UHP or ultra high pure hydrogen is _____ % hydrogen with less than 1 ppm total hydrocarbon contamination.

15. True or false? FIDs will operate properly in severely oxygen deficient atmospheres.

PROBLEM SET

1. A 55 gallon drum of used paint remover and paint residue has been pierced by a fork lift truck. FID readings are 400 units at entry, and 900 units within the spill area. The foreman reports that at least a third of the contents of the drum is water. Should the presence of water in the spilled material affect your FID readings?

2. At the same incident, you suspect that some of the spilled paint remover and residue may have entered a small drain. A reading of 500 units is obtained when the sample probe is placed directly over the drain. The probe is then extended into the drain itself; the meter immediately flames out. What could have caused this effect?

3. An FID used at a site investigation keeps flaming out and background readings keep falling to zero. What could be the problem?

4. At a gasoline spill, the local fire department requests that you check the nearest sewer for gasoline vapors. An FID gives a reading of 120 units. A charcoal filter is inserted into the sample hose and a reading of 10 units is obtained. What does this indicate?

5. An FID is turned on and electronically zeroed in a clean area. The flame is then ignited and a reading of 2.4 units is obtained. The operator zeros the meter using hydrocarbon-free air; the zero-air bag is removed and the meter calibrated with 30 ppm methane. After calibration, a reading of 1.2 is obtained in the clean area. Is this procedure correct? What portion of the initial background reading was removed?

8

Multi-Specific Gas Detectors

LEARNING OBJECTIVES

1. List the major components of a gas chromatograph.
2. Describe how a gas chromatograph works.
3. Describe some conditions that can affect columnn retention time.
4. List the major components of an infrared analyzer.
5. Discuss some limitations of infrared analyzers.

Instrument users are still searching for the elusive "black box" which will instantly measure and identify all contaminants present in air. Unfortunately, such an instrument exists only in the mind or on Star Trek; it is not possible with the currently available technology. There are, however, several different types of instruments which can measure multiple compounds simultaneously, or sequentially after a few instrument adjustments.

A **gas chromatograph (GC)** separates multiple contaminants in an air sample into discrete peaks on a chromatogram, which facilitates subsequent quantification or identification. An **infrared (IR) spectrophotometer** provides real-time concentration data for one contaminant after another, with minor adjustment to its optical filter.

Portable gas chromatographs, equipped with flame or photo-ionization detectors, are used for ambient air sampling at hazardous waste site investigations and remediations, on-site soil and water sample contamination surveys, in situ soil gas analysis, emissions testing, industrial hygiene surveys, and community health hazard assessments. IR spectrophotometers are more commonly encountered at fixed facilities, where they are used for occupational/industrial hygiene monitoring, leak detection, and emissions testing. Portable IRs have also been used for fume hood testing and monitoring and verifying concentrations of chemicals used in controlled laboratory studies.

PORTABLE GAS CHROMATOGRAPHS

Portable GC units are sufficiently small and portable to be set up and used in the field. There is a difference between **movable** and **portable**; a unit that must be carried by two people, transported on a luggage carrier, or kept stationary in a field trailer should not be considered portable. Truly portable GCs can be taken on site and set up to perform actual analysis under field conditions. If the unit cannot be picked up and carried by one handle, it is not a field portable GC. Some authors distinguish between person-portable and vehicle-portable instruments[1].

A typical portable GC unit consists of a GC column, an injection system for introducing samples into the column, and a detector; samples may be introduced into an injection port using a gas-tight microsyringe or a vacuum pump may pull a standard volume of ambient air directly into the column. A recording device is usually employed in order to retain an analysis record.

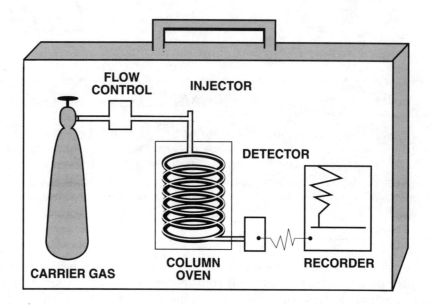

Figure 8-1: Components of a typi-cal portable GC unit.

The injection port is where the sample air is introduced; the sample may be automatically injected by the instrument, or manually injected by the user with a gas-tight syringe. At low ambient temperatures, low volatility materials with high boiling points may condense inside the injection port. A heated injection port prevents condensation and facilitates initial separation and subsequent detection.

After a sample is introduced into the column, it is carried through the column by a carrier gas; within the column, the individual components are separated before entering the detector. Separated components are represented as peaks on the recorder. Carrier gas is usually a high purity inert gas such as nitrogen or helium; portable FID GC units typically use ultra high pure hydrogen.

GC columns contain a packing or coating which attracts organic contaminants; each contaminant has a different affinity for the packing material. Compounds with little or no affinity will rapidly pass through the column; other materials will pass through more slowly and have a longer retention time. Under optimal conditions, all components are separated and elute from the column and into the detector individually, forming discrete, easily recognizable peaks. Figure 7-8 is an example of good chromatographic separation of five contaminants with different retention times.

Retention time of a particular contaminant is dependent on the type of packing material, length of the column, flow rate of the carrier gas, and temperature of the column. Retention time decreases with increased carrier gas flow rate or column temperature.

There is a variety of column designs to accommodate different types of contaminants and anticipated sample size. Improper column selection or separation conditions can result in inadequate peak discrimination, broad peaks, tailing, or shouldering. **Packed columns** are packed with a fine inert

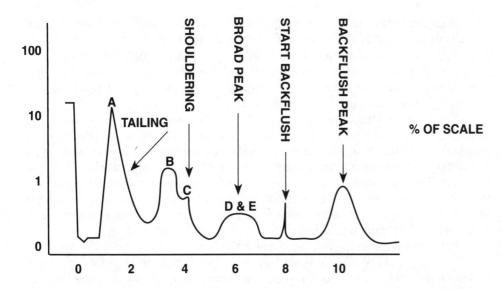

Figure 8-2: An example of inadequate separation of multiple contaminants. Compare this chromatogram with Figure 7-7.

solid material (the support phase) which is coated with a known volume of a non-volatile liquid (the stationary phase or liquid phase). In some packed columns, there is no liquid phase, the solid material acts as the adsorbent. A **capillary column** is a hollow tube; the walls of the tube serve as the support phase, or a fine layer of the inert support material is applied to the walls; the support material is then coated with the liquid phase.

Solid support materials are usually chemically inert, stable, and offer a large surface area for holding the thin film of the liquid phase. Commonly used solid supports include Chromosorbs, which are diatomaceous earth supports treated to make them more chemically inert; Poropaks or porous polymer beads, Teflon, carbon or silicon molecular sieves; porous silica or glass beads are also used. Liquid phase materials may be Carbowax or UCON 50 HB (both polyethylene glycol polymers), diisodecyl phthalate, dioctyl phthalate, organosilicones, or polysiloxanes. New, improved GC column designs and packing materials are always being introduced; it is best to consult the manufacturer before selecting a column.

Figure 8-3: A Foxboro OVA-128 with GC and strip chart recorder; the circular GC column has been twisted into a figure-8 (arrows) and should be discarded. (Author photo.)

Manufacturers usually supply selection guides or charts to assist the user. Column length varies from 6 inches up to 30 feet long. Columns are usually coiled or looped to accommodate their length into the small space provided. Regardless of column length and design, they should be handled carefully. Twisting or bending a column disrupts the interior packing and irrevocably alters separation capabilities.

Since more than one column may be necessary for a specific field application, a variety of columns and column lengths should be readily available from the manufacturer. The user should have an idea of what type of compounds are anticipated in order to select the appropriate column and detector. Most columns must be conditioned prior to actual analysis; conditioning instructions are provided by the manufacturer. It is always a good idea to have a spare, conditioned column available in case the one in use is damaged or becomes unusable.

Portable GCs with unheated columns are more difficult to use since column temperature affects retention time; materials with relatively high boiling points will have inordinately long retention times when the column is cold. Variability in column temperature during the day also necessitates more frequent calibration.

Most portable GCs have an operating temperature range of 10 to 40°C (50 to 104°F); operating temperature is not the same as column temperature. Heated columns offer stable temperatures and reproducible results within a range of ambient conditions. GCs which offer **isothermal operation** have the column enclosed in an oven which is capable of maintaining a fixed temperature; isothermal ovens are usually adjustable in increments of 1°C. The smaller the difference between operating temperature and ambient temperature, the faster the column will reach thermal equilibrium, and less battery power will be required.

Most GC users fret about cold temperatures which can limit operating temperature and reduce battery life. Maximum isothermal oven temperatures may be limited by ambient temperature. For example, maximum oven temperature may be only 25°C above ambient; if ambient temperature is 10 °C, then the maximum column temperature attainable is 35°C. Separation conditions must often be changed to accommodate cold weather conditions.

High ambient temperatures, however, can also be a problem. If the temperature difference between ambient and column temperatures is too small, ambient temperatures can increase above column temperature. When this occurs, column conditions are no longer isothermal and separation will be affected. Column operating temperature should always be at least 5°C above ambient temperature.

Backflushing is used to clean the column of residual contamination from eluted peaks or materials with long retention times that have not yet cleared the column. In GC units with a single column, contaminants cleared from the column usually go to the detector; the size of the peak can give some information regarding the relative amount of backflushed material. The duration of backflushing is usually 1.5 to 2 times the analysis interval.

Many GCs actually have two columns, a short precolumn and a much longer analytical column. The precolumn is designed to trap slow, heavy, less volatile materials with high boiling points that have long retention times; more volatile materials pass through quickly and are then separated on the analytical column. After the volatile compounds have cleared the precolumn, carrier gas flow is reversed and remaining contaminants are flushed out, away from the detector. Backflushing only the precolumn can significantly shorten the interval between analyses.

The air sample and carrier gas enter the detector after eluting from the column. Separation of sample contaminants is of little use unless the detector is sensitive to them. The most commonly used detectors employed are PIDs and FIDs. Most portable GC units accommodate only one detector. HNU Systems, Inc. offers a field GC with a variety of detectors; two detectors can be operated separately or in series. These GCs are less portable and more amenable to benchtop use.

Other detection methods that can be employed with a GC include electron capture (EC), far-UV (FUV), thermal conductivity (TC), and flame photometry (FPD). FUV detectors (FUVDs) use a lamp which emits light in the far UV range (120-180 nm); as contaminants pass through the light, they absorb some of its energy. The decrease in light energy is detected at a photodiode; this change is then amplified and translated into a recorded signal on the strip chart. FUVDs are considered universal detectors since they respond to nearly all materials except the noble gases.

In flame photometry, light emitted from substances heated to an unstable, excited state is measured. The greater the temperature, the greater the amount of light energy emitted; light spectra may be in the UV, visible, or infrared range. In FPDs, a hydrogen/oxygen-fed flame is used as the heat source. A photomultiplier tube converts light energy into an electrical signal which is then recorded.

Electron capture detectors (ECDs) employ a low level radioactive source to supply a constant beam of beta particles which are attracted to a collecting electrode. Contaminants can capture electrons, attract the beta particles and decrease the intensity of the beam reaching the collecting electrode, which is translated into an electrical signal.

Improvements in detector design and versatility are continually made. Detailed information on the theory, operation, use, limitations, and maintenance of GC columns and detectors should be obtained from the manufacturer and carefully reviewed **prior** to purchase. In most cases, a single column and detector are not sufficient to detect the entire range of contaminants suspected to be present. Users should recognize and understand the limitations of their GC prior to actual field use. Some manufacturers will provide individualized instruction on portable GC units prior to purchase; prospective buyers should take advantage of such an opportunity. There are also specialized courses available in field sampling and GC analysis[2].

Truly portable GCs have an internal battery capable of powering the unit for at least 4-6 hours; optional battery packs are usually available to extend field use to a full day. Portable GCs may also be powered off 115V AC power supply; some can be connected to a vehicle 12V DC battery through the cigarette lighter. Most GCs can be operated on AC power while simultaneously recharging the internal battery.

A refillable internal high pressure cylinder serves as a reservoir for the carrier gas. The carrier may be air, hydrogen, helium, nitrogen, or another inert gas. Only high purity gas (99.999%), with **less than 0.1 ppm** total hydrocarbon contamination, should be used; such gases may be called Ultra Zero, Zero Zero, or Ultra Zero Pure. A large cylinder of carrier gas should be available to refill the GC reservoir. The high purity refill regulator should be clean and free from oil, particulates, or any other contamination. Teflon or stainless steel tubing should be used for all connections. Vinyl, rubber, or tygon tubing will contaminate the carrier gas and should never be used.

GCs equipped with microprocessors allow operator control of a variety of parameters including column temperature, duration of analysis, backflushing, detector sensitivity, and peak height integration using stored calibration data. Microprocessor-controlled data storage, retrieval, and evaluation of analytical results make field analyses and contamination assessment faster, easier, and more reliable. While these instruments offer a variety of flexible features which improve field analyses, they also require more training and experience on the part of the user.

USING A PORTABLE GAS CHROMATOGRAPH

Gas chromatography is often viewed as a means to **identify** contaminants. GC is best applied in situations with a limited number of **known** contaminants. Under these conditions, the operator must have the relative retention time and peak height characteristics of each contaminant of interest under the conditions of field analysis. This information can be obtained only by introducing a known concentration of each contaminant individually or as a calibration gas mixture containing known amounts of each material. The need for calibrant gas mixtures is an important consideration that is often overlooked.

Calibration gases are commercially obtainable from specialty gas suppliers. A single gas, such as n-butane or isobutylene should be used to ensure the GC is functioning properly. A gas containing a known concentration of one or more of the anticipated contaminants should then be introduced; relative retention times and sensitivities or peak heights should also be recorded. Calibration should be repeated if separation conditions are changed (type of column, column temperature, carrier gas flow rate, autoinject vs. syringe), or if maintenance procedures have been performed (cleaning the detector, purging the column, replacing inlet filters).

Instrument response to materials as they are eluted from the column is usually recorded or displayed as a series of peaks on a strip chart recorder, LCD display, or monitor. Microprocessor-controlled GCs can store calibration data and integrate peak size and determine the concentration of each contaminant present. Other methods to estimate approximate concentrations include measuring relative peak height, calculating the area under the peak (1/2 base x height = area) for peaks that are approximately triangular in shape, counting squares under each peak, or cutting out and weighing each peak of the strip chart recording.

When manually injecting low concentration standards or samples, a glass syringe with a stainless steel plunger should be used; the sealing ring on the plunger should be Teflon or a similar material with little or no affinity for volatile organic contaminants. The stainless steel needle may be screwed on or bonded to the syringe. Syringes should be regularly checked for leaks or blockages. To check for leaks or blockages, draw air into the syringe and then submerge it in water; watch for bubbles which indicate leakage. If no bubbles are noted, push the plunger down slowly; if no blockage is present, bubbles will be seen emerging from the tip of the needle until the plunger is fully depressed.

Syringes should be carefully purged with clean air after use; it is recommended that a syringe used to inject calibration gas or a standard gas mixture not be used for other purposes. Inadequate cleaning of syringes can result in cross-contamination between samples.

Field GC analyses can be a powerful analytical tool; the usefulness of field data, however, is directly linked to the QA/QC (quality assurance/quality control) program which ensures that data generated are accurate and reproducible. The QA/QC program should address sample collection as well as analysis. Such a program ensures that samples collected are representative of actual conditions and minimal chemical or physical change occurs as a result of the collection or analytical process. Overall adherence to good laboratory practices (GLPs), as well as specific methods of calibration and analysis, including duplicate measurements, standards, and checks, are also used to assure proper sample analysis and evaluation[3].

Interpretation of Results

An important limitation often overlooked by novice users is that each peak may represent **multiple** contaminants with the same retention time. There is not a unique retention time for each organic contaminant. **The presence of a peak at a particular point in the analysis does not guarantee its identity**. It is also not unusual to detect peaks which cannot be identified. At sites where unknowns are present, the identity of contaminants should be confirmed by laboratory analysis.

Similarly, the absence of peaks does not automatically indicate the lack of contamination; it does, however, suggest that there are no compounds present within the limits of the analytical method used. A change in one or more parameters (type of column, column temperature, detection limit, duration of analysis, carrier gas flow) may change results. At sites containing unknowns, the lack of contamination should also be confirmed. Laboratory analysis is especially important to detect low volatility compounds that are backflushed and never reach the detector.

Photovac 10S Plus

The Photovac 10S Plus is currently the industry standard for portable field GCs. The 10S Plus is the successor to previous 10S models (10S-30, 10S-50, 10S-70); many of the limitations of older versions have been remedied. The 10S Plus is relatively lightweight (28 lbs), can utilize a variety of carrier gases, and is equipped with a photoionization detector, computer, and modem. This instrument can be operated for 7 hours using the internal, rechargeable battery; battery packs and an AC adaptor are also available.

Figure 8-4: The Photovac 10S Plus field portable gas chromatograph is easily carried and adaptable to field or fixed facility use. (Courtesy Photovac.)

The on-board computer utilizes function/application and memory cards which allow the user to program specific parameters, store methods and results, view completed analyses, add notes, magnify small peaks, view or print reports of individual analyses, or provide a summary of results for a sampling interval.

Although the unit comes with only a photoionization detector, it is capable of detecting contaminants as low as 1 ppb. The standard UV lamp is 10.6 eV; optional 8.4, 9.5, 10.0, and 11.7 eV lamps are available. The detector can be adjusted to accommodate high and low concentrations in the same analysis.

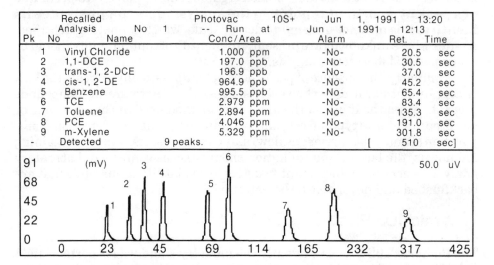

Figure 8-5: An example of a 10S Plus Analysis Report which includes chromatogram and compounds detected with concentrations. (Courtesy Photovac.)

Individual user-created libraries are stored on memory cards; each card may contain information, including analysis methods and chromatogram, for up to 25 compounds. Information regarding measured response versus concentration is obtained during multiple compound calibration and then stored; the response/concentration ratio or "sensitivity" for each compound is also calculated and stored. Retention time and sensitivity ratios for three concentration ranges can be maintained for each compound; Information is updated after each calibration event. During sample analysis, the computer compares detected peaks with those stored on the memory card. Matched peaks are identified by retention time; the appropriate sensitivity ratio is then selected and used to calculate the actual concentration present.

Frequent calibration is required to compensate for PID output, changes in sample flow, and UV lamp window cleanliness[4]. While calibration with standard gas mixtures containing all known or suspected contaminants ensures the greatest accuracy, it not always possible to obtain a gas standard with all compounds of interest. Surrogate calibration, using one or a few compounds, can be used to set the sensitivity of the instrument under specific operating conditions and of update sensitivity settings for other designated compounds within a library.

Photovac provides a wealth of technical and support information to 10S Plus users or prospective buyers; they also offer individualized instruction on maintenance and use of this instrument. They do not, however, provide calibration gas mixtures which must be obtained from a specialty gas supplier. The user must program the library memory cards for future use.

Photovac Snapshot™

The Snapshot™ is a hand-held, factory programmed GC designed to detect a limited number of specific contaminants in air. The unit weights only 8 lbs; a pump draws a sample for automatic injection into the GC column. Identity and concentration of contaminants are displayed on the screen and can be stored in the datalogger. The unit maintains isothermal control of column temperature and precolumn backflushing.

Snapshot users must change Application Modules to accommodate specific monitoring requirements. Each Module is designed to detect a limited number of contaminants. Modules are available to detect chlorinated hydrocarbons, aromatic hydrocarbons, linear hydrocarbons, hydrogen sulfide and mercaptans. Some modules are also designed to meet industrial hygiene needs of specific industries such as rubber manufacture (styrene, acrylonitrile, and 1,3-butadiene) and sterilization/fumigation (ethylene oxide, methyl bromide).

The Snapshot should not be used, however, as a monitoring instrument at unknown sites. **The unit will provide information on only a few compounds; other contaminants will not be identified and the user will not be alerted to their presence.** For instance, when using the Application Module for chlorinated hydrocarbons, other contaminants, regardless of concentration, will not be detected. The Snapshot is designed for use in situations where specific contaminants are known or suspected to be present. Under these conditions, it can also be an invaluable tool for rapid screening of environmental samples.

HNU Systems, Inc.

The HNU Model 301P Portable GC can be used in the field but is more suited to stationary use. The 301P is designed to accommodate two detectors (PID and FID) which can be used separately or in series; other detectors (ECD, TCD, FUVD) are also available[5]. The 301P has a heated injection port and isothermal oven which can hold two packed columns or one capillary column. A separate temperature control module is required to monitor and control oven and injection port temperature. The Field Pak allows limited field use; the unit will operate for 10 hours with the oven at ambient temperature. The instrument should be considered field-deployable rather than portable; the GC weighs 25 lbs; a separate Battery Support Pack weighs another 16 lbs.

The HNU Model 311 Portable GC weighs 55 lbs and requires an AC power source. This unit is designed for bench top use and is suitable for a stationary field lab; despite its name, it is **not** portable. The 311 employs a PID (8.3, 9.5, 10.2, 11.7 eV UV lamps are available[6]). The GC has an isothermal oven and heated injector block; the unit has a keypad and is microprocessor-controlled but does not have integration capabilities. A built-in printer/plotter provides hard copy of analytical conditions and results.

The Foxboro Company

The Century™ OVA model 108 or 128 offer a rugged, portable field GC option with limited capabilities at an affordable price. The GC option on these FID models include an exterior GC column, manual inject and backflush valves, and activated charcoal scrubber assembly. The inject valve is used to select either the survey mode (valve out) or the GC mode (valve in). When the inject

Figure 8-6: The Photovac Snapshot™ is a hand-held analyzer that incorporates advanced GC technology into a powerful 8-lb package. (Courtesy Photovac.)

valve is depressed, hydrogen is routed through the GC column and 0.25 ml of sample air is injected. An optional septum adaptor may be used for making syringe injections directly into the column.

During survey mode operation, sample air containing contaminants is drawn into the sample inlet and routed directly to the combustion chamber. During GC operation, contaminants are carried with hydrogen gas and separated in the column. The hydrogen-fed flame still requires air during GC analysis; ambient air drawn into the sample inlet is first routed to the charcoal scrubber which removes hydrocarbon contaminants (except for methane and ethane), before reaching the flame.

After sample injection, the user watches the analog display to determine the number of peaks eluting from the column; an optional strip-chart recorder can also be used. The operator must know the identity of sample air contaminants and their retention times under current field conditions. Retention times can be affected by hydrogen pressure (should be no less than 12 psi) which influences carrier flow rate. The activated charcoal must be regularly replaced to prevent contamination of air feeding the hydrogen flame. After the peaks of interest have eluted from the column, the column

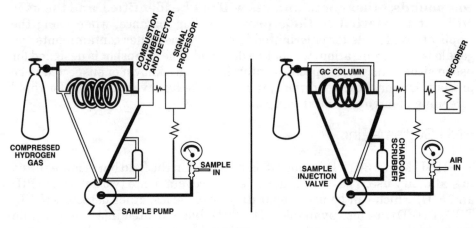

Figure 8-7: Schematic of OVA operation during survey mode (left) and GC mode (right).

is backflushed and size of the backflushed peak is observed and recorded. A large backflush peak indicates the presence of significant concentrations of unknowns or unidentified compounds.

A portable isothermal pack (PIP) is available to maintain the GC column at a uniform temperature (0, 40, or 100°C) using a temperature control slug inserted into the center of the pack. Also available is an instrinsically safe portable strip-chart recorder designed to be carried with the OVA. The recorder can be used to record chromatograms obtained during GC mode use, or it can provide a continuous read-out of total hydrocarbons detected during use in the survey mode. Previous recorders designed for stationary use were not intrinsically safe.

Some users experience flame-out when operating an OVA in the GC mode. Flame-out can occur if the inject or backflush valves are only partially activated and are not completely in or out. This often happens when the backflush valve is activated, because of its hard-to-reach position under the GC column. When neither valve is completely activated, both air and hydrogen flows are blocked; always activate valves by quickly and firmly pushing in or pulling out. Insufficient hydrogen pressure, often caused by loose or worn column connections, a blocked sample inlet, or a kinked column which obstruct hydrogen flow, can also cause flame-out when the instrument is in the GC mode.

TRACE AIR GAS ANALYSIS UNITS

Trace air gas analysis (TAGA) units combine chromatographic separation with mass spectrometry; mass spectrometry (MS) is a powerful analytical tool which provides information about the atomic and molecular composition of organic and inorganic materials. MS can identify unknowns and confirm the presence of contaminants known or suspected to be present.

The need for on-site analytical data has prompted the development of the TAGA unit. A typical TAGA mobile analytical laboratory unit contains two tandem MS analyzers and at least one GC housed in a large self-propelled recreation-type vehicle. Also on board is a powerful computer with an analytical library containing information on thousands of compounds, as well as equipment and space necessary to prepare samples and calibration standards[7].

TAGA units are capable of direct sampling of ambient air; rapid mobile analytical surveys allow a relatively large area to be surveyed and are often

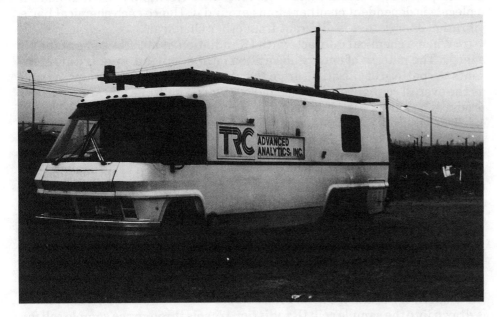

Figure 8-8: A mobile TAGA unit. (Author photo.)

undertaken as a result of odor complaints, tracking the source of fugitive emissions, or after a hazardous materials emergency. Alternatively, samples may be collected and injected manually; this technique was successfully used to allay fears regarding potential dixoin exposure in a community adjacent to a burning landfill.

Equipping the unit with additional sample preparation facilities allows rapid analysis of soil and water samples as well and facilitates rapid screening of large areas for volatile and semivolatile contaminants; hot spots are quickly located and specific contaminants identified.

A less expensive alternative to the TAGA unit are field deployable GC/MS analyzers that are now available. Viking Instruments now offers a 120 lb SpectraTrak™ 600 Transportable GC/MS with features typically found in laboratory instruments. The SpectraTrak 600 can accommodate prepared soil and water samples; automated air sampling inlets allow for real time monitoring and trace analysis.

While mobile TAGA units and transportable field GC/MS are expensive to maintain and use, they offer the opportunity for rapid and reliable sample

Chemical Groups	Absorption Band (microns)
Alkanes (C-C)	3.35 - 3.65
Alkenes (C=C)	3.25 - 3.45
Alkynes (C≡C)	3.05 - 3.25
Aromatic HCs	3.25 - 3.35
Substitued Aromatic	6.15 - 6.35
Alcoholic (-OH)	2.80 - 3.10
Acids (C-OOH)	5.60 - 5.90
Aldehydes (COH)	5.60 - 6.00
Ketones (C=O)	5.60 - 5.90
Esters (COOR)	5.75 - 6.00
Chlorinated (C-Cl)	12.80 - 15.50

Table 8-1: Specific infrared absorption bands for hydrocarbons (HCs)[7].

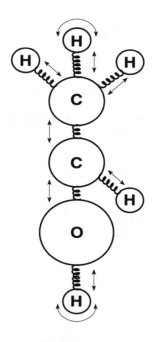

Figure 8-9: Bonding forces which hold a molecule of ethanol together can be visualized as springs which allow each atom to bend, twist, and rotate and impart a vibration energy pattern to the entire molecule.

analysis. Such real-time data offers the opportunity to change procedures and focus efforts on areas known to be contaminated. Increased work site flexibility and decreased use of contract laboratory analysis inevitably results in overall cost savings.

PORTABLE INFRARED SPECTROPHOTOMETERS

The infrared region of the electromagnetic spectrum extends from the red end of the visible light spectrum to the microwave region. The most-used spectral or wavelength range is the mid-infrared region (2.5-15.4 microns). Infrared spectroscopy examines the bending, twisting, and vibrational movements of atoms in a molecule.

Atoms or groups of atoms in a molecule are always in motion. The bonding forces that keep molecules together can be thought of as springs; the springs keep the molecule linked together and balanced. Different chemicals are composed of different atoms; the motion or vibration of each atom in a molecule is dependent on other atoms around it. Hence each chemical has a vibration energy that is different from other chemicals.

When a chemical interacts with IR radiation, it absorbs some of the IR energy. The pattern of energy absorption or **absorption spectrum** represents the pattern of vibrations from the atoms and functional groups, as well as the overall configuration of the molecule. The multiple vibration pattern produces a characteristic absorption spectrum that is unique to each chemical. This absorption spectrum can be considered a fingerprint of the chemical.

Each chemical demonstrates optimal absorption, or a unique absorption pattern within a relatively small IR wavelength range. For example, the absorbance spectrum for ammonia in Figure 8-10 indicates optimal absorption occurs between approximately 10.5 and 11.5 microns. Absorption ranges or analytical bands for classes of chemicals have been defined and are usually quite small; this allows the user to filter out all but a small portion of the IR spectrum and still measure the concentration of the material known to be present.

A typical portable IR analyzer is a single beam spectrophotometer and consists of a sample inlet, an IR radiation source, a filter to control the wavelength of radiation delivered, a sample cell, and a detector. Sample air is drawn into the sample cell; IR radiation travels through the sample cell for a specific distance or **pathlength** before reaching the detector. IR absorbance by a chemical over a given pathlength is proportional to its concentration. The longer the pathlength that the IR radiation travels before it reaches the detector, the greater the sensitivity and the lower the detection limit (since there is more time for sample contaminant to interact with the IR radiation and absorb some of its energy). Mirrors are used to reflect or fold the IR radiation within the sample cell to achieve long pathlengths (i.e. 10-50 feet). Instruments may have fixed or variable pathlengths.

Some analyzers are factory-calibrated to measure only one material; these instruments have one or two fixed pathlengths (to accommodate a wide concentration range) and a single filter. Other IR units have variable settings and can detect many different contaminants; these analyzers are usually microprocessor-controlled and convert absorbance to concentration in ppm or % by volume. Concentrations as high as 100% by volume and as low as 0.10 ppm can be measured.

Fixed IR monitors are used to detect methane, carbon dioxide and oxygen in monitoring boreholes around landfills and methane extraction systems within landfills. IR analyzers are utilized because high carbon

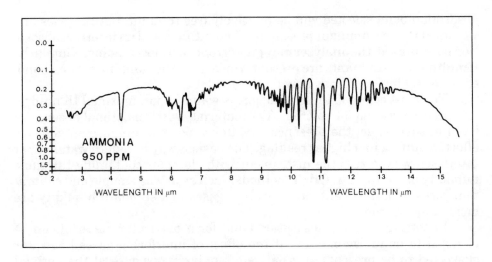

Figure 8-10: Infrared absorbance spectrum of ammonia. (Courtesy The Foxboro Company.)

dioxide concentrations will not affect methane readings. Geotechnical Instruments offers a small, IR monitor which measures landfill gases.

Using Portable Infrared Analyzers

A newly-purchased IR spectrophotometer should be calibrated with multiple concentrations of span gas as soon as possible after delivery. This is to ensure that the sample cell and mirrors have not been damaged or alignments changed during shipping. Manual calibration can be performed using known concentrations of gas. Calibration can also be conducted using a closed-loop system into which a pure gas or liquid sample is introduced with a syringe. A separate pump then circulates the calibration gas into the sample cell.

IR analyzers are designed to measure the amount of one or more **compounds known to be present**; they are not capable of identifying the material. These units are designed for situations where only one or a few contaminants are present. Some analyzers can measure only one compound of interest at a time; if multiple contaminants are known to be present, they must be measured sequentially. The user adjusts the IR wavelength and the sample cell pathlength for each contaminant of concern. Some analyzers automatically monitor for multiple contaminants.

The optimal absorption band for a particular chemical may be shared or overlapped by other compounds; it is extremely important that the

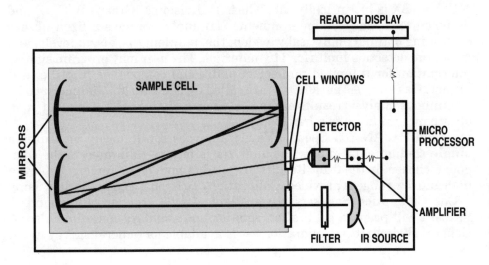

Figure 8-11: A simplified functional diagram of an IR analyzer. The mirrors within the sample cell reflect the infra-red radiation several times to achieve a long pathlength.

Figure 8-12: The Miran 1B2 (top) is bulky and awkward to carry. (Photos by Steve Napolitano.)

absorption band selected will be relatively free from interfering gases and vapors. If the contaminant present is misidentified, and an incorrect absorption band is used, the analyzer may give erroneously low readings. Similarly, if multiple contaminants are present which have the same absorption band, erroneously high readings may be obtained.

The most common interfering gas is water vapor, maximal IR absorption occurs between 5-8 microns. For materials with an optimal absorption band in this range, the presence of water vapor will produce an additive effect, resulting in a higher reading. The easiest way to reduce water vapor interference is to zero the instrument with clean air containing the same humidity as sample air. Other methods, such as drying sample and calibration gases with desiccant, or saturating gases to a standard humidity are more cumbersome.

In some cases, the absorption band for a particular material can be changed to minimize or prevent the effect of interfering gases known or suspected to be present. Alternatively, analyzer response at the optimal absorption band and a secondary band can be compared.

Analyzers may be zeroed using clean ambient air known to be free of the contaminant of interest; if that is not possible, a zero air gas standard can be used, or zero gas cartridges. Zero gas cartridges remove IR absorbing contaminants and are supplied by the manufacturer. Special zero gas cartridges may be required for some contaminants such as nitrous oxide, sulfur dioxide, formaldehyde, carbon monoxide, carbon dioxide, and ammonia.

IR analyzers are equipped with particulate filters to keep dust out of the sample cell; dust will scatter the IR radiation and decrease sensitivity. Dust on mirrors and sample window surfaces will also decrease sensitivity and instrument efficiency. The particulate filter should always be in place when the analyzer is in use.

The sample windows which transmit IR radiation into and out of the sample cell are usually made of polished flat discs of sodium chloride, silver bromide, or potassium bromide[8]. Windows can be damaged by water condensation or by exposure to high concentrations of some gases and vapors. This information should be supplied by the manufacturer.

MIRAN® Portable Analyzers

The MIRAN®(MIniature IR ANalyzer) series of IR analyzers manufactured by The Foxboro Company. Probably the most powerful and versatile models are the MIRAN 1B2, and the 1BX. Both are microprocessor-controlled; the MIRAN 1BX is intrinsically safe (Class I, Division 1, Groups B, C, D) and designed for hazardous environments. 1B analyzers have a fixed library containing optimal analytical wavelengths, pathlengths, alarm levels, and minimum detection limits for 116 materials. The user may program similar information for a limited number of additional compounds into the user library. Gas analysis parameters are called up using an alphanumeric code. Continuous analysis results are indicated on a digital LED display in ppm, absorbance units, or as a bar graph[9].

The MIRAN 203 Specific Vapor Analyzer continuously measures a **single** contaminant in air. The analyzer is factory-calibrated to the compound chosen by the user. Results are read in ppm on an analog display; a high and low range is provided. Calibration can be changed in the field using a Gas Calibration Set; sets are available for 28 standard compounds[10]. Foxboro will prepare calibration sets for non-standard materials also. A General Hydrocarbon Calibration set is available for general survey work.

MIRANs are limited to no more than four hours continuous use when operated off the internal battery. Battery recharge requires 14-16 hours. Spare battery assemblies are available; analyzers may be operated on AC power while simultaneously recharging interval batteries.

An often-overlooked limitation is the size, weight, and bulk of MIRAN analyzers. The MIRAN 1BX, for example, weighs nearly 30 lbs, and is awkward to carry. The alignment of mirrors in the sample cell can be inadvertently changed if the unit is handled roughly. MIRANs are best suited for situations where they can be placed on a stationary surface during sampling.

Monitoring Landfill Gases

Fixed IR monitors are used to detect methane and carbon dioxide in monitoring boreholes around landfills and in methane extraction systems within landfills. IR analyzers are utilized because high carbon dioxide concentrations will not affect methane readings. These IR monitoring systems are capable of measuring up to 100% by volume concentrations of methane and carbon dioxide. It is important to recognize, however, that such systems are specific for these two normally-occurring landfill gases; other gases and vapors will not be detected.

Geotechnical Instruments offers a small, handheld IR monitor which also measures landfill gases. Methane and carbon dioxide concentrations are calculated by infrared absorption; oxygen is measured by an optional electrochemical sensor. This analyzer is designed to detect relatively large concentrations; because of the short pathlength of the infra-red beam, concentrations less than 0.5% methane and carbon dioxide cannot be accurately measured. The oxygen sensor operates at an extremely high resistence to inhibit poisoning of the electrolyte by carbon dioxide. The analyzer has an in-line water trap and replaceable particulate filter; it also has a data logger which can download data for analysis. The instrument carries no intrinsic safety approvals at this time.

Figure 8-13: The GA 90 Infra-red Gas Analyser (sic) detects percent by volume concentrations of carbon dioxide and methane. Current models are not certified as intrinsically safe. (Courtesy Geotechnical Instruments.)

PHOTOACOUSTIC SPECTROSCOPY

Photoacoustic spectroscopy (PAS), also called optoacoustic spectroscopy, uses sound and UV or IR radiation to measure the concentration of air contaminants. When atoms or molecules vibrate, they vibrate at a particular frequency, called resonance frequency. The number and type of atoms determine the resonance frequency; each chemical therefore has a unique resonance frequency. Frequency is measured in units called Hertz (Hz); one Hz equals one vibration per second. The resonance frequency of most molecules is around 10^{13} Hz, or 10^{13} (10 trillion) vibrations per second; IR radiation has a frequency within the same magnitude. IR radiation will transfer energy to a molecule if the molecule has exactly the same frequency as the radiation. When radiant energy is absorbed, the affected molecules gain energy and vibrate more vigorously. The excess energy is quickly transmitted to the surrounding medium in the form of heat. The increase in heat produces an increase in pressure which is detected by a microphone.

A typical photoacoustic analyzer consists of a radiation source which is emitted in pulses, a photoacoustic cell filled with optically transparent gas (usually air or helium), a sample cell, a microphone, amplifier, and recorder/readout display. Detection thresholds are typically between 0.001 and 1 ppm.

Photoacoustic analyzers are typically used as portable or fixed location monitors in situations where a limited number of known contaminants are present.

Interpretation of Analyzer Results

IR and PAS analyzers are designed to measure the concentration of a limited number of gases or vapors **known** to be present. These analyzers are frequently used in hospitals, laboratories, and manufacturing facilities where a limited number of chemicals are used. Under these conditions, the analyzer is used to detect only one or a few compounds. A lack of a reading indicates that the concentration of the contaminant of concern is less than the detection limit of the analyzer. For instance, at a hospital sterilizing facility, an IR analyzer can be used to detect very low concentrations of ethylene oxide (EO). The lack of a reading indicates the absence of significant amounts of EO; it does not indicate the absence of **other** contaminants that may also be present.

Despite the appearance and obvious level of sophistication of these analyzers, they are not "black boxes" capable of identifying any and all contaminants present. It is impossible to identify all contaminates present at uncontrolled and uncharacterized sites by interpreting analyzer readings.

REFERENCES

1. Fingas, H.F. and Bobra, A.M. 1985. Environment Canada's Project for Spill Analytical Systems. Proceedings of the Hazardous Materials Management Conference/West '85, pp.147-150. Wheaton, IL: Tower Conference Management Company.
2. Customized environmental sampling/sample analysis training programs are offered by Nielson Ground-Water Inc., Galena, OH.
3. American Society for Testing and Materials. 1985. Quality Assurance for Environmental Measurements, Special Publication 867, PCN 04867000-16. Philadelphia: ASTM.
4. 10S Plus Digital Gas Chromatograph User's Manual (Revision A). 1991. Photovac International Inc., Deer Park, NY.
5. Instruction Manual for Model 301 Gas Chromatograph. 1986. HNU Systems, Inc. Newton, MA.
6. Model 311 Portable GC Operator's Manual Version 1.0. 1989. HNU Systems, Inc.
7. Shushan, B.I., DeBrou, G., Mo, S.H. and Webster, W. 1990. Mobile field monitoring of volatile organics and toxic air pollutants using a mobile tandem mass spectrometer system. In Monitoring Methods for Toxics in the Atmosphere, W.L. Zielinski, Jr. and W.D. Dorko (Eds.), pp.75-91. Philadelphia: American Society for Testing and Materials (ASTM).
8. National Institute of Occupational Safety and Health. The Industrial Environment, Its Evaluation and Control, NIOSH: Cincinnati, OH, 1973.
9. Instruction Manual MI 611-098 for the MIRAN® 1BX Portable Air Analyzer. The Foxboro Company, April 1991.
10. Instruction Manual MI 611-070 for the MIRAN® 203 Specific Vapor Analyzer. The Foxboro Company, March 1990.

CHAPTER REVIEW

1. List three components of a portable gas chromatograph: _____

_____

2. The gas that pushes the sample through the column is called the _____ gas.

3. List three variables that can affect column retention time: _____

_____

4. Describe the difference between a precolumn and an analytical column:_____.

5. List the recommended specifications for GC carrier gas purity:_____.

6. The most common detectors found in portable GC analyzers are the _____.

7. List three components of a portable infrared analyzer: _____.
_____.

8. True or false? Within the sample cell, infrared radiation is reflected off mirrors to achieve long pathlengths.

9. True or false? The longer the IR pathlength the lower the detection limit for the compound of interest.

10. When using an IR analyzer, the most common interfering gas present is usually _____.

9 Radiation Detectors

We are surrounded by many kinds of radiation, all of which are a form of energy. Some types of radiation we can see (visible light) or feel (infrared radiation can be felt as heat); we can see the effects of exposure to UV radiation (a sunburn). Many forms of radiation, however, we cannot see or feel (cosmic rays, radar, television, radio and microwaves).

Radiation can be broadly classified into two categories, ionizing and non-ionizing. High levels of ionizing radiation can cause significant injury to biological tissues. It is not unusual to encounter ionizing radiation at hazardous waste sites or during emergency responses. It is important, then, that the responder or site worker determine if radiation is present at levels which pose an immediate health hazard.

Radiation is a form of electromagnetic energy that can be described as a series of waves which differ in length, frequency, and energy. The spectrum of wavelengths extends over a broad range, from less than 10^{-12} cm to greater than 10^{10} cm; ionizing radiation has a much shorter wavelength than non-ionizing radiation.

WAVELENGTH IN MICRONS (10^{-6} METER)

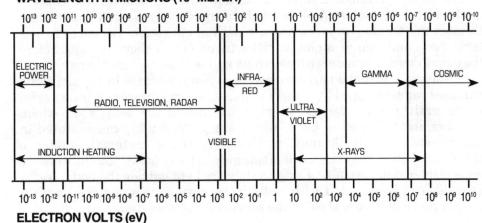

ELECTRON VOLTS (eV)

Figure 9-1: The electromagnetic spectrum.

157

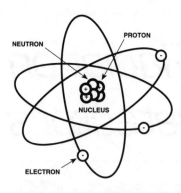

Figure 9-2: Basic model of an atom; each electron occupies an orbit around the nucleus.

Ionization is a process that changes the normal electrical balance in an atom. Each atom has a heavy core or **nucleus** containing positively charged particles called **protons**, and neutral particles called **neutrons**. Light-weight, negatively charged particles called **electrons** orbit the nucleus. When an atom loses an electron, it becomes a positively charged **ion**; if a free electron becomes attached to another atom, a negatively charged ion is formed. Ionizing radiation is capable of producing ions, either directly or indirectly. Direct ionization occurs when a charged particle interacts with other ions; indirect ionization occurs when uncharged particles or energy waves induce the emission of charged particles.

Radioactive materials emit atomic particles and/or high energy electromagnetic waves. Radioactive atoms are unstable because they contain excess energy; excited or radioactive atoms release energy and return to their normal or stable state by emitting subatomic particles and high energy electromagnetic radiation (gamma rays). The most commonly encountered types of ionizing radiation are alpha, beta, and neutron particles, and x-rays and gamma rays.

Radioactive materials or **radionuclides** undergo spontaneous transformation or **decay**. This process is also called **nuclear disintegration**; when a radioactive atom disintegrates, it emits a particle from the nucleus. The remaining atom or **decay product** is now a different element and is often called a **daughter product**.

As decay takes place, the number of disintegrations occurring in a given amount of radioactive material per unit time decreases exponentially. The amount of time for half the disintegrations to occur is called the **half-life**; put another way, the half-life is the amount of time required for the material to lose half of its radioactivity. Each radionuclide has a specific half-life; they range from less than a second to billions of years, during this time radiation is emitted and a new radionuclide is formed.

Activity of radionuclides is commonly expressed by the number of disintegrations occurring per minute (dpm) or per second (dps). A **curie** (Ci) is equivalent to 3.7×10^{10} dps.

TYPES OF IONIZING RADIATION

Alpha Particles

Alpha radiation is not composed of waves, but rather particles which are ejected from the nucleus during nuclear disintegration. An alpha particle consists of two protons and two neutrons and have the same structure as the nucleus of a helium atom. Alpha particles are positively charged; as they pass through matter they produce a dense path of ionization. Because of their large mass and charge, alpha particles travel only a short distance before they slow down, combine with electrons and become helium atoms.

The range of alpha particles is at most about 10 cm (4 in.) in air; alpha particles can be stopped by a sheet of paper, a film of water, or intact skin. Alpha **emitters** are dangerous when taken into the body; some alpha emitters are deposited in bone, others are preferentially concentrated in organs such as the kidney or liver. As these materials undergo disintegratration, they release damaging alpha particles. Alpha emitters are not considered hazardous as long as they are kept outside the body. Alpha particles can be detected with radiation survey meters equipped with a special probe held very close to the source or alpha emitter.

Beta Particles

Beta particles are **negatively charged electrons** ejected from radioactive atoms during disintegration. Beta particles have a velocity near the speed of light; their range in air may be up to 18 m (60 ft); commonly encountered beta particles have ranges less than 9 m (30 ft). Beta particles are smaller, faster moving, and have a greater range than alpha particles. Beta particles can penetrate wood, clothing, and human skin. High energy beta particles can penetrate up to 1.5 cm (0.5 in.) into the human body; skin burns can result from overexposure to beta radiation. Deceleration of a beta particle produces secondary radiation (x-radiation) which also causes tissue damage. Beta emitters are considered internal and external radiation hazards.

Neutrons

Neutrons are particles that are released during disintegration and have no electrical charge. Neutrons radiation is released by those radionuclides capable of spontaneous fission which may be found around the core of nuclear reactors, nuclear fuel, and nuclear waste. Neutrons are capable of penetrating human skin. Neutrons collide with and disrupt atoms into high-energy fragments. As they repeatedly collide with atoms, the neutrons slow down and lose energy; in the process, excess energy is released in the form of protons, beta or alpha particles, or gamma rays. The secondary radiation released produces additional tissue damage. Neutron radiation is not commonly encountered; specialized equipment must be used to detect neutron emitters.

Positrons

A **positron** is a particle similar to an electron, except that it has a positive charge. Positrons immediately combine with negatively charged electrons in a process called **annihilation**; during this process gamma radiation is released, which can be detected.

Gamma Rays

Gamma rays are high energy electromagnetic radiation with a very short wavelength. Gamma radiation is emitted from the **nucleus** of the atom during annihilation. High energy gamma rays travel at the speed of light, have an extended range, and can penetrate several inches of steel. Gamma rays readily penetrate the body, where they are scattered, reflected, or absorbed; this in turn produces secondary radiation in the form of particles or x-rays. Specific radionuclides emit gamma radiation of characteristic energy and wavelength. Because of its energy and penetration capabilities, gamma radiation is considered the most hazardous form of radiation.

X-Rays

X-rays are a form of penetrating electromagnetic radiation similar to gamma radiation except it does not originate from the nucleus. Man-made x-rays are produced by accelerating electrons from a heated tungsten filament into a metal target. The accelerating electrons interact with orbiting electrons in the target and quickly decelerate; this sudden deceleration results in the emission of x-rays. X-rays are also produced when beta particles interact

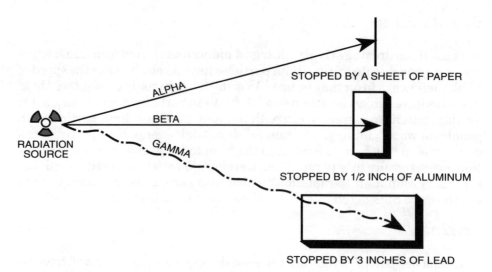

STOPPED BY A SHEET OF PAPER

STOPPED BY 1/2 INCH OF ALUMINUM

STOPPED BY 3 INCHES OF LEAD

ALPHA

BETA

GAMMA

RADIATION SOURCE

Figure 9-3: Relative penetration of alpha, beta, and gamma radiation.

with orbiting electrons in the body. "Soft" x-rays have a relatively long wavelength and are less penetrating than "hard" x-rays which have a very short wavelength and can penetrate several inches of steel.

RADIATION EXPOSURE TERMINOLOGY

All forms of ionizing radiation transfer energy to matter; in living organisms, the amount of energy imparted per kilogram of irradiated material is defined using a unit called a **gray (Gy);** the conventional unit of absorbed dose is the **rad** (radiation absorbed dose); 1 Gy equals 100 rads, 1 centigray (cGy, 10^{-2} Gy) equals 1 rad.

The biological effect of equally absorbed doses, however, differs depending on the type and energy of the radiation. Quality factors (Q) are used to account for these differences and to determine dose equivalency; x-rays, gamma rays, and beta particles have a quality factor of 1.0, neutrons 10.0, and alpha particles 20.0. The dose equivalent (H) is the absorbed dose multiplied by the quality factor. One dose equivalent unit is the **sievert (Sv)**; the conventional unit is called the **rem** (radiation equivalent for man). One rem is equivalent to 10 millisieverts (mSv). The **roentgen (R)** is a unit of exposure for gamma radiation or x-rays; since the quality factor for gamma radiation is 1.0, one R equals one rem. Dose in R, rem, or millirem (mR) is expressed on a per hour time basis (mR/hr).

Naturally occurring radiation has always been present on earth; cosmic rays, a form of gamma radiation, is another source of natural or background radiation. Background radiation levels will vary depending on the type of building materials or natural geologic formations. Typical natural radiation background levels are between 0.01 and 0.05 mR/hr. In the United States, the annual dose equivalent from natural radiation is estimated to be about 125 mR.

RADIATION DETECTORS

There is a variety of detectors used for radiation monitoring or measurement. Detectors are normally associated with a readout which displays radiation on a scale marked R/hr, rems/hr, mR/hr, or cpms (counts per minute). **Dosimeters** are used to determine the dose accumulated by the wearer during an exposure interval. Radiation **survey instruments** indicate the amount of radiation present at the time of sampling.

The measured value is accurate **only for the calibrating radiation**. Significant errors can occur when the type and activity of the radiation source is different than the calibrating radiation. While users should be aware that the errors can occur, it should not be cause for alarm; exposure limits and action levels are sufficiently low to safeguard responders from radiation overexposure.

Users should also be aware that a detector intercepts and registers only a fraction of the radiation emitted from a radioactive source. Information on detector efficiency is provided by the manufacturer and will vary by the type of source; for example, a detector may have efficiencies of 10% for C^{14}, 60% for Cs^{137}, and 40 for Sr^{90}.

A common misconception is that cpm and mR/hr readings are the same for all detectors. Although they may be indicated on the same meter display, cpm and mR/hr readings are **not** interchangeable; the manufacturer must supply information on how to translate a cpm response to mR/hr. The conversion factor varies between different detectors and manufacturers. For example, an extremely sensitive detector may show a reading of 900,000 cpm in response to a Cs^{137} source of 1 mR/hr, while another may only register 2000 cpm. The cpm to mR/hr conversion should always accompany the instrument; it is impossible to convert cpm to mR/hr without this information.

Calibration

Calibration of detectors and dosimeters should only be performed by qualified personnel using carefully prepared calibration sources. Calibration is best performed by the manufacturer on a yearly basis. Radiation detectors should have a radioactive **check source** to ensure that the meter is still functioning. The detector should give a response when placed directly over the check source. The manufacturer should supply information regarding the type of radiation emitted as well as anticipated instrument response at the surface of the check source.

GAS-FILLED DETECTORS

Geiger-Mueller counters, ionization chambers, and proportional counters are gas-filled detectors; the gas may be air, nitrogen, or an inert gas such as helium or argon. Radiation interacts with the gas and produces ions; electrodes establish an electric field across the gas volume. Negatively charged ions are collected at the positively charged electrode or anode, while positive ions are collected at the cathode.

Ionization Chambers

The ionization chamber detector measures the ionization current flow in air between two electrodes in response to a beam of radiation. These detectors directly measure ionization and can respond to beta particles, x-rays, and gamma radiation. If the "window" or covering over the chamber is thin, alpha radiation can also be detected.

Geiger-Mueller (GM) Counters

The GM counter or tube is capable of detecting very small amounts of radiation; it is one of the most commonly used radiation survey instruments at hazardous materials sites. GM tubes are versatile, easy to operate, inexpensive, sensitive, reliable, and respond to x-rays, gamma and beta radiation. The GM counter also uses a gas filled ionization chamber, but the inert gas is usually supplemented with a small concentration of a halogenated hydrocarbon and a greater voltage is applied between the two electrodes. A single ionization event within the chamber causes secondary ionization of all the gas molecules. Each ionization event is counted and often displayed as counts per minute (cpm).

A popular type of GM survey meter uses a detector with a cylindrical tube shape in a protective metal casing that is open on one side; a metal shield can be slid over the detector tube to absorb some types of incoming radiation. During the initial survey the shield is removed (the window is open) to allow all radiation to reach the detector. When a reading above background is obtained, the shield is slid over the detector (the window is closed); in this configuration, only x-rays and gamma radiation can reach the detector.

GM detectors require a recovery time of 10^{-4} or 10^{-3} seconds after each pulse or ionization event; if the amount of radiation present does not allow sufficient recovery time, subsequent ionizations will not be detected. This is not a problem when relatively low levels of radiation are present. When high

Figure 9-4: A GM tube detector encased in a metal probe. When the window is open (left), x-rays, gamma, and beta radiation can reach the detector. When the shield is slid over the detector (right), beta particles are deflected or absorbed and cannot reach the detector. (Photos by Steve Napolitano.)

levels of radiation are present, the rate of ionization exceeds the recovery rate and counts are lost. If the ionization rate becomes excessive, the counter can become saturated or "paralyzed" and no counts are recorded. When this occurs, the meter will display a zero reading regardless of the amount of radiation present. If saturation is suspected, the detector can be exposed to a check source. A meter condition of no response or zero cpm in response to a radiation check source should considered an indication of an excessively high level of radiation, or meter failure. In either case, personnel should leave the area and replace the GM detector with another type of detector. The potential for meter paralysis or failure is a good reason why all personnel involved in radiation detection should always wear personnel dosimeters.

Proportional Counters

Proportional counters are usually cylindrical. The gas-filled chamber in a proportional counter has a central wire which serves as a high voltage anode; the inner conducting surface of the chamber serves as the cathode. The higher the voltage, the greater the current gradient around the central electrode. If the voltage is carefully maintained, the size of each ionization current pulse will be proportional to the original amount of ionization.

Current pulses are sorted, recorded electronically, and analyzed according to size. The amplitude of the pulse peak is equivalent to the amount of energy absorbed by the gas; properly calibrated proportional counters can be used to actually identify specific radionuclides and have been used to measure alpha and neutron radiation.

LIGHT-EMITTING DETECTORS

Scintillation Detectors

Scintillation detectors are based on the phenomenon that many substances emit visible light when exposed to ionizing radiation. Radiation interacts with solid, liquid, or gas phosphors. Commonly employed phosphors include zinc sulfide, sodium iodide, and cesium iodide crystals. These crystals are produced with a tiny amount of impurities to improve their scintillation properties. The light signal is converted to an electrical pulse by a photomultiplier tube, these signals are amplified and counted. Light output is proportional to the radiation energy absorbed in the scintillator.

Luminescent Detectors

Luminescent detectors utilize solids which store radiation energy; when processed, the solid material emits a quantity of light proportional to the original amount of energy. Commonly used luminescent solids are metaphosphate glass, calcium fluoride, and lithium fluoride. These detectors must be heated or treated with UV light to obtain a measurement.

Thermoluminescent detectors (TLDs) utilize small chips of lithium fluoride crystals within a thin plastic casing; after exposure, the chips are heated and the amount of light released measured. The amount of light released is equivalent to the total amount of radiation energy received. Since the reading cannot be repeated or verified after initial heating and reading,

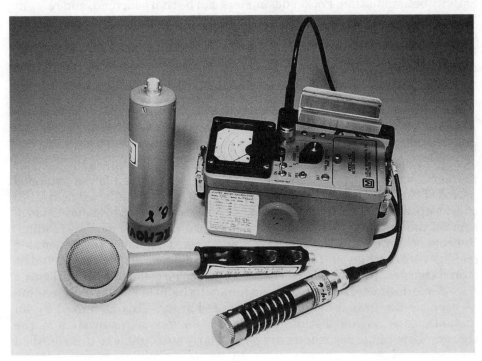

Figure 9-5: A radiation survey instrument with three different probes. The box contains readout display, battery, and scale range. The detector inside the metal housing or probe is attached by a flexible cable; (from top left) scintillation detector, pancake probe, GM detector. (Photo by Steve Napolitano.)

Figure 9-6: A thermoluminescent dosimeter (TLD) resembles a credit or ID card. (Photo by Steve Napolitano.)

each TLD usually contains at least two chips. TLDs are typically used as cumulative exposure personal dosimeters for x-rays, gamma, and beta radiation; they have also been used for neutron exposure evaluation. TLDs are often worn in conjunction with pencil-type dosimeters which can be read by the wearer.

RADIATION DOSIMETERS

Film Badges

A film badge consists of a small piece of photographic film wrapped in a light-tight cover and supported with a metal backing. Radiation interacts with the silver ions in the photographic emulsion which "exposes" the film in a manner similar to visible light. The more radiation the badge is exposed to the more metallic silver is left on the film after developing, and the darker the film.

A densitometer focuses a fixed amount of light through the badge and onto a photodiode; the amount of current generated is proportional to the optical density of the film. The radiation dose is proportional to the density. The amount of light passing through the film is used to determine the amount of radiation present.

Film badges are worn on outer clothing to monitor exposure to x-rays and gamma; beta/gamma radiation badges have a small opening to allow beta particles to reach the film. Neutron badges are also available. Badges are usually worn clipped onto clothing; wrist and ring badges are used when localized hand exposure is anticipated.

Pocket Dosimeters

Direct reading pencil-shaped pocket dosimeters are clipped to clothing and measure x-rays and gamma radiation; some units are also sensitive to high activity beta radiation. Pocket dosimeters can be charged, read, and reset by the wearer. The pocket dosimeter consists of a quartz fiber, a scale, a lens to observe fiber movement across the scale, and an ionization chamber. A battery-operated charger is used to charge and reset the dosimeter before and after use.

The pocket dosimeter has a viewing end (clip end) and a charging end. To charge, the dosimeter is pressed firmly into the charging receptacle; the user looks into the clip end and adjusts the zero control until it reads zero. After use, the user simply points the dosimeter towards a light and looks through the clip end to read the accumulated dose.

A frequently overlooked limitation of pocket dosimeters is the need for frequent charging. Dosimeters that have not been charged for an extended period of time (or have been stored uncharged) will normally not hold a full charge upon initial charging or re-charging. This is caused by penetration of some of the charge into the insulation and results in an erroneous reading. Repeated charging over several days may be necessary before an acceptable leakage rate is achieved. To prevent excessive leakage, dosimeters should be stored charged; they should be checked and recharged on a regular basis.

Some dosimeters also display an annoying habit of displaying a reading as soon as the unit is removed from the charger. This is caused by an instantaneous partial discharge as the dosimeter is removed from the charger. Low-range dosimeters are particularly susceptible to this so-called

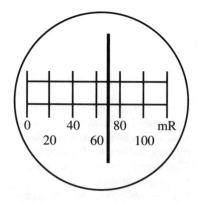

Figure 9-7: Example of a pocket dosimeter reading.

Figure 9-8: Charging and resetting a pocket dosimeter is accomplished in one step. (Author photo.)

"capacitance jump"; users should always check the dosimeter after charging to ensure it still reads zero. If necessary, the dosimeter can be adjusted to a point below zero which is equal to the capacitance jump.

All dosimeters leak a small amount of stored charge, causing the quartz fiber to drift and show a radiation reading. The leakage rate during use and storage should be determined in terms of radiation units per unit time. This can be done by charging and zeroing the unit, storing in a radiation-free area for 24 hours, and then reading the calibrated scale.

Personal alarm dosimeters are now available. These pocket-size units respond to radiation over a pre-set threshold level, usually between 1.0 and 2.5 mR/hr. Alarm dosimeters provide real-time personal radiation detection and dosimeter protection and can respond to alpha, beta, x-ray, and gamma radiation. Most alarm dosimeters have visible and audible alarms and utilize gas-filled GM tube detectors.

RADIATION EXPOSURE GUIDELINES

Occupational and non-occupational radiation exposure guidelines have been recommended by the Nuclear Regulatory Commission (NRC) and the National Council on Radiation Protection and Measurement (NCRP). The recommended maximum whole-body radiation dose is currently 5 rems per year (see Table 9-1).

NCRP[2] and NRC[3] guidelines recommend that workers under the age of 18 be limited to one-tenth of the adult permissible radiation dose exposure limit. Because of the sensitivity of the unborn child, it is recommended that the dose equivalent of the expectant mother be limited to 500 millirems (0.5 rem) for the entire pregnancy.

In the past, an action level of 2 mR/hr was recommended for hazardous waste workers[3]; this dose was derived from the NRC annual permissible exposure limits. Although the 2 mR/hr level is still used by Department of

Figure 9-9: This personal alarm dosimeter runs on a 9-volt battery and responds to x-ray, gamma, and strong beta radiation. Weak beta and alpha particles can be detected when the circular black cover over the GM detector is removed. (Author photo.)

Defense agencies, an EPA action level of 1 mR/hr is currently recommended to provide protection against the effects of overexposure to sources of radiation[4]. This is an extremely safe level; an individual would have to be continuously exposed to 1 mR/hr for 14 hours per work day for an entire year before the maximum recommended annual dose limit of 5 rem is reached.

At radiation levels of 1 mR/hr or less, workers can remain on site with continued radiation monitoring. A potential radiation hazard is considered present when radiation survey instruments show levels greater than the action level of 1 mR/hr above normal background.

Annual Maximum Permissible Dose Equivalent Exposures

1. Occupational Exposures

Whole body	50 mSv	(5 rems)
Lens of eye	150 mSv	(15 rems)
Hands (per quarter)	250 mSv	(25 rems)
Forearms (per quarter)	100 mSv	(10 rems)
Other organs	150 mSv	(15 rems)
Cumulative exposure	age x 10 mSv	(age x 1 rem)

2. Public Exposures

Frequent exposure	1 mSv	(0.1 rem)
Infrequent exposure	5 mSv	(0.5 rem)

3. Emergency Life Saving Exposure (one per lifetime)

Individual	1 Sv	(100 rems)
Hands and forearms	2 Sv (additional)	(200 rems)
Total overall exposure	3 Sv	(300 rems)

4. Emergency Non-Life Saving Exposures

Individual	250 mSv	(25 rems)
Hands and forearms	750 mSv	(75 rems)
Total overall exposure	1 Sv	(100 rems)

5. Embryo-Fetus Exposures

Total dose equivalent limit	5 mSv	(0.5 rem)
Monthly limit	0.5 mSv	(0.05 rem)

Table 9-1: Recommendations on maximum exposure dose equivalents to ionizing radiation for occupational and non-occupational exposure. Taken from NCRP Report Nos. 43[1] and 91[2].

When using radiation detectors, background radiation levels should be determined prior to entering the suspect area. In general, there is little or no background for alpha and beta radiation. Gamma radiation background is caused by cosmic radiation, as well as naturally occurring radiation in soil, rock, and building materials.

In many cases, the radiation survey meter employed will be a GM tube detector. Always perform radiation monitoring with the detector shield or window open; when the shield is open, beta and gamma radiation can be detected. Remember to turn the detector so that it faces all directions; beta radiation can only enter through the opened window, beta cannot penetrate the metal housing which surrounds the detector tube. When the meter reads radiation levels above background, the shield should be closed to determine if the reading is caused by beta or gamma radiation.

When radiation levels above the action level are encountered, minimizing the **duration** of exposure will limit the potential for biological injury. The longer the exposure, the greater the absorbed dose, and the greater likelihood of injury. There is a direct relationship between time and exposure; doubling the time of exposure doubles the radiation dose received. Thus an individual exposed to a 1 mR/hr source for an 8 hour day receives a total dose of 8 mR. A worker who comes upon a 50 mR/hr radiation source, leaves the area within 10 minutes, and does not encounter any additional radiation for the remainder of the day also receives a total dose of approximately 8 mR.

This can be readily calculated; remember than minutes must be converted to hours, one minute is equivalent to 0.0167 hour (1 minute divided by 60 minutes in an hour = 0.0167).

$$\begin{aligned} Total\ dose\ &=\ Source\ x\ time\ (hours) \\[6pt] &=\ 50\ mR/hr\ x\ 0.167\ hour \\[6pt] &=\ 8.35\ mR \end{aligned}$$

Most health and safety plans call for evacuation of the site when the radiation action level is reached. This may not always be necessary. Exposure to radiation can be reduced or limited by using protective measures including **distance** and **shielding**. The relationship between distance and dose follows the inverse square law; the intensity of radiation decreases by the square of the distance from the source. Doubling the distance from a 4 mR/hr source decreases exposure by one-fourth, or to 1 mR/hr.

A radiation source has been rated at 4 mR/hr at a distance of 1 meter; if a worker is standing 2 meters away from the source for two hours, what is his total dose?

$$\begin{aligned} Total\ dose\ &=\ \frac{source}{distance^2}\ x\ time\ (hours) \\[6pt] &=\ \frac{4\ mR/hr}{2^2}\ x\ 2\ hours \\[6pt] &=\ 1\ mR/hr\ x\ 2\ hours \\[6pt] &=\ 2\ mR \end{aligned}$$

An individual standing 2 meters away from the source would accumulate a dose of 2 mR; a worker standing 4 meters away would receive one-fourth the radiation dose, or 0.5 mR.

The effect of distance on exposure is an important concept to remember when encountering a source of radiation. Moving a short distance away from a source can dramatically decrease the radiation dose. It is not always necessary to leave the entire site when low levels of radiation are encountered; it is often possible to prevent radiation exposure simply by restricting personnel from the immediate vicinity of the source.

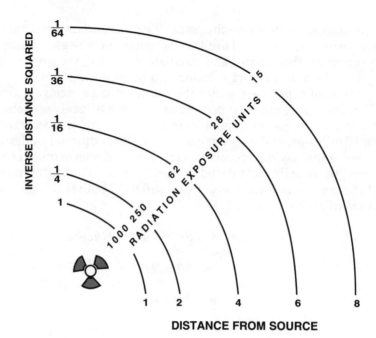

Figure 9-10: The effect of distance on radiation exposure. As the distance increases, exposure decreases by the square of the distance from the source.

Similarly, placing shielding between responders and the source can also reduce exposure. Shielding is commonly used to protect against X-rays, gamma radiation, and high energy beta radiation. The more mass placed between the source and a person, the less radiation will reach the person. Lead is commonly used to protect medical workers against X-ray exposure. In an emergency, haz mat responders can employ earthen berms, water-filled drums, sandbags, concrete blocks, or steel or iron plates to reduce exposure to high levels of radiation.

It is important to realize the potential limitations of instrumentation and dose reduction methods when dealing with significant levels of radiation. Always consult a health physicist if you have health and safety questions regarding radiation exposure.

Figure 9-11: Snow fencing was used at this landfill site to isolate a radiation source and prevent inadvertent radiation exposure. (Author photo.)

REFERENCES

1. National Council on Radiation Protection and Measurements (NCPR). 1975. Review of the Current State of Radiation Protection, NCPR Report No. 43. Bethesda, MD: NCPR.
2. National Council on Radiation Protection and Measurements (NCPR). 1987. Recommendations on Limits for Exposure to Ionizing Radiation, NCPR Report No. 91. Bethesda, MD: NCPR.
3. National Institute for Occupational Safety and Health, Occupational Safety and Health Administration, U.S. Coast Guard, U.S. Environmental Protection Agency. 1985. Occupational Safety and Health Guidance Manual for Hazardous Waste Activities.
4. U.S. Environmental Protection Agency. 1992. Standard Operating Safety Guides. Washington, DC: Office of Emergency and Remedial Response, Hazardous Response Support Division, Environmental Response Team.

CHAPTER REVIEW

1. List 3 types of ionizing radiation: _____.

2. List 3 common sources of non-ionizing radiation: _____.

3. Alpha radiation is difficult to monitor because it_____.

4. Three ways to reduce exposure to radiation are _____.

5. What is the normal range for radiation background readings?_____.

6. Which type of radiation is most penetrating?_____.

7. Which types of radiation pose an ingestion hazard?_____.

8. Describe the difference between a radiation survey instrument and a radiation dosimeter:_____
   _____
   _____.

9. A GM detector indicates a reading of 0.8 mR/hr when the shield is open; when the detector is covered by the shield, the reading decreases to 0.03 mR/hr. What type of radiation is present?_____.

10. The distance from a 400 mR/hr source is doubled; the radiation exposure at the doubled distance is now_____mR/hr.

11. The recommended action level for hazardous materials workers is _____mR/hr. For pregnant workers, the recommended total pregnancy exposure dose is _____mrems.

12. The annual recommended permissible radiation dose exposure for occupational exposure is _____.

13. A worker exposed to a radiation source of 0.08 mR/hr for 8 hours receives an accumulated radiation dose of _____mR.

14. True or false? Pocket dosimeters can be charged, read, and reset by the user.

15. True or false? Thermoluminescent dosimeters can be charged, read, and reset by the user.

16. List some materials that can be used to shield against gamma radiation._____
    _____.

10

Field Applications

LEARNING OBJECTIVES

1. Given field situation, select the appropriate instruments to characterize the materials present.
2. Correctly interpret individual and multiple meter responses.
3. Recognize inconsistencies in responses from different types of monitoring equipment.
4. Identify limitations of instrument use under field conditions.

HAZARDOUS WASTE SITE CONSIDERATIONS

Air monitoring is used at hazardous waste sites to establish control zones, locate sources of contamination, evaluate the extent and migration of ground-water contamination, identify areas in need of remediation (hot spots), and determine the level of personal protective equipment required to safeguard on-site workers.

The selection of personal protective equipment (PPE) is based upon the need to protect workers against potential exposure to hazardous air contaminants. The EPA has developed recommendations for four levels of protection for PPE ensembles (Level A through Level D).

Level A is the highest level of respiratory, skin, and eye protection. The Level A ensemble consists of a one piece, gas/vapor tight, totally encapsulating chemical protective suit. Respiratory protection consists of a full facepiece positive-pressure self-contained breathing apparatus (SCBA), or a full facepiece positive-pressure airline with emergency positive-pressure escape SCBA. Level B requires the same level of respiratory protection as Level A but less skin protection. In most cases, respiratory protection equipment is worn over protective clothing.

Level C allows the use of air-purifying respiratory protective equipment instead of positive-pressure supplied air. Level C air purifying respirators (APRs) may be full-face or half-face. APRs utilize filters or sorbent materials to remove specific types of contaminants from air. Purifying agents are contained within a cartridge or canister. Level D consists of a normal work outfit without respiratory protection. A typical Level D ensemble consists of steel-toed footwear, hardhat with faceshield or safety glasses, and work clothes or coveralls.

A hazardous waste site health and safety plan (HASP) is a document which specifies safety, occupational, workplace, and health precautions to be employed by personnel working at the site. The initial draft of the HASP must identify each anticipated health and safety hazard for each work activity, and describe how those hazards will be eliminated or controlled. The HASP should also clearly specify appropriate air monitoring procedures as well as

Figure 10-1: Level A (top) offers the highest level of respiratory and skin protection. Level B (bottom) provides a high level of respiratory protection but less skin protection. (Author photo.)

Figure 10-2: Level C is defined by the use of an air-purifying respirator. This worker is wearing a full-face respirator equipped with a chin canister. (Author photo.)

Trichloroethylene
Lead and compounds
Toluene
Benzene
Polychlorinated biphenyls
Chloroform
Tetrachloroethylene
Phenol
Arsenic and compounds
Cadmium and compounds
Chromium and compounds
1,1,1-Trichloroethane
Zinc and compounds
Ethyl benzene
Xylenes
Methylene chloride
trans-1,2-Dichloroethylene
Mercury and compounds
Copper and compounds
Cyanides (soluble salts)
Vinyl chloride
1,2-Dichloroethane
Chlorobenzene
1,1-Dichloroethane
Carbon tetrachloride

Table 10-1: The twenty five most frequently identified substances at 546 Superfund sites, in rank order.[4]

the level of PPE required for initial site characterization. After site hazards have been analyzed, the HASP is revised to reflect any new information collected. Air monitoring procedures should also be described to ensure that new hazards are readily identified. Periodic monitoring should also be conducted when there is any indication that air contamination is present, when site conditions change, or when work activities are initiated. The HASP will also describe the minimum acceptable level of PPE, which is then upgraded or downgraded according to a variety of criteria, including instrument readings.

In most cases, the level of PPE parallels recommendations from the EPA-issued Standard Operating Safety Guides[1] (SOSG) and Order 1440.2[2] contained specific guidance on criteria for selection and use of protective equipment. Level C respiratory protection (defined by EPA as a full-face air-purifying mask equipped with an organic vapor/particulate canister) is recommended when readings on portable total hydrocarbon (THC) meters are greater than background but less than 5 ppm-equivalents. If the presence of IDLH conditions, including highly toxic substances and oxygen deficiency, have been reliably ruled out, Level C respiratory protection is acceptable as long as instrument readings remain below 5 ppm-equivalents above background. Every effort should be made to identify the contaminates present. Level B respiratory protection (positive-pressure supplied air) is recommended when meter readings are 5 ppm-equivalents or greater.

The EPA readily acknowledges that PID and FID survey meters do not respond to all hazardous materials, nor do they respond identically to gases and vapors. PIDs and FIDS can underestimate or overestimate the concentration of unknown gases and vapors present. The EPA clearly recommends that the protection level should be based not only on survey meter readings but also on potential exposures and hazard characteristics of the materials suspected to be present.

The criteria for the use of Level C always made us (and others[3]) uncomfortable because of the way they were used and interpreted. First of all, there is no such instrument as a **"total** hydrocarbon meter", yet all too often a single meter (usually a PID instead of both a PID and FID) was used to determine the level of respiratory protection needed. Many workers were in the wrong level of respiratory protection, or were not using any protection at all, because of the inadvertent reliance on only one type of instrument. Secondly, the presence or absence of highly toxic gases is frequently interpreted in terms of easily detectable toxic gases such as carbon monoxide or hydrogen sulfide; the potential for other highly toxic air contaminants that may not be detectable by a single instrument is often largely overlooked.

Another concern is that EPA guidance documents are not compatible with the NIOSH criteria for use of Level C respiratory protection. Air purifying respirators may be used only when specific NIOSH criteria are satisfied for each contaminant present. If all criteria cannot be met, then Level C respiratory protection should not be used.

NIOSH criteria require that the identity and approximate concentration of each contaminant is known. A cartridge or canister of sufficient capacity must be available for each contaminate. Individual contaminants must have good warning properties to alert the wearer when the service life of purifying elements are exceeded, allowing contaminates to breakthrough the cartridge or canister. Good warning properties include odor, or nasal or eye irritation; warning properties must also be manifested at non-hazardous concentration (i.e. below the PEL/TLV/ WEEL). Air purifying respirators cannot be used when IDLH conditions, including oxygen deficiency, are known or anticipated to be present.

EPA-issued guidance documents were superseded by the OSHA rule on Hazardous Waste Operations and Emergency Response (HAZWOPER, 29 CFR 1910.120)[4]. The rule requires that direct-reading air monitoring instruments be used during site characterization to identify hazards and select personnel protection methods. An on-going air monitoring program is also required after site characterization has established the site is safe for initial start-up and subsequent operations. The rule makes no mention of the previously-used EPA criteria for selection of respiratory protection.

At uncharacterized sites covered in the OSHA rule, air monitoring is required for specific hazards including radiation, oxygen deficiency, toxic substances, explosive/flammable gases, and IDLH conditions. Thus the use of multiple types of instrumentation is mandated under the OSHA rule. A review of the 25 most-commonly encountered contaminants at Superfund sites (Table 10-1) reinforces the argument that more than one type of air monitoring device is required to adequately characterize sites with unknown hazards. In addition, not all contaminants are volatile; particulate-borne contaminates require dust monitoring. In some cases, portable direct reading instruments may not be available to detect some types of contamination.

Figure 10-3: These air-purifying respirators are equipped with cartridges instead of the larger-capacity canister. NIOSH/OSHA does not recommend the use of Level C to protect workers from unknown breathing zone contaminants at hazardous materials sites. (Author photo.)

Action Levels for Health and Safety

Protecting personnel at uncontrolled hazardous waste sites can be difficult, since the identity and concentration of contaminates present are usually unknown. Action levels that can apply to uncontrolled sites include those for radiation (1 mR/hr), combustible/flammable gas (20-25% LEL meter reading), oxygen deficiency (19.5%), and oxygen excess (23.5%). Action levels for specific toxic gases and vapors are usually one-half their exposure limit value; for instance, a commonly used action level for hydrogen sulfide is 5 ppm, which is one-half the OSHA exposure limit of 10 ppm. At uncontrolled sites, however, action levels for specific contaminates may not apply, since the identity of materials present are unknown.

Since direct reading, real-time instrumentation is largely inadequate to identify and quantify the extent of air contamination, and it is usually impossible to guarantee against the potential for IDLH conditions, the use of Level B PPE is usually required during initial investigations at uncharacterized hazardous waste sites. However, when there is extensive site history, and previous soil and water analysis has demonstrated that the level of contamination present is very low, Level C may be an option, since in this case the NIOSH criteria can be satisfied. At sites containing ppb levels of contamination, Level D should be considered. For example, if workers are collecting water samples from previously sampled wells known to contain ppm concentrations of toluene, ethyl benzene, and xylene, then Level C is an appropriate choice; Level D should be considered if low ppm or ppb levels are known to be present. On the other hand, Level B should be employed during collection of water samples from wells containing unknown contaminates. Whenever Level C or D is used, however, air monitoring should be conducted continuously to alert workers of IDLH conditions.

Confined Spaces

A confined space is defined by OSHA in the permit-required confined space rule[5] as a space that is large enough to enter, is not designed for human occupancy and has limited or restricted means for entry and exit. A permit-required confined space may also contain either a hazardous atmosphere, a material that could engulf a worker, or internal parts that could trap a worker, or other unrecognized hazards such as energized electrical wires or moving, unguarded equipment. Commonly encountered confined spaces include sewers, culverts, wells, large diameter pipes, pipe galleries, above and below ground storage tanks, tank trucks, rail tank cars and barges, and electrical conduits and vaults.

Confined spaces can pose physical, mechanical, electrical, and chemical hazards, oxygen deficiency or excess, and poor visibility. While air monitoring is required prior to entering a confined space, the use of instrumentation is only one of the tools available to ensure safety during confined space entry.

Initial testing of a confined space atmosphere should be performed remotely, from outside the space. Testing of a confined space atmosphere should always start with oxygen deficiency; after ensuring adequate oxygen levels, the atmosphere should be tested for combustible gases, then potential toxic contaminates. Toxic gas monitoring may be fairly straightforward such as for hydrogen sulfide and carbon monoxide, or for the chemical which was stored in the tank or line. More insidious contaminants may include combustion by-products, transformation or degradation products, and radiation.

Air monitoring should continue as the entry proceeds; this is especially important in horizontal confined spaces such as conduits, ducts, and culverts, since the entire space cannot be tested prior to entry. In vertical confined spaces such as tanks, air from multiple levels should be tested prior to entry; monitoring should continue after workers reach the bottom of the space. Continuous monitoring is warranted in confined space entry, since concentrations can change dramatically. Increased concentrations can result from leakage through cracks or micropores, enhanced vapor release as a result of disturbing contaminated sediment or sludge, or exposing enclosed contaminated spaces to heat and light.

Action levels for confined spaces have been defined in the OSHA permit required confined space rule. The action level for combustible gases and vapors is 10% LEL; the action level for oxygen deficiency is 19.5%, and 23.5% for oxygen excess. These action levels were assigned for situations when the displacing contaminates are known to be non-toxic and physiologically inert. For example, although the action level is 10% LEL, a space containing 9% LEL should not be entered unless the combustible gas is known to be methane, which is non-toxic. An instrument reading of 9% LEL could represent a contaminate concentration of 900 ppm (with unknowns, each 1% LEL meter reading should be considered to represent at least 100 ppm). Similarly, the action level of 19.5% oxygen will not protect a worker who enters a confined space containing 20% oxygen when the displacing agent is unknown. A drop in oxygen levels from 20.9% to 20% indicates that approximately 45,000 ppm of an unknown, potentially toxic contaminate is present (oxygen represents only 20% of the air envelope, 0.9 or 9,000 ppm x 5 = 45,000 ppm).

After air monitoring has ensured the atmosphere is safe for entry, confined space permits and safety procedures must be followed to ensure continued safety during confined space operations.

FIELD MONITORING ACTIVITIES

Instrument monitoring during investigatory and remedial activities is usually unexciting and thankless work. Instrument operators are usually the first to arrive on site since they must warm up and field calibrate each instrument, and then take initial background readings. Instrument users

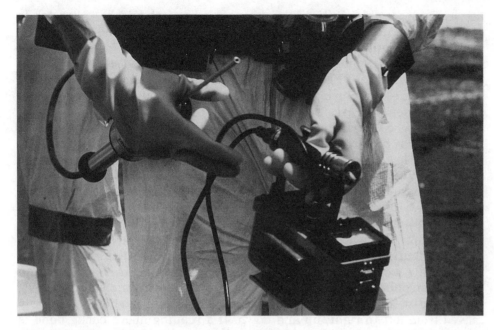

Figure 10-4: Workers should not be encumbered by carrying multiple instruments. This worker was assigned a GM radiation detector and an HNU PI-101. (Author photo.)

are the last to leave after all other personnel have finished their activities. Finally, at the end of the day, all instruments must be recharged and readied for the next day; field logs and summary sheets must be completed and checked. It is not surprising that new-hires are frequently targeted as "instrument bearers" regardless of their training or experience!

The use of air monitoring at investigations and remediations is often misunderstood. Meters should not be used once or twice a day; nor should they be used only when someone smells something or starts to feel ill. Meter use must be carefully specified in the health and safety plan and a supervisor or manager should ensure that plan specifications are followed. The instruments designated in the plan must be available to field personnel. All too often field personnel are sent out without adequate equipment. We have rarely encountered a project that required only one instrument!

The role of air monitoring should be clearly defined in the health and safety plan. In most cases, the primary role of air monitoring is to protect site workers; gathering information for the client is secondary. When investigating an uncharacterized site covered by the OSHA rule, air monitoring is required for specific hazards, including radiation, oxygen deficiency, toxic substances, explosive/flammable gases, and IDLH conditions. The use of multiple types of instruments is mandated under the OSHA rule; it follows that similar monitoring procedures be used at all hazardous waste sites, regardless of their status under OSHA.

Prior to on-site monitoring, non-contaminated areas surrounding the site should be assessed; such perimeter monitoring should be conducted before site activities begin to determine background readings for each instrument used. Whenever possible, background monitoring should encompass the same types of environments that will be encountered on site (i.e. asphalt or blacktop, grassy areas, heavily vegetated areas, sand and bare soil).

Background meter readings should also be obtained for any equipment that is to be used on-site. Vehicle engines, drilling rigs, compressors, pumps, and heaters are all potential sources of volatile emissions from fuels, engine exhaust, oils, greases, hydraulic and other fluids. Equipment should be kept as clean as possible. Initial instrument readings should be taken while equipment is running to assess the "mechanical background that will be present when equipment is used.

Initial investigatory activities at a non-characterized site are typically conducted in Level B PPE ensemble with good skin protection. The primary hazards to screen for during initial site monitoring should include radiation, oxygen deficiency or excess, toxic gases, and combustible/flammable gases. Obtaining a good site history is often helpful in determining the type of toxic gases that may be encountered. If no dangerous conditions are identified, the presence of low-concentration contaminates can then be addressed using more sensitive survey instruments such as a ppm-CGI, FID, PID, and IR; a second radiation survey is also recommended to detect the presence of alpha and low low-activity beta radiation. Sites known to be contaminated with radioactive waste materials should be assessed by radiation specialists.

Air monitoring activities conducted during initial and subsequent site characterization should not be rushed. Remember that volatile and semivolatile contaminates are not stationary in the air; contamination moves! Personnel with instruments should walk slowly and sample above, below and within their breathing zone. Sample probes should be slowly moved from left to right, up and down. If a reading above background is obtained, the user should stand still and determine if the meter response fluctuates or remains stable.

If the instrument response is localized to one area, the user can try to determine the source, look for discolored soil, drums or other containers nearby, or idling heavy equipment. In some cases, the source of contamination is a good distance from the location of initial meter response; it is not always possible to determine the source of instrument readings.

After site work has been initiated, monitoring should be conducted whenever new conditions are encountered, when work begins at a different section of the site, or a new activity is started (i.e. drilling, opening drums, sampling obvious areas of contamination). Remember that these are new conditions or activities; that means that all instruments should be ready and available for use. Monitoring should be repeated with several instruments using different methods of detection. Lack of a response from one instrument is not sufficient evidence to consider the area free of contamination.

Figure 10-5: Types of perimeter monitoring (left to right): long term tube connected to sampling pump, passive organic vapor badge attached to fence, direct-reading FID at fence-line. (Author photos.)

Continuous perimeter monitoring is often performed to determine if any volatile contaminants have left the site during the work day. Passive dosimeters or badges, long term tubes with low-flow sampling pumps, or direct-reading instruments with recorders or data loggers are positioned at a fenceline or other non-contaminated, perimeter locations and left in place for the entire work day. At the end of the day, equipment is collected and results analyzed. Perimeter monitoring is recommended at sensitive sites where there are off-site exposure concerns.

Screening Environmental Samples

Portable, direct-reading instruments are frequently used to rapidly discriminate soil or water samples according to the presence or relative absence of contamination. At hazardous waste sites, data from real-time field instruments are used to refine sampling plans and estimate the extent of contamination. In other cases, screening results are used for health and safety purposes or to determine which samples will be sent for laboratory analysis and how sampled materials will be handled.

The EPA recommends the use of a "portable total organic vapor monitor" for screening in its Compendium of Superfund Field Operations Methods[6]; FIDs and PIDs are included in this group. The "selective" response of PID meters is considered "advantageous" since they do not respond to methane or non-ionizable materials. FIDs are mentioned only as non-selective organic vapor meters; there is no mention of specific advantages or disadvantages.

The compendium goes on to state in Section 7.6.2.1 that a "Zero instrument response is a definitive result; it indicates an undetectable amount of organic vapors...within the range of the instrument's ionizing lamp." This, of course, is not completely correct. It is true that the lack of instrument response may indicate the lack of detectable concentrations of ionizable vapors within the range of the UV lamp; however the lack of response may also be due to the presence of excess concentrations of methane, water vapor, or other non-ionizable gases. Regardless of the application, no one instrument should be used to determine the presence or relative absence of contamination. Because of the many limitations associated with its use, a PID should not be the only instrument employed to detect contamination in field samples.

We are not saying that a PID can never be used alone to screen samples, however results of such screening are **not** definitive. That is, the absence of a meter reading cannot be interpreted as a lack of contamination. We recommend the use of both a PID and FID when screening field samples. FID and PID response should be compared and a charcoal filter can be used on the FID if the presence of methane is suspected. Another screening tool that is becoming more popular is the Miran 203, an IR air analyzer which can be operated in a general petroleum hydrocarbon sensing mode.

Although we are opposed to such practices, some municipal and state agencies allow the use of only a PID for field screening of environmental samples, if the user can demonstrate that the PID can respond to a reference material in a manner similar to an FID. If the PID gives a lesser response, the user can provide a "relative to FID response factor. The response factor is determining by comparing the FID and PID responses to a reference mixture.

We cannot recommend the use of a PID as an "FID-equivalent meter". There is an enormous difference between meter efficiency in

Figure 10-6: Sampling headspace contamination levels in collected soil samples. The operator is holding a readout assemby of a Foxboro OVA. A Photovac MicroTIP is in the background. (Photo by Steve Napolitano.)

Figure 10-7: Screening potentially contaminated soil by the bucket method. The operator is using an HNU PI-101. (Author photo.)

response to known concentrations of reference gases in dry air, and meter response under field conditions. Remember that PID response is markedly reduced in the presence of humidity, methane, and other non-ionizable gases and vapors. Meter efficiency is also reduced when the lamp is dirty, or when dust or particulates are drawn into the ionization chamber.

When screening soil or water samples, a pint or quart jar with a ring lid is commonly used to hold each sample which is placed in the jar and then covered with aluminum foil. The ring lid holds the foil in place. The jar should be half-full; if the jar is too full there will be insufficient head-space for sampling. We recommend saving samples until a sufficient number is collected; all samples should be stored in the shade and allowed to reach a uniform temperature. Insert the probe through the foil and sample the headspace.

Containers of water samples are frequently agitated, and soil sample containers are placed in warm water (25-30°C) for 5-15 minutes before sampling to drive semi-volatile materials into the headspace. Such methods are not always successful. Users should be aware that semi-volatiles may quickly condense within a cooler sample probe and not be detected. Heating or agitation can also increase the concentration of water vapor in the headspace, which decreases PID efficiency.

Another method used for rapid soil screening is the so-called bucket method employed during large-scale clean-up operations. A back-hoe bucket scoops out soil and the air immediately above the soil is sampled; in most cases the probe is held no more than 6 inches away from the soil. A better method is to use a solid rod and make several 1 to 2-inch diameter holes in the soil of each bucket; the probe is then placed directly over the top of each hole for sampling.

Soil gas and water headspace contaminants can be detected and often identified using field portable GCs. Another field screening method which uses IR absorbance of hydrocarbons after Freon extraction is also available using a Foxboro Miran 1FF/1ACVF analyzer. This method requires sample collection and several simple preparation steps. Detection limits for total petroleum hydrocarbons are 2 ppm in soil and 0.1 ppm in water.

Landfill Operations

Landfills are an operational nightmare. Landfills were originally considered final depositories; they were never designed to be dug up, drilled through, sampled, or re-located. Landfills have the potential to contain every category of hazardous materials. Additional hazards include unstable terrain, physical and biological hazards; the slope or toe of the landfill can collect heavier-than-air vapors and should be considered a potential "semi-confined" space.

Landfills also generate methane gas, carbon dioxide, and heat. Natural composting of organic materials in landfills can produce up to 50-60% by volume methane and 40-60% carbon dioxide by volume. Escaping moist air and gases can reach temperatures up to 180°F. Steam is formed when warm escaping gases come in contact with cooler ambient air. The presence of methane may pose a flammability hazard. More often, methane gas complicates air monitoring protocols and increases the need for comparative instrumentation. Recall that FIDs respond well to methane; PIDs, on the other hand, cannot ionize methane. IR analyzers designed to monitor landfill gases do not detect other types of contamination.

PIDs have always been the meters of choice at landfills, since they do not respond to methane. However, the presence of methane as a non-ionizable gas dramatically decreases PID efficiency. As discussed in Chapter 6, methane concentrations of approximately 25% LEL (1.25% by volume) reduced meter response to a known concentration of toluene by approximately 50%; at higher methane concentrations, the PIDs tested gave low, negligible, or negative readings[7]. Water vapor or steam produced during composting can also decrease PID efficiency. It is dangerous to rely only on a PID at landfills; both a PID and an FID should be used; an IR analyzer can be used to monitor methane when relatively large concentrations are

Figure 10-8: Steam produced as a result of natural composting of organic materials can complicate use of air monitoring equipment. (Author photo.)

Figure 10-9: Workers preparing to open a well casing. The headspace is first sampled with a Bacharach combination CGI / oxygen deficiency meter; the CGI is used in the ppm range (up to 10,000 ppm hexane equivalents) before low-concentration survey meters are used. (Author photo.)

present. Readings should be compared and a charcoal filter used to discriminate between methane and non-methane instrument readings. A GC-equipped FID can also be used to separate methane from heavier hydrocarbon contaminants.

During drilling or excavation, naturally-occurring materials may be encountered which give meter readings. For instance, lemon, orange, and grapefruit rinds contain volatile oils; conifers such as pine, hemlock, and spruce contain resins and turpenes. When high readings are obtained, look at the excavated soil for evidence of naturally-occurring organic materials. Unfortunately, there is no method available to readily discriminate between naturally occurring organic materials and similar materials that have been contaminated.

Installation of methane collection wells at landfills require a variety of strategies in order to comply with action level guideline of 25% LEL. The presence of high concentrations of methane gas, carbon dioxide, and heat makes selection of instruments and interpretation of meter readings difficult. It must be noted that the action level of 25% LEL is designed to trigger an action to reduce combustible gases to a more acceptable level in the workzone; remember that downhole concentrations will be significantly greater.

The concentration of combustible gases in the workzone can be reduced by the use of muds, drilling foams, and employing double casing; intrinsically safe fans can also be used to blow gases away from workers. In some cases, the hole may be inerted with dry ice, carbon dioxide gas, or nitrogen.

The Level B PPE ensemble worn by personnel at such locations should be upgraded to include fire-protective coveralls, undergloves and hood worn under chemical resistant outerwear. Monitoring for oxygen and combustible gas should be conducted continuously in the workzone; periodic monitoring should also be performed under equipment and vehicles, and in nearby low lying areas. If carbon dioxide is known to be present, the oxygen sensor should be periodically checked to ensure it is functioning properly.

Drilling Operations

Drilling at uncontaminated sites is dangerous because of the physical and mechanical hazards involved; at hazardous materials sites, the dangers are compounded many-fold. The use of personnel protective equipment, while decreasing the risk of chemical overexposure, markedly increases the risk of physical or mechanical injury. Interestingly, we often find personnel at drilling operations reluctant to use air monitoring instruments. This is unfortunate, since many (often unrecognized) accidental overexposures could have been avoided if instruments were used.

Vehicle engines, drilling rigs, compressors, pumps, and heaters are all potential sources of volatile emissions from fuels, engine exhaust, oils, greases, hydraulic, and other fluids. Equipment should be kept as clean as possible. Initial meter readings should be taken while equipment is running to assess the "mechanical background" that will be present when drilling activities begin.

Air monitoring instrumentation should be employed to alert the user and fellow workers of the presence of immediately hazardous conditions such as oxygen deficiency or flammable and toxic gases in the workzone. Instruments can also be used to monitor down-hole contamination; monitoring levels of down-hole contamination gives an indication if the hole is advancing into more- or less-contaminated soil. Downhole contaminants can be forced out of the hole and into the workzone during drilling. Always monitor initially for oxygen deficiency and combustible gases; if CGI readings are low or negligible, lower-concentration survey meters (PID or FID) should be employed. At sites where the only hazardous materials present are flammable or combustible (i.e. gasoline stations, petroleum product underground storage tanks) a CGI which measures in ppm-equivalents can be used instead of the PID/ FID combination.

Remember that down-hole oxygen concentrations are usually low; a %LEL CGI must be used in concert with an oxygen deficiency indicator. Repeated flame-out on an FID, erratic response on a PID, and no response on a %LEL CGI suggests insufficient oxygen for proper meter function **or** the presence of extremely high levels of contamination. If such responses occur, move the meter probe to a point **just above** the annulus of the hollow stem auger or casing and resample; there should be sufficient dilution from ambient air to supply adequate oxygen for meter function. Wind or air currents can be expected to produce fluctuations in meter readings but this will give some indication of the concentration at the very top of the hole. The same procedure can be used to assess headspace contaminants of finished monitoring wells.

*Figure 10-10: Monitoring headspace concentrations using a HNU PI-101. To avoid serious injury, drilling tools **MUST** be stationary when monitoring is conducted. **NEVER** attempt to monitor close to moving equipment. (Author photo.)*

Headspace readings can be derived from volatile gases and vapors that gradually collect over a period of time, or from contaminants that are under a slight positive pressure and maintain a continuous reading after the casing is removed. Initial reading may be high as the casing is opened, then immediately decrease. To determine how rapidly the headspace contaminants collect, place a sheet of aluminum foil over the well and secure with a rubber band; wait ten minutes, then resample through the foil. If readings are similar to that obtained when the casing was first removed, the soil or water contamination levels are high and/or contaminants are being forced out of the soil and into the headspace.

Drums and Containers

This section could also be entitled "putting it where the sun don't shine". Instrument users are constantly doing just that—putting the meter probe or sample hose into a drum or container without any thought whatsoever. Never thrust sample hose into a container or drum; it may contain a corrosive material or be filled to the top with a liquid; excessive concentrations of vapors well beyond the range of the instrument can damage a low-concentration survey meter. Exposing a meter of excessive concentrations can damage an analog meter display or at least take the meter out-of-service for several minutes until it is allowed to sample clean air and cleanse itself. After exposure to very high concentrations, residual contamination may linger within the sample hose and in filters and continue to off-gas for several minutes after sampling.

Always use a high-range instrument before the lower-range survey meters. Remember the 1300 rule—a solvent or liquid with a vapor pressure of just 1 mm Hg can be expected to have approximately 1300 ppm in the headspace of a closed container at 68°F. Most liquids have vapor pressures greater than 1 mm Hg. A drum containing residual amounts of liquid methylene chloride (VP 350 mm Hg) could have a headspace concentration in excess of 450,000 ppm! PID or FID survey meters cannot handle such concentrations.

We always sample drums with a %LEL CGI meter before using a low-concentration survey meter. The CGI responds to any material that has an

Figure 10-11: This worker is preparing to conduct headspace monitoring with an HNU HW-101; the drum contains unknown contaminants. (Author photo.)

Figure 10-12: Unsafe use of a survey meter during drum removal operations; the operator is in jeopardy of falling into the trench! Potential readings from the drum can be obtained after it is removed from the trench; the surrounding soil can be monitored using the bucket method. (Author photo.)

LEL, not just to materials classified as flammable or combustible, which is determined by flash point. Most organic materials will burn and have an LEL. Methylene chloride, for example, forms flammable vapor-gas mixtures at temperatures well above 200°F but is readily detected by a CGI.

We do not recommend indiscriminate opening and headspace sampling of drums. Opening of drums should always be carefully planned; techniques for opening drums include remote opening using non-sparking (usually brass, bronze, or copper-beryllium) spike attached to a backhoe, hydraulically operated drum piercer, or pneumatic drills. Manual procedures for opening drums should safeguard workers from explosion and exposure to drum contents. In most cases, personnel working with drums should be behind a protective shield and be in Level B PPE; air monitoring should be conducted during drum opening and sampling. Fire extinguishing equipment should be readily available close by.

Whenever drums are moved or opened, there is an increased risk of worker and environmental contamination; drums may rupture, leak product, or release vapors. Drums should be examined for leaks, signs of deterioration, or excess pressure prior to opening. Drums should be opened and sampled by trained personnel for a specific reason, such as waste characterization, compatibility testing, preparation for bulking or disposal.

Once opened, we do not, however, shove the sample hose into the drum! If possible, determine if the drum is empty, full, or contains some liquid; this information can usually be obtained by checking the assigned drum number against the field log compiled during drum-opening operations. If the container is full, keep well clear of the liquid level. Always take initial readings above the bung hole or opening; if no readings are evident, carefully place the probe just into the opening and note the instrument reading. If there is minimal response, advance the probe into the opening another inch or two and take another reading. Never agitate a container while the probe is inside it; never place the probe where you cannot see where the tip is going! If you cannot see the end of the probe it should not be there!

It is dangerous to sample close to the surface of any liquid; a sudden movement can inadvertently immerse the probe tip and liquid will be drawn into the meter. Even when the meter is equipped with a liquid trap, the probe, sample hose, and trap must be thoroughly decontaminated or replaced before sampling can resume.

Drum or container monitoring results can only be used as yes-no information. A yes meter response indicates the presence of some volatile material still present in the container. A lack of a meter response does not indicate the absence of contamination, only that the meters used did not give a reading. The container may contain non-volatile organic contaminants (such as many pesticides), or a variety of inorganic contaminants may be present.

In situations which suggest that inorganic materials may be in the container (i.e. plastic carboy or drum, corrosive/oxidizer label, evidence of corrosive damage to container), wetted pH paper or detector tubes for acids and alkalis are used to ensure that corrosive vapors are not present which could damage meters.

Figure 10-13: Corrosive materials can damage sensitive air monitoring equipment. Suspect corrosive contents when investigating plastic, glass, or stainless steel containers. (Author photo.)

Remediation Activities

Attention to detail frequently suffers during remediation; this is probably because most of the uncertainty experienced during the initial investigatory phase is missing. Instruments may be used infrequently and not calibrated; instrument logs may not be kept or regularly updated. This is not the time to become sloppy! Meter use should be carefully reviewed and specific monitoring procedures developed for the remediation health and safety plan; supervisors should ensure that these procedures are followed.

We often find that only one or two meters are used during remediation activities. In many cases the meter employed is not the optimal one, based on site conditions or contaminants identified. For instance, in Figure 10-12, a PID is used to assess trench contamination; the presence of standing water in the trench, however, suggests that a PID is not the instrument of choice.

Cement and other dry materials are used in construction of slurry walls, during installation of monitoring or recovery wells, and for stabilization of

Figure 10-14: Meters without sample inlet filters should not be used around cement, crushed stone, or other particulates. (Photo courtesy of SEC-Donohue.)

drum contents. These are often very fine particulate materials which can clog inlet filters. When such materials are used, filters must be regularly cleaned or replaced; meters without filters should not be used.

COMPARATIVE INSTRUMENTATION

When we "attack" a new site, we always attempt to discover as much as possible regarding the type of contaminants we are likely to encounter, as well as their potential concentration ranges. In many cases, we are not sure how our instruments will respond to these unique combinations of contaminants. In other cases, as when we purchase a new instrument, we like to assess its response to known contaminants, relative to other instruments. We then compare the response of each of instruments, in order to get an idea of what to expect from each meter in the field.

This does not always entail the purchase of expensive calibration gases. In most cases, we use a plastic gallon jug as a sample container and add one or two microliters of the liquid or solid of interest. The jug is then capped and the vapor is allowed to reach equilibrium over a short interval. The jug's atmosphere is then successively sampled by each meter. Instrument users are always surprised at how easy and effective this method is to determine the best instrument mix for use on sites with known contaminants.
This is also an good way to compare relative meter response to detector tubes.

In cases where the material of interest is a gas, or when we anticipate relatively low concentrations on site, we use known concentrations of vapor or gas from specialty gas suppliers. For instance, for a site known to be contaminated with hexane and vinyl chloride, we assessed the relative response of several survey meters after calibrating each meter with the factory calibrant or span gas. We then compared each meter's response to sample bags containing hexane or vinyl chloride in air. Both PIDs, the MicroTIP (10.6 eV UV lamp) and HNU (10.2 eV UV lamp), responded with efficiencies of about 10%; i.e. both gave a reading of about 1.0 unit in response to 9 ppm hexane in air. The FID, which gave a reading of 4 units, was considered the best instrument to detect hexane. In contrast, the MicroTIP was clearly the superior instrument to detect vinyl chloride, with an efficiency of approximately 50%; the OVA-128 and HNU HW-101 each efficiencies of only about 25%.

The primary survey instruments selected for the site were the MicroTIP and OVA-128; meter readings on the MicroTIP were interpreted as being caused by vinyl chloride or other non-hexane contaminants. FID readings were interpreted as caused primarily by hexane or other non-vinyl chloride contaminants. Vinyl chloride and hexane detector tubes were also used to confirm the presence of chlorinated and petroleum hydrocarbons.

It is important to remember that meter response to known concentrations of gases and vapors represent an optimal response. Humidity levels are low and there are no interfering gases and vapors present. Meter efficiency obtained under such artificial conditions should not be used to calculate a conversion factor for use in the field. Similarly, the use of a PID under oxygen deficient conditions is not recommended unless the displacing gas is known to be nitrogen. Displacing gases such as methane or ethane, which absorb or scatter UV light, can severely decrease meter efficiency and produce false negative responses.

The use of multiple types of instrumentation at hazardous materials sites can best be envisioned by using field scenarios. The remainder of this chapter could also be entitled "putting it all together" because that is what we hope will happen as you read and work through these examples. The most important decision is determining what instruments to use. Interpretation of meter readings is also important, but such interpretation will be of little value if the wrong instrument is selected and no meter response is obtained.

Before you begin these examples, we strongly recommend that you review the materials in the Appendices. We also encourage you to follow along as each example is discussed. Try answering each question yourself before reading our solution.

Example 1:

A 55-gallon drum of decontamination wash, containing n-hexane and water has been spilled at a remediation site. The temperature is 86°F. What types of hazards are present?

Flammability and toxic overexposure. Hexane is a flammable liquid with a flash point of −7°F. It is relatively insoluble in water, has a high vapor pressure, and is potentially toxic by inhalation; the OSHA/ACGIH 8-hour TWAs are 50 ppm.

What specific instruments could be used to monitor for these hazards?

A %LEL CGI can be used to characterize the flammability hazard; since only one volatile organic compound has been spilled, a conversion factor (CF) or response curve can be used to estimate the actual concentration present. A detector tube calibrated to hexane can be used to estimate the approximate concentration present. Instruments such as an FID or PID can be used to verify the presence of contamination but cannot be used to characterize the hazard in terms of actual concentration present.

The LEL of hexane is 1.1% (11,000 ppm). Suppose an MSA 360 gives a reading of 2% LEL in the spill area; what actual concentration is present and how does it compare to the TWA of 50 ppm? (Hint: check the MSA 360 conversion factor in Appendix C.)

The LEL of hexane is 1.1%, therefore, 100% of the LEL = 11,000 ppm, 1% of the LEL = 110 ppm. Using the meter reading of 2% and a CF of 1.3 for hexane from Appendix C, an actual concentration of 2.6% of the LEL is obtained. By rounding up to 3% of the LEL, the actual concentration present can easily be estimated to be 3 x 110 ppm or 330 ppm. This concentration is well over the TWA of 50 ppm.

What other instruments can be used to confirm this interpretation?

A detector tube calibrated to n-hexane or a ppm-equivalent CGI with a response factor for n-hexane could be used. An FID would be anticipated to give a response within the same order of magnitude (i.e. over 100 units).

A PID with a 10.2 eV lamp is also brought to the spill area; would you anticipate this instrument would give a reading comparable to the FID?

According to the list of IPs in Appendix D, hexane has an IP of 10.18; a PID with a 10.2 eV lamp would be anticipated to have a very low ionization efficiency. In addition, the spilled material also contained water; the presence of water vapor would decrease meter efficiency. In this situation, a PID would not give a response comparable to an FID; a significantly lower reading (i.e. probably less than 50 units) should be anticipated.

Example 2:

A dry cleaner wants your clean-up company to assess the extent of soil contamination that has occurred as a result of improper disposal of waste chlorinated solvents in the rear of his establishment. The most commonly used dry cleaning solvent used was tetrachloroethylene (perchloroethylene, perc, or perk) which has the following physical/chemical characteristics:

Vapor pressure (VP): 14 mm Hg
Flash point (FP): not flammable
Ionization potential (IP): 9.32
Water solubility (sol): 0.02%
OSHA TWA: 25 ppm
IDLH: 500 ppm; NIOSH occupational carcinogen

What are the potential hazards to investigatory personnel?

The primary hazard to personnel is from inadvertent exposure to concentrations above recommended or permissible exposure levels during investigatory activities.

Would you use the following instruments: %LEL CGI, FID, PID (10.6 eV lamp), detector tubes. Why or why not?

CGI: yes, even though perc is not flammable, there is always the possibility that other, flammable solvents have also been dumped at the same location. FID: yes, the FID will respond to chlorinated as well as non-chlorinated organic materials.

PID: yes, there is sufficient energy in a 10.6 eV lamp to ionize tetrachloroethylene with an IP of 9.32. The lack of a meter response however, does not rule out the presence of other contaminants with IPs greater than the lamp eV capacity.

Detector tubes: yes, tubes can be used to confirm the results obtained with other instruments. While there are tubes which can discriminate between chlorinated hydrocarbons and other types of organic contaminants (such as the MSA Halogenated Hydrocarbon tube), they cannot identify the actual contaminant present.

During the assessment, a hand auger is used to make bore holes into the soil. When the sample probe is placed within the borehole, readings of 1% to 3% LEL are obtained on a combustible gas indicator. What does this suggest? How would you proceed?

Low LEL readings on a CGI may suggest that other flammable materials may have been disposed of on site, or decomposition of perc or other organic materials is generating flammable gases such as methane or vinyl chloride. The presence of low molecular-weight materials such as methane and vinyl chloride can be confirmed by using a charcoal filter; low molecular weight materials pass through the charcoal filter, and the instrument reading will not be affected. Higher molecular weight materials, however, are retained in the filter; the instrument response will be less than the initial reading after the filter is attached.

The CGI instruction manual should be checked to determine if the catalytic filament sensor is inordinately sensitive to chlorinated hydrocarbons; if it is, the meter may be giving a false positive result. If necessary, obtain %LEL readings with a CGI from another manufacturer that does not exhibit chlorinated hydrocarbon sensitivity. A detector tube specific for chlorinated hydrocarbons that does not cross-react with other non-chlorinated contaminants, such as the MSA Halogenated Hydrocarbon tube, can also be used to determine if halogenated materials are present.

If all information suggests that other non-methane contaminants are present, review the site's history with the dry cleaner. Get information, if possible, about how the property was used before it became a dry cleaning facility.

A Dräger Polytest tube is used within a freshly augered hole; after 5 pump strokes (ps), a 3 mm-wide ring with multiple colors is obtained; color noted include brownish-green and purple. (Hint: see Dräger Polytest detector tube instructions in Appendix A).

Obtaining multiple colors in the indicating layer suggests the presence of multiple classes of contaminants (i.e. chlorinated hydrocarbon, petroleum hydrocarbon, oxygenated hydrocarbon). The Dräger Polytest tube instructions state that the indicating layer changes color as a function of the substance in the sampled air. While the colors observed cannot be used to determine the number or chemical classes of contaminants present, the result is evidence that more than the one anticipated class of contaminant (i.e. chlorinated hydrocarbons), is present.

Could the Polytest color change be caused by methane?

No, careful reading of the instructions (under Measurement and Evaluation) reveals that methane, ethane, and carbon monoxide cannot be detected by the Polytest tube.

Example 3:

You have been asked to assess the extent of environmental contamination from a furniture stripping shop. Materials used in the stripping process include aromatic petroleum solvents such as xylene and toluene, chlorinated hydrocarbons such as methylene chloride, alcohols such as methanol and isopropyl alcohol, and corrosives such as sulfuric acid, sodium hydroxide, and hydrochloric acid. Current practice is to brush a mixture of stripping materials onto furniture, then use of high pressure water spray or steam to clean each piece. Water runs into a floor drain and then out to a dry well. A small-bore hollow stem auger has been used to drill into the immediate area of the dry well. How would you conduct initial monitoring over the bore hole?

This is definitely one of those "where do I begin?" situations. Start out with wetted pH paper or detector tubes for acids and alkalis to determine if significant concentrations of corrosives are present; this is necessary in order to minimize the potential for corrosive damage to instruments. If corrosive levels appear low or negligible, bring in a combustible gas indicator to determine if high concentrations of flammable vapors are present. If CGI readings are minimal or not present, a survey meter such as an FID can be used to get an idea of relative amounts of organic contaminants present in the head space of the dry well.

What problems do you foresee?

Direct-reading instruments will not discriminate between the various organic contaminants present. Detector tubes will be useful only to give relative concentrations of various classes of contaminants (i.e. chlorinated hydrocarbons, alcohols, aromatic hydrocarbons, acids). Since water spray or steam is used to wash stripping materials off furniture, water and water vapor will also be present in the well. High humidity conditions can be expected to lower PID efficiency.

Inside the bore hole, a Dräger Methylene Chloride tube gives a reading of approximately 100 ppm after 10 ps. Is it appropriate to conclude that 100 ppm methylene chloride are present? Why or why not? (Hint: check tube instructions in Appendix A.)

Definitely not! Section 7 of the instructions indicates that besides methylene chloride, other organics will also be indicated, including alcohols and benzine hydrocarbons (benzine is German for petroleum). This tube is very susceptible to interfering materials and often gives false positive readings. A positive tube reading cannot be used to determine the actual concentration of methylene chloride unless there are no cross-specific, interfering compounds present. In this case, alcohols (methyl and isopropyl) and petroleum hydrocarbons (toluene and xylene) are known to be present. MSA offers a halogenated hydrocarbon detector tube that does not suffer from such interferences.

Example 4:

You are notified by the plant manager that a small quantity of gasoline has been spilled in the parking lot. Some gasoline may have entered a storm sewer which travels under the parking lot and when flowing, empties into a small stream. The plant manager claims he smells gasoline at the storm sewer outfall. There is approximately 3 to 6 inches of stagnant water and decaying leaves in the sewer. The probe of a hexane-calibrated Scott-Alert meter is slid through the hole in the cover of a manhole in the middle of the parking lot. Readings of 0% LEL and 20.9% O_2 are obtained. Is this sufficient evidence that there are no gasoline vapors in the sewer? Why or why not?

*The presence of normal oxygen concentrations effectively rules out the presence of UEL concentrations of gasoline or other flammable materials and indicates that there was sufficient oxygen present for the meter to operate properly. The LEL of a typical gasoline is about 1.4% or 14,000 ppm; to elicit a meter response on an LEL meter, around 0.5% to 1% of the LEL, or **at least** 70 to 140 ppm must be present. The lack of a response on a %LEL CGI indicates that if gasoline vapors are present, their concentration is below the instrument's limit of detection. Another instrument must be used to determine if low concentrations of gasoline vapor are present in the sewer headspace. An HNU with a 10.2 eV lamp is brought in; an ambient background reading* of 1.4 units is obtained in an area of the lot where there are no vehicles. A reading of 1.0 units is obtained when sewer air is sampled. Does this indicate a lack of contamination?

A reading of less than ambient background can be caused by moving the instrument into a less contaminated area, or by the presence of an increased concentration of non-ionizable gases, including water vapor or methane gas. In this case, since there is water in the sewer and humidity may be a significant factor, the lack of an HNU response could be interpreted as a water vapor effect; the lack of a positive PID response, however, does suggest that significant concentrations of gasoline vapors are not present in the sewer.

What other instruments could be used to determine if low concentrations of gasoline vapors are present within the sewer's atmosphere?

An FID, CGI with a ppm range, or low concentration range colorimetric detector tubes calibrated to petroleum hydrocarbons.

An FID is used to sample the sewer atmosphere through the manhole and a reading of 1.2 units above background is obtained. A charcoal filter is attached; the meter reading does not change. What does this suggest?

FIDs are sensitive to low concentrations of organic hydrocarbons, including petroleum hydrocarbons and methane. The meter reading did not change when a charcoal filter was used; since methane is not retained in the charcoal filter, this result indicates that a small concentration of methane is present in the sewer, probably produced from decomposition of naturally-occurring organic materials. It also indicates there is no detectable concentration of gasoline or other hydrocarbon vapors in the sewer. This does not necessarily mean there is no gasoline contamination, only that concentrations are below the limit of instrument detection, which for fresh gasoline is between 0.2 ppm and 0.6 ppm. The odor threshold for gasoline in some individuals is much lower. It is not unusual for an individual to smell gasoline at concentrations that are non-detectable by direct-reading instruments.

Does the FID response of 1.2 units represent an immediate health hazard?

Remember that all meter readings were taken through the holes of a manhole cover and represent readings from the sewer, not the workzone. The results of such remote sampling does not automatically represent a health hazard to the meter user standing in the parking lot.

Example 5:

Office workers are complaining of eye and throat irritation after new carpeting was installed over a 3-day weekend. The chemical culprit may be formaldehyde from the carpet, or the glue used during installation. The glue consisted of a powdered resin polymer mixed with a volatile petroleum distillate.

It is not unusual for municipal Health Departments, Environmental Protection Departments, or Hazardous Materials Response Teams of Fire Departments, to deal with such problems. In many cases, it takes at least several hours, sometimes days, to obtain the professional assistance of an industrial hygienist or ventilation specialist; it can also take several days to get results back from laboratory analysis. In the interim, the local agency must handle the issue with the resources it has on hand, and determine if a building should continue to be occupied.

What instruments could be used to determine if formaldehyde and petroleum hydrocarbons are present?

The FID can detect petroleum hydrocarbons but will not respond to formaldehyde. FID readings can be considered to represent relative response to other organic contaminants present, including petroleum hydrocarbons. Formaldehyde has an IP of 10.9; a PID with an 11.7 eV lamp could be used to detect both petroleum hydrocarbons and formaldehyde contaminates. For both the PID and FID, the meter reading cannot be interpreted as the ppm level of contamination, the actual ppm concentration present is greater than the meter unit reading. An IR analyzer would be of value to measure specific concentrations of formaldehyde. A minimum detection limit for formaldehyde is 0.2 ppm. A field portable GC could also be used.

Will the Dräger Formaldehyde 0.2/a tube be of any value in determining the extent of formaldehyde contamination?

The Dräger Formaldehyde tube used has a limit of sensitivity of 0.2 ppm; the tube will also cross react with other contaminates, including diesel fuel, and perhaps petroleum distillates. The tube will not be useful for measuring formaldehyde concentrations of less than 0.2 ppm and therefore cannot be used to rule out formaldehyde as a potential source of irritation.

What problems may be encountered when using survey instruments (PID and FID) in an office situation?

Offices and office personnel typically use a variety of materials that can elicit a meter response, including perfumes, deodorants, hair sprays, colognes, food and drink products, room deodorizers, white-out type correction products, solvents, glues, and cleaners.

Suppose all the instruments previously discussed were used and failed to give any readings; what can you say about the relative presence or absence of contamination?

Lack of instrument response indicates that the level of contamination is less than the detection level of each instrument used. Under optimal conditions, the minimum detection level is 0.2 ppm; under actual use conditions, the detection level may not be as low. The ACGIH Ceiling limit for formaldehyde is 0.3 ppm; the OSHA TWA is 0.1 ppm. Since exposure limits are at or below the limits of detection, the instrument results obtained cannot be used to rule out the presence of formaldehyde as the cause of worker complaints. In this case, the absence of instrument readings cannot be interpreted as the absence of potentially irritating concentrations of formaldehyde. The building should be considered to contain formaldehyde concentrations that are potentially greater than the OSHA TWA. Until air samples are collected and analyzed to confirm or deny this conclusion, workers should not remain in the affected areas of the building.

Example 6:

While installing a slurry wall around a former landfill, workers uncover a small package with the remnants of a radioactive label. A radiation meter with a GM pancake detector is on site. How should the package be surveyed as a possible source of radiation?

The meter should be turned on and the background level of radiation measured with all shielding removed. Approach the package slowly and determine if there is an increase in meter response. If there is no change in meter reading, continue to approach the container, bringing the pancake detector as close to the box as possible (without touching it) to detect very weak beta and alpha particles. If an increase is noted, place the shield between the box and detector and note if the meter response changes.

About 20 feet from the container, the meter reading changed from 0.02 mR/hr to 0.04 mR/hr; at 10 feet, the meter response is 0.10 mR/hr. What meter response should be obtained at 5 feet from the source, according to the inverse square rule?

The initial background was 0.02 mR/hr, and at 20 feet the reading increased to 0.04 mR/hr, for a radiation source reading of 0.02 mR/hr. When the distance was halved, the reading increased four-fold (0.02 x 4 = 0.08 + 0.02 background = 0.10 mR/hr). This follows the inverse square rule, which is used most often to compute dose as the distance increases from the source. As the distance is doubled, the dose is multiplied by the inverse square of the distance. In this case, the distance is halved from 10 to 5 feet, so the dose is multiplied by 4. If the distance is halved again, the dose from the source would increase to 0.32 mR/hr (0.08 x 4 = 0.32), for a total meter reading of about 0.34 mR/hr (0.32 + 0.02 background = 0.34).

A stainless steel shield is placed between the detector and source; there is no change in meter response. Should the worker approach any closer to the source? What should the radiation meter read if the distance to the source was only 2.5 feet?

Halving the distance would increase the radiation dose four-fold; the meter should read approximately 1.3 mR/hr. Since the action level recommended is 1 mR/hr, the worker should not approach any closer to the source.

Figure 10-15: Initial radiation monitoring should always be conducted with an unshielded detector (left). A pancake detector must be placed close to the source to detect alpha radiation (right). (Detrick Lawrence photos.)

Workers decide to leave the package where it is; they mark off the area and call a health physicist for advice. In the meantime, they continue to work keeping a distance of at least 10 feet from the package. What total radiation dose will workers receive if they are exposed to the package for 8 hours at 0.08 mR/hr?

Workers will receive a total dose of 0.64 mrems (0.08 mR/hr x 8 hr = 0.64).

Example 7:

A contractor at a nail polish manufacturing plant is removing a stationary above-ground 10,000 gallon storage tank that was used to store ethyl acetate. The tank interior was hosed down with water and the wash water allowed to drain out. Approximately 100 gallons of wash water remain in the tank. A cutting torch will be used to dismantle the tank. What instrument would you use to determine if it is safe to use the torch?

Ethyl acetate is a flammable liquid with a flash point of 24°F; it has a solubility of about 10% in water and a specific gravity of 0.90. Since the unsolubilized solvent will float on water, the liquid remaining in the tank may be mostly ethyl acetate. There is a potential for a significant concentration of ethyl acetate within the confines of the tank. A %LEL/oxygen deficiency indicator should be used to determine the concentration within the tank.

Ethyl acetate has a vapor pressure of 74 mm Hg. Use the 1300-rule to determine the approximate maximum headspace concentration within the tank at 68°F. How does this concentration compare to the flammable range (2%–11.5%) for ethyl acetate?

An approximate maximum headspace concentration would be about 96,000 ppm; this is within the flammable range; it is well above the LEL of 2% or 20,000 ppm, and below the UEL of 11.5% or 115,000 ppm.

The vapor density of ethyl acetate is 3.0. Where in the tank would you anticipate the highest readings?

A vapor density of 3.0 indicates that the vapor is 3 times heavier than air; the highest vapor concentrations should be found at the bottom of the tank.

An LEL meter gives the following readings as the sample hose is lowered into the tank from the top: 10% LEL at 2 feet, 15% LEL at 4 feet, 40% LEL at 6 feet, 80% LEL at 10 feet, 100% LEL at 15 feet, greater than 100% between 16 feet and the bottom of the tank. How would you proceed?

The concentration of flammable vapors must be lowered before cutting can begin. The liquid within the tank must be removed and the tank washed again; purging with an inert gas or fresh air, followed by venting, may also be necessary.

After removal of the liquid residue at the bottom of the tank, the contractor has decided to purge the tank with carbon dioxide by dropping 50-lb blocks of dry ice into the tank. An LEL meter gives a reading of 0% LEL, but a %GAS CGI gives a reading of 10%. What is affecting the meter readings? Is it safe to start cutting?

It is not safe to start cutting. Carbon dioxide from the dry ice may have displaced sufficient oxygen to affect the function of the catalytic filament sensor. The LEL reading should not be accepted unless an oxygen reading is taken simultaneously. Carbon dioxide is a well-known interfering gas for %GAS CGIs because it cools the thermal conductivity filament as well or better than many flammable gases. It is not possible to determine the hazard based on the meter readings obtained. In this case, the carbon dioxide must be evacuated from the tank before valid CGI readings can be obtained. Another alternative is to use a detector tube to determine the approximate concentration of ethyl acetate present. Sensidyne offers an ethyl acetate tube with a measurement range of 0.04 to 1.5% (400 to 15,000 ppm); the Dräger ethyl acetate tube has a measurement range of 200 to 3000 ppm.

The contractor washes tank once more with water and completely drains the tank. An MSA 361 gets a reading of 20.7 oxygen and 25% LEL at the bottom of the tank, a Scott-Alert gives a reading of 30% LEL. A Dräger ethyl acetate tube gives a reading between 2000 and 3000 ppm after 10 pump strokes. Are these results consistent? (Hint: check CGI conversion factors and detector tube instructions.)

The ethyl acetate conversion factors for the MSA and Scott-Alert CGIs are 1.1 and 0.9, respectively. The approximate actual %LEL present based on the MSA is 1.1 x 25% LEL = 27.5 or 28% %LEL. The actual concentration present based on the Scott-Alert is 0.9 x 30 %LEL = 27% LEL. The information from the two CGIs is consistent.

The Dräger ethyl acetate tube requires 20 ps before concentration can be read directly off the tube; since only half the required number of ps was used, the results must be multiplied by 2. The approximate concentration present, based on the tube results, is therefore between 5000 and 6000 ppm. The LEL of ethyl acetate is 2% or 20,000 ppm; the concentration present, based on CGI readings, is approximately 27% of the LEL, or 5400 ppm (20,000 ppm x 0.27 = 5400 ppm). All instrument results are therefore consistent with each other.

Example 8:

You are supervising repair work in an underground service vault at a tank storage farm. Tanks currently contain #6 fuel oil. The vault is checked every morning prior to entry by a safety officer employed by the facility; an instrument which measures %LEL and oxygen deficiency is used. After several hours in the vault workers complain of headache and nausea; episodes of dizziness have also been reported. Do you think the safety officer's monitoring procedures are sufficient?

Worker symptoms suggest that there is a problem within the vault. The %LEL/oxygen deficiency meter cannot be used to rule out the presence of hydrogen sulfide, carbon monoxide, carbon dioxide, or low concentrations of other contaminates. The monitoring practice of once per day is obviously not sufficient to protect workers inside the vault.

What questions would you ask of the facility health and safety officer?

Continuous monitoring is required when personnel are working within permit-required confined spaces. Why is monitoring conducted only once per *day? Is the monitoring instrument used working properly and when was it last calibrated? What is the limit of CGI sensitivity to #6 fuel oil? (With a flash point of up to 270°F, most LEL meters will be relatively insensitive.) What is the level of H_2S in the fuel oil? What was previously stored in the tanks? Has the vault ever been contaminated, and if so, with what? Have there been similar complaints from other workers, and if so, where and when?*

What other instruments should be used when evaluating this confined space atmosphere? Be specific.

Toxic gas sensors or detector tubes to detect carbon dioxide, carbon monoxide, and hydrogen sulfide. A PID (10.6 eV lamp) or FID survey meter could also be used to detect very low levels of petroleum hydrocarbon contamination. Low concentration colorimetric detector tubes for petroleum hydrocarbons can also be used. Air samples can also be collected throughout the day for laboratory analysis to identify the specific air contaminant responsible for worker symptoms.

Do you think it is a good idea to evaluate the vault only once per day? Why or why not?

It is not a good idea! It is also not safe and does not adhere to OSHA regulations. Vault conditions may change throughout the day in response to activities inside and outside the vault. A better protocol would be to continuously monitor whenever personnel are inside the vault.

What other factors may be involved in causing worker symptoms?

The vault probably has poor air circulation; worker activity within the vault may cause a build-up of carbon dioxide and a decrease in oxygen. The temperature within the vault may increase during the day, causing temperature stress. The size of the vault may also be a factor; some personnel find it stressful to work in a confined space.

Positive pressure ventilation, which pushes fresh air into the vault, can be used in concert with continuous air monitoring, in an attempt to alleviate worker symptoms and identify the source of worker complaints. Using Level B positive pressure supplied-air respirators should be viewed as the last resort, since the OSHA permit-required confined space rule requires engineering controls wherever and whenever possible, and continuous use of Level B confined spaces can be physically demanding and psychologically stressful.

Figure 10-16: Extra care is required when using air monitoring instruments around heavy equipment. (Author photo.)

REFERENCES

1. U.S. Environmental Protection Agency. 1984. Standard Operating Safety Procedures. Edison, NJ: USEPA Office of Emergency and Remedial Response, Hazardous Response Support Division. (Revised 1988 and 1992).
2. U.S. Environmental Protection Agency. 1981. Health and Safety Requirements for Employees Engaged in Field Activities, U.S. EPA Order 1440.2. Edison, NJ: USEPA Office of Management Information and Support Services.
3. Levine, S.P., Turpin, R.D., and Gochfeld, M. 1991. Protecting Personnel at Hazardous Waste Sites: Current Issues. Occup. Environ. Hyg. 6: 1007-1014.
4. Kerfoot, H.B. and Mayer, C.L. 1989. The use of industrial hygiene samplers for soil-gas measurement, EPA Report 600/4-89/800. Las Vegas, NV: Environmental Monitoring Systems Laboratory Office of Research and Development, U.S.EPA.
5. 29 Code of Federal Regulations, Part 1910. Permit-Required Confined Spaces for General Industry.
6. U.S. Environmental Protection Agency. 1987. A Compendium of Superfund Field Operations Methods. Washington, DC:USEPA Office of Emergency and Remedial Response.
7. Nyquist, J.E., Wilson, D.L., Norman, L.A. and Gammage, R.B. 1990. Decreased sensitivity of photoionization total organic vapor detectors in the presence of methane. Am. Indus. Hyg. Assoc. J. 51:326-330.

PROBLEM SET

1. A hollow stem auger is used to put in test borings adjacent to a former landfill. Down-hole monitoring produces the following readings: MicroTIP (10.6 eV lamp): 870 units, OVA Model 108: 1200 units, CGI: 15% LEL, 20.7% O_2. What would you do to further characterize meter response?

2. You are assigned as health and safety officer for an initial investigation of a 20 acre site suspected to contain hazardous materials. You arrive to find 5 workers ready to don Level B respiratory protection (SCBA) and protective clothing. Three workers are assigned as the entry team, two will be available as a back-up team to assist the entry team if an emergency occurs. One entry team member will map the site, carry a radio, and record instrument readings. The following instruments are available to use: %LEL/ppm CGI, %LEL/O_2 combination meter, %LEL/O_2/H_2S combination meter, GM (beta/gamma) radiation detector, PID, FID, Matheson Leak Detector, and colorimetric detector tubes. Which instruments should the two entry team members carry? Be specific.

3. A number of 55-gallon drums have been found in a field adjacent to a shopping mall. State environmental officials have requested your firm to examine the drums and determine if they contain hazardous materials. Most drums appear rusty but intact and are not leaking. Describe the role of air monitoring in this situation.

NOTES

11 *Emergency Response*

OPERATIONAL CONSIDERATIONS

At a hazardous materials emergency, it is irresponsible and often dangerous to make decisions and take action before sufficient information has been obtained and the scene is controlled. An important aspect of scene control is the establishment of specific areas or **control zones**. Within designated zones, specific tasks are performed by responders wearing levels of PPE appropriate to the hazard. Delineation and use of control zones helps ensure that responders are properly protected against known or anticipated hazards, the spread of contamination by personnel is limited, and additional equipment and personnel are staged and ready for use.

While control zones receive the most attention by emergency responders, they are also used at hazardous waste sites. Three frequently used zones are the exclusion zone, the contamination reduction zone, and the support zone. The **exclusion zone** is the immediate area where contamination exists as well as the area surrounding the contamination which may become contaminated or be affected by the hazardous materials release. The exclusion zone extends far enough to reasonably ensure that a contamination release will not adversely affect less-protected personnel outside the zone. The exclusion zone is also called the **hot zone** and **red zone**. The outer border of the exclusion zone is called the **hotline**.

The hotline is the border between the exclusion zone and the **contamination reduction zone** (CRZ), also called the **warm zone**, **yellow zone**, and **limited access zone**. The contamination reduction zone is a transition area between the contaminated exclusion zone and non-contaminated areas not affected by the hazardous materials release. Within the CRZ, personnel and equipment are decontaminated. The width of the CRZ is determined by the hazard posed by contaminates within the exclusion zone and the type of required decontamination procedures, and the number of personnel and type of equipment to be decontaminated. In addition to the

197

Polychlorinated biphenyls	23.0%
Sulfuric acid	6.5%
Anhydrous ammonia	3.7%
Chlorine	3.5%
Hydrochloric acid	3.1%
Sodium hydroxide	2.6%
Methyl alcohol	1.7%
Nitric acid	1.7%
Toluene	1.4%
Methyl chloride	1.4%

Table 11-1: Top ten hazardous materials encountered, and percent of incidents involved, at incidents involving releases other than vehicle fuel.

decontamination stations located in the contamination reduction corridor, the CRZ may also contain support personnel and equipment to assist responders in the exclusion zone.

The **support zone** is a non-contaminated area where administrative, communications, and support functions are performed. The command post, medical station, equipment and supply centers, and field laboratory are usually located in the support zone. The border between the contamination reduction zone and the support zone is called the **contamination control line**. Personnel returning from the exclusion zone must enter the support zone through the contamination reduction corridor. The support zone is also called the **cold zone**, **green zone**, and **clean zone**. The **primary staging area** is usually located within the support zone.

Levels of protection within control zones are determined by the known or anticipated hazards of the contaminates released. In general, Level B with appropriate skin protection is the minimum level of protection used in the exclusion zone. Structural firefighting clothing does not offer chemical protection and is not recommended for use as an overgarment at hazardous materials incidents where there is a potential for vapors or splash. Most emergency responders wear coveralls made of fire-retardant material or structural firefighting clothing under their chemical protective outer garments. Selection of chemical protective clothing should be based on the knowledge of chemical resistance and limitations of the garment and the type of hazardous materials present.

Responders in the contamination reduction zone should wear the level of PPE required to ensure their safety. Levels of PPE can vary depending on the task of responders within the CRZ. For instance, backup and support teams ready to assist responders in the exclusionary zone should wear the level of protection required in the exclusionary zone. Workers manning temporary rest areas and equipment resupply areas at the edge of the CRZ may require a lower level of PPE. Personnel in the support zone may wear Level D protection, or normal work clothing.

HAZARD AND RISK ASSESSMENT

The first step in preparing to confront a hazardous materials incident is to identify the hazards—the types of materials released, their source, and the probable impact area. There are about 10 million chemicals known to man and the number is growing; nearly 75% are considered "intermediate chemicals" used to produce a final chemical product. Intermediate chemicals are short-lived are often created and then quickly consumed. That still leaves approximately 2.5 million chemicals that a responder could conceivably encounter; fortunately, only a small number (around 50,000!) are in common use, and not all are hazardous.

A short-list of hazardous materials has been compiled by the USDOT in the Emergency Response Guidebook[1] to assist first-on-scene emergency responders. The USEPA has developed guides for a more extensive list of over 2500 hazardous chemicals[2]. The Association of American Railroads has developed the Emergency Action Guides[3] for about 150 hazardous materials transported by rail. NIOSH regularly publishes the Pocket Guide to Chemical Hazards[4] which contains valuable information on nearly 400 hazardous substances which have OSHA exposure limits.

These response guides, and others like them, are of little use unless the released material can be identified. Hazard assessment is the process of

Figure 11-1: Terrorist or trickster? In this case, the package label which indicated the presence of highly radioactive "plutonium detonators" turned out to be a hoax. (Author photo.)

identifying the chemical, evaluating the hazard and risk it presents. Sources of information which assist in hazard identification include placards and labels, warning and marking systems such as NFPA 704M[5] and HMIS[6], Material Safety Data Sheets (MSDS), and shipping papers or manifests. In some cases, the configuration of containers, location of the release (such as a fuel depot, manufacturing plant, dry cleaner, or tank farm) can suggest the type of material involved.

A 1985 survey of nearly 7000 hazardous materials incidents involving substances other than commercial vehicle fuel (gasoline and diesel) revealed that nearly 75% of the releases were at fixed facilities while the remainder were in-transit incidents[7]. Nearly half of the nearly 7000 incidents involved only 10 chemicals. Chlorine, ammonia, hydrochloric acid and sulfuric acid had signifiant impact on human health; these four substances were involved in more than 25% of the incidents which caused human injury or death.

While real-time air monitoring cannot identify the material released, it can identify specific types of hazards, such as flammability, oxygen deficiency, and radiation exposure. If the material is known, then instruments can assist in determining more specific risks to responders and community receptors.

EMERGENCY INSTRUMENTATION

Initial air monitoring should assess the scene for hazards that can be immediately life-threatening: flammability, radiation, and depending on the situation, oxygen and toxic gases. Air monitoring equipment used for initial assessment should be easy to operate and calibrate, and require little maintenance. Typical %LEL CGIs, oxygen deficiency indicators, radiation detectors, and many toxic gas sensors can be considered emergency response instruments because they can literally be turned on and used immediately with very little adjustment.

Responders performing initial air monitoring assessment should not be encumbered with every instrument the team possesses! Not only can this result in injury to the responder, but meter readings are often missed or mis-

Figure 11-2: This responder is burdened with three instruments and would have difficulty reaching his SCBA regulator or reacting to an emergency. (A Detrick Lawrence photo.)

read. Responders can be so busy keeping track of meters that other hazards are not readily recognized.

Calibration

It is often impossible to check calibration of instruments prior to initial use at a hazardous materials emergency. There should be a regular schedule for recharging or changing batteries and instrument calibration. We recommend monthly recharging and calibration for instruments that are not subjected to potential mishandling. Meters stored in busy response vehicles can be bounced around a bit and often need more frequent adjustments (i.e. weekly or even daily); meters kept in little-used vehicles require less frequent (i.e. monthly) calibration checks. A calibration/maintenance log should be maintained for each instrument.

During a hazardous materials emergency response, if time permits, it is prudent to take each meter aside in turn and check its calibration. If meter calibration is not possible before or during a response, it should be performed at termination to ensure that meters were working properly and information collected and used during the response was valid.

Initial Response

In many cases emergency response personnel have a good idea of the material involved, or the nature of the hazard, when they arrive on the scene. This allows the Haz Mat Sector commander to select the most appropriate instruments available for use at the incident. When using instruments, it is important to remember that the absence of a meter reading does not automatically indicate the absence of contamination. The meter may not be sufficiently sensitive to detect low concentrations, it may not be working properly, or it may be unable to detect the contaminant present.

Whenever possible, check instrument readings with another device to verify the response obtained. For instance, a CGI meter reading may be verified by a detector tube, another CGI, or if readings were negligible or very low, a ppm-equivalent CGI or a survey instrument such as a PID or FID. At the site of an obvious release, start with high concentration instruments first and then progress to lower range meters as needed.

Figure 11-3: When a vapor cloud is present, a meter is not necessary to warn responders of significant gas concentrations. (Author photo.)

We have been describing air monitoring instruments designed to measure **gases and vapors** in air; they are **not** designed for, and are not accurate in, the presence of aerosols or mists; this includes so-called "vapor clouds" resulting from the release of cryogenic or cold gases such as hydrogen or propane. The visible cloud is formed by the freezing of water vapor in the air; as the mixture of ice crystals and flammable gas is drawn into the instrument, the ice crystals thaw and liquid water may be deposited within the instrument and affect the function of the sensor or flashback arrestors. In situations where a vapor cloud is visible, air monitoring should be conducted at the edge of the cloud and beyond to determine the concentration and extent of the contamination.

Instruments designed to monitor gases and vapors should not be used in the presence of explosive dusts such as flour, grain dust, or coal dust. High concentrations of non-explosive dusts or particulates can also clog filters, sensors, and flashback arrestors and affect their function. Instruments must have dust filters at responses involving explosions or building collapse where high concentrations of particulates are present; filters should be frequently checked and changed as necessary.

In addition to the potential for injury from an unstable structure and smoldering rubble, evaluating a fire scene for known or potential hazardous materials includes other variables or conditions which makes instrument use difficult or impossible. These include the presence of smoke, heat, steam, products of combustion, and extinguishing agents such as foam and dry chemical. In most cases, monitoring with instruments should be limited to the periphery of the scene until the fire has been controlled and the scene is made safe for personnel and equipment. Air monitoring in the immediate area surrounding the fire scene, as well as down-wind locations, can play an important role in assessing the relative amount of contamination released and identify areas affected.

Many gases and vapors are heavier than air and can collect in low-lying areas; materials released as a result of fire will often behave like a lighter-than-air gas because they are heated and carried up on thermal air currents created by the heat of the fire. If not consumed in the fire, these gases and vapors will eventually cool and sink. Instrument users should be careful to take readings from above as well as below waist level; a good rule is to move the sample hose or probe up or down with every other step.

Instruments can be very useful in determining if there has been a significant vapor release in an enclosed area. Always try to check the interior of a building or enclosure before entering. In some instances, it is prudent to allow the vapor or gas to stay within the enclosed area rather than opening a door and allowing a highly hazardous material to escape.

Combustible Gas Indicators

The %LEL CGI should always be used first as a safety meter to determine the extent of the hazard present; the %LEL meter reading is accurate only for the calibrant gas, all other readings are relative to the calibrant. Conversion factors or response curves significantly increase the usefulness of the CGI when dealing with known materials. Curves and response factors can be used ONLY when there are no other interfering gases or vapors AND the identity of the contaminant is known.

Factors and curves represent the best fit or average value of many trials; the accuracy is often ± 10% or even 20% of the actual %LEL calculated. It is not necessary to use a calculator! Do not worry about getting an exact

Figure 11-4: Check drains and other low areas for heavier-than-air gases and vapors which collect in low areas. (Author photo.)

answer down to the decimal point; the instrument itself is not that precise! For example:

A pentane-calibrated MSA 361 is used at the site of a methyl ethyl ketone spill. The meter reading is 10% LEL. The LEL of MEK is 1.4%, the NIOSH IDLH is 3000 ppm. What is the approximate actual concentration present in %LEL and ppm? Has the IDLH been exceeded? (Hint: see Appendix C for MSA conversion factor for MEK.)

The MSA conversion factor for MEK is 1.6; the actual concentration present is therefore 16% LEL. The LEL of MEK is 1.4% or 14,000 ppm; thus

100% LEL	=	*14,000 ppm*
50% LEL	=	*7,000 ppm*
10% LEL	=	*1,400 ppm*
5% LEL	=	*700 ppm*
1% LEL	=	*140 ppm*

Approximately 15% LEL is actually present, which is equivalent to 1,400 ppm plus 700 ppm, or 2100 ppm; the IDLH of 3000 ppm was not exceeded at the time and location of the reading. The addition of the "extra" 140 ppm makes little difference in the overall hazard assessment.

The presence of low concentrations of gases and vapors can be detected with a ppm-equivalent CGI. These meters usually require a lengthy warm-up and should be turned on as soon as possible after arrival on scene. Meters which demonstrate significant drift after turn-on should not be used until readings stabilize.

Figure 11-5: Frequent calibration checks should be performed when operating a catalytic filament sensor CGI around firefighting or vapor suppression foam. (Author photo.)

The catalytic filament in CGI sensors can be poisoned by the silicone and silicate compounds used to stabilize firefighting and rapid response foams. CGIs operated around foam should have frequent calibration checks. It is possible for a sensor to fail very quickly, especially when assessing the effectiveness of vapor suppression foams by drawing sample air from directly above the foam blanket.

Figure 11-6: Radiation is often found in laboratories; suspect radiation at illegal chemical processing, storage or disposal operations. (Author photo.)

Survey Instruments

Some industrial and municipal response teams are equipped with low concentration survey instruments including PIDs, and FIDs. These are extremely valuable instruments which can be used for odor complaints, to locate the source of low-volume chemical releases, or for tracking the extent of contamination from larger releases. While the usefulness of these instruments cannot be discounted, they should not be placed in the same category of emergency instruments as CGIs and detector tubes that are available for immediate use. Survey meters require significant warm-up time and should be calibrated before each use. They are designed to measure low concentrations of materials and are unusable in the immediate area of a significant release. Most importantly, these meters can be incapacitated by exposure to excessively high concentrations.

Survey meters can, however, be valuable tools to assess the extent of a release, evaluate the potential health risk to the unprotected public, and estimate the extent of environmental contamination. Survey instruments are typically used at the edge of a release, in areas where the concentrations are no longer detectable by CGIs. Survey instruments can also be valuable when the material released is non-flammable, and does not elicit a response on a CGI. Prior to the introduction of disposable chemical protective outer garments, instruments where used to assess the effectiveness of decontamination procedures. Although radiation survey instruments are very useful to detect residual radioactive contamination on personnel, the utility of survey meters to detect residual chemical contamination on protective equipment after decontamination is completed, is questionable.

Figure 11-7: PID and FID survey meters are not true emergency response instruments and should not be used when high concentrations are known to be present. (A Detrick Lawrence photo.)

Radiation Detectors

Radiation detectors should always be used at responses involving hospitals, diagnostic or treatment centers, research facilities, university, industrial, or medical laboratories, medical waste storage or disposal facilities, and incidents involving unknown materials. The pocket-sized alarm rate-meters can

Figure 11-8: This alarm rate meter is worn by the responder and can detect gamma, x-rays, and high energy beta radiation. The device will alarm if the measured radiation dose exceeds 1 mR/hr.

be worn by responders and require no attention by the wearer. These devices can detect X-rays and gamma radiation as well as beta particles. Some rate meters can also be adapted for alpha particle detection. Dosimeters should also be worn when radiation is suspected.

The presence of very low levels of radioactive contamination is often not discernible by radiation detectors. This is because the level of radiation emitted is below the detection limit of the instrument or the type of radiation present (alpha particles and some beta particles) cannot reach the detector. A wipe test is often used to document the lack of surface contamination. The test is performed by wiping the surface with filter paper or other absorbent material and then submitting it for counting in a gamma counter or scintillation counter.

Some GM detectors can be overwhelmed by high readings. The CD V-700, for example, is designed for low range beta/gamma radiation; rates over 50 mR/hr will produce off-scale readings. Exposure rates in excess of 50 mR/hr may jam or saturate the meter, when this occurs the reading may be zero or less than full scale.

Detector Tubes

There are occasions when a survey meter or CGI cannot measure the contaminate released. In these instances, colorimetric detector tubes are often the only means to determine the approximate or relative concentration; this is especially true when the contaminates are inorganic. Detector tubes are often used to measure airborne concentrations of ammonia, chlorine, hydrogen sulfide, and cyanides. A well-equipped response team should have high and low concentration tubes available. High range tubes are valuable to determine the hazard in the immediate area of the release and to evaluate efforts to control the release. Low concentration tubes are necessary to estimate the extent of the release, set up control zones, and assess the health hazard to the unprotected public.

Detector tubes are often the last device considered for use at an incident when other types of real-time instrumentation is available. This is unfortunate, since tubes offer a valuable means to confirm readings from other instruments. In some cases, a detector tube may be available for a known contaminant; comparing CGI reading to the tube response taken at the same time and location is a potential method of establishing a relative conversion factor for the material. For example:

Figure 11-9: Responders use a single MSA rack with 4 tubes to sample the interior of a truck prior to entry. (Author photo.)

A CGI is monitoring cyclohexane vapors, and gives a reading between 25% LEL and 30% LEL; there is no conversion factor. An MSA petroleum hydrocarbon tube gives a reading of 4000 ppm on the cyclohexane scale. The LEL of cyclohexane is 1.3% or 13,000 ppm; the detector tube gave a reading of 4000 ppm, which is about 25% of the LEL.

A comparison of the detector tube and CGI results suggests that at the concentration tested, the CGI is giving a reading that is approximately the actual LEL, i.e. the conversion factor is 1.0. Evaluation of tube results at the same time and location where lower CGI responses are obtained to determine if the conversion factor is valid at lower %LEL readings.

Detector tube results can also be used to determine relative response efficiency of survey meters to known contaminants. The detector tube and survey meter draw a sample at the same time and location and compare

Figure 11-10: Three tube racks are used to hold all 12 tubes which comprise the MSA hazmat kit. All 12 tubes can be drawn at the same time (left). This design can make it difficult to sample through drains, sewer grates, and other restricted areas (right). (Author photos.)

results. A "response factor" is then used to give an indication of the relative concentration present. The response factor should be verified using multiple tubes at different locations. This method works fairly well with FIDs but only when detecting concentrations within a limited range. Response factors can change dramatically as concentrations increase or decrease.

Remember that detector tubes vary in their accuracy; always add at least a **two-fold safety factor** when using detector tube readings to establish response factors. The dual use of survey meters and detector tubes to approximate actual concentrations should never be used in place of laboratory analysis or other air monitoring instrumentation used by health professionals.

Detector tube manufacturers offer hazmat kits that can be used to determine the type or class of contamination present. Most kits use racks which hold and sample four or five tubes at a time. Users should be aware, however, of the potential for cross-reactivity and false positives. Detector tube kits cannot be used for identification, but can be used to assist in verification of a substances identity.

For example, suppose several leaking drums labelled DOWPER are found on the side of the road. Chemtrec reports that DOWPER is the trade name for tetrachloroethylene, a chlorinated hydrocarbon. Physical testing of the liquid indicates it is insoluble in water, and heavier than water. There is no response from the CGI, even when the sample probe is placed directly over the spilled material. Survey instruments (PID with 10.2 eV lamp and FID) give good responses. The 12 tubes of the MSA hazmat kit are sampled directly over the spilled material. The Unsaturated Hydrocarbon tube gives a slight positive response; the Halogenated Hydrocarbon Group B tube gives a strong positive. These results can be used to confirm the suspicion that the material in the drums is an unsaturated chlorinated hydrocarbon. The detector tubes responses cannot, however, confirm that the specific chlorinated hydrocarbon present is actually tetrachloroethylene.

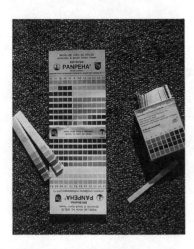

Figure 11-11: Multi-colored pH test strips are useful when dealing with colored liquids. (Author photo.)

Other Detection Devices

Toxic gas sensors may also be useful when the hazardous materials known or suspected to be present is not detectable by survey meters or CGIs. Responders at hazardous materials incidents involving inorganic materials such as chlorine gas, ammonia, sulfur dioxide, and carbon monoxide will be limited in the choice of direct reading real-time instrumentation available. Toxic gas sensors calibrated to the released material can be of value, however, it is important not to overwhelm the sensor with concentrations far in excess of its maximum detection range. Exposing a chlorine gas detector with an upper response limit of 30 ppm to a chlorine vapor cloud will disable the sensor and probably ruin the entire instrument. Toxic gas sensors with relatively low detection ranges can still be extremely valuable, however, in assessing the outer boundaries of the release, establishing the position of control zones, and determining if there is immediate hazard to the public.

In other cases the hazardous material involved is not easily detectable with air monitoring equipment. Polychlorinated biphenyls (PCBs), sodium and potassium hydroxide, many pesticides and pharmaceuticals, toxic metals, explosives, and a variety of materials with very low or negligible vapor pressures fall into this category. Other devices can be used to assist the responder in detecting and assessing the hazard present. Individual and multiple test strips and kits are available to identify the presence of oxidizers, peroxides, chlorides, fluorides, acids, and bases.

There is a variety of kits available to detect the presence of chlorinated materials in soil and oils. Most kits actually detect chlorine ions liberated during the test. Test kits are susceptible to false positives, however, PCB-free soils containing road salts, chlorine salts from animal urine, or chlorine-containing pesticides often give strong positive results. When dealing with liquids suspected to contain PCBs, try to obtain a sample from the container rather than scooping up a sample from the road or ground. If a false positive result is suspected after a soil test, run a negative control using a non-contaminated soil sample.

The HAZCAT® Chemical Identification System[8], in the hands of an experienced user, can be used to categorize an unknown liquid or solid. This system is valuable in classifying difficult to recognize compounds such as pesticides and oxidizers; multiple hazard characteristics can also be recognized. Perhaps the most beneficial aspect of HAZCAT is that it is capable of confirming the presence of non-hazardous materials, saving time and unnecessary clean-up costs.

Perhaps the simplest and easiest to use detection device is pH paper. Simple pH paper can be wetted and used to detect the pH of solids; it can also be used to detect acid or alkaline vapors in air. Unwetted paper can be used to test the pH of liquids.

COMPARATIVE INSTRUMENTATION

Comparing instrument readings is often easier for emergency responders because they often deal with only one, known hazardous material. The principles developed in Chapter 10 are applicable however, not only to field operations, but to all types of instrument use.

Emergency responders can be easily placed in hazardous and life threatening situations. It is important that instrument use and interpretation of readings are performed correctly. Responders should trust their instruments; injuries have resulted when meters were not used, or readings ignored. If there are doubts concerning the reliability of an instrument it should be taken out of service. **Never use a meter that cannot be zeroed or calibrated; never ever use an instrument with insufficient battery power.**

Once again we will ask you to work through a series of response scenarios, many based on actual incidents. Cover the printed answer and attempt to solve each problem; when you have finished see if you agree with our solution.

Example 1:

An emergency response team arrives at a sanitary waste treatment plant which has reported explosive levels of gas at the intake, based on its stationary CGI. Responders enter the intake area in full turnouts with SCBA and a methane-calibrated portable %LEL CGI. After re-setting the treatment facility's methane-calibrated stationary meter, it gives a reading of 75% LEL. The responders place the sample hose of their meter next to the stationary CGI inlet and get a reading of 20% LEL. Both CGIs are calibrated to methane. Are the readings consistent? What do they suggest?

These results are definitely inconsistent. Both meters are calibrated to methane, yet they show significantly different readings. Since both meters are methane-calibrated and were drawing sample air at the same time and location, they should give similar readings if the incoming material is methane. The dissimilar readings suggest that one may be malfunctioning, or the material eliciting the alarm condition is NOT methane.

What could you do to help determine the reason for the inconsistent meter readings?

*A charcoal filter can be used to determine if the material is naturally-occurring sewer gas (methane) or a heavier hydrocarbon. Retention of organic vapors and gases in a charcoal filter increases as the molecular weight of the material increases. Methane and ethane, as very low molecular weight organic materials, readily pass through the activated charcoal; the meter reading not affected by adding a charcoal filter. Higher molecular weight hydrocarbons, such as propane, butane, pentane, and more complex organic materials, are retained on charcoal. For instance, if a CGI is reading an actual propane concentration of 30% LEL, the addition of an in-line charcoal filter will immediately reduce the response to 0% LEL; the reading will remain at 0% LEL for at least one minute. In this case, if the material entering the treatment facility is sewer gas, there will be **no** change in the reading when a charcoal filter is used. If the entire response is due to a another material, the CGI reading will fall to 0% LEL. If a combination of methane gas and one or more heavier hydrocarbon is present, the reading will be decreased but not entirely eliminated.*

A charcoal filter is placed in the sample line and there is no decrease in the meter reading on the portable CGI. What should be done now?

The charcoal filter results indicate the incoming gas is a low molecular weight gas similar to sewer gas or methane; one meter may be malfunctioning. A calibration check should be quickly performed on the portable CGI, and, if possible, on the stationary meter. If a calibration check is not possible, another CGI should be used, either calibrated to methane or with a methane conversion factor.

Example 2:

A 9,000 gallon capacity methanol (methyl alcohol) tanker has rolled over under an overpass at a busy highway interchange. A large volume of methanol has pooled under the overpass. What hazards are present?

Methanol vapors can collect under the overpass; the primary hazard is flammability. Toxicity by vapor penetration of the skin is also a hazard, but should be a secondary concern to responders wearing appropriate PPE. The alcohol may also enter the storm-water drainage and cause problems at locations distant from the initial release. Environmental contamination of water and soil at the drainage discharge must also be considered.

What instruments should be used? Be specific.

The initial meter used should assess the flammability hazard; a %LEL CGI equipped with an oxygen deficiency indicator.

Emergency clean-up crews are at the spill site; you have been asked to respond to nearby home owners who are complaining of headaches and an "alcoholic" odor. The MSA 360 gives a response of 1-2% LEL. The MSA Alcohol tube shows a 12 mm stain after 3 ps. Are these readings consistent? The LEL of methanol is 6%. (Refer to Appendices for MSA 360 conversion factors and MSA detector tube instructions.)

The MSA 360 conversion factor for methyl alcohol is 1.0; the meter reading of 1-2% LEL is equivalent to an actual concentration; this corresponds to approximately 600 to 1200 ppm methanol. A 12 mm stain after 3 ps on the MSA Alcohol corresponds to approximately 800 ppm. The results obtained are therefore consistent.

The OSHA PEL for methyl alcohol is 200 ppm, the NIOSH IDLH is 25,000 ppm. Is there a hazard to unprotected residents in the neighborhood where the tube reading was obtained? How would you proceed?

*Methanol concentrations are **at least** 3 to 4 times the 8-hour OSHA exposure limit. Remember to add a two-fold safety factor to detector tube results and concentrations calculated from CGI readings when using the results to determine if an area is safe for unprotected personnel. In this case, there is a very real danger of overexposure to unprotected neighborhood residents. Appropriate officials should be notified regarding the potential need for evacuation. Outdoor monitoring should continue with an instrument capable of detecting low concentrations of methanol; monitoring should be conducted inside homes also. Residents should be advised to remain inside with windows closed until further notice.*

What other instruments could you use to confirm the presence of methanol air contamination?

FID, ppm-equivalent CGI, PID with 11.7 eV UV lamp, IR analyzer.

What color change is anticipated on the MSA Alcohol tube in the presence of alcohol?

The tube should change from yellow to pale green stain when alcohol is present.

Example 3:

A response team arrives at a detached garage of a residence which reportedly contains a leaking 20 lb propane tank. Readings at the windows and under the garage door are as follows:

MSA 260: 20% LEL, 20.9% O_2
Scott Alert: 30% LEL, 20.8% O_2
Dräger Hydrocarbon 0.1%/b detector tube: after 15 ps, pale brown-gray stain which is initially the lighter than the comparison layer but after 10 minutes becomes much darker.
MSA Petroleum Hydrocarbon tube: full stain after 1 ps.

Are these results consistent? What approximate concentration of propane is at the windows and the door? The LEL of propane is 2.2%.

Figure 11-12: Whenever possible, check windows, doors, or other openings prior to entering structures where a release has occurred. (A Detrick Lawrence photo.)

The MSA 260 response curve indicates the actual concentration present is 15% LEL. The Scott-Alert conversion factor for propane is 0.56, which can be rounded up to 0.6. The approximate actual concentration is about 18% LEL. The CGI readings are consistent with each other.

The LEL of propane is 2.2% or 22,000 ppm. The Dräger tube instructions states that 15 ps to match the comparison layer indicates the presence of 0.5% propane or 5,000 ppm. The indicating layer did not match the comparison layer, it was lighter; this means the concentration present was less than 5000 ppm. The maximum range of the MSA tube is 1000 ppm; full stain for the entire indicating layer after 1 ps demonstrates the presence of concentrations well above 1000 ppm. The detector tube results are consistent with each other and the CGI readings.

Could the detector tube and CGI readings be caused by another contaminant such as gasoline, kerosene, or a petroleum-based solvent?

Yes, both detector tubes respond to other petroleum products, not just propane.

Examine the sensitivity/cross specificity/limitation and corrections sections for other likely detector tubes in Appendix A. Is there a tube that would be useful in distinguishing between light hydrocarbon gases (such as methane, ethane, and propane) and heavier hydrocarbons such as gasoline or petroleum-based solvents?

Yes, the Dräger Natural Gas tube responds to methane, ethane, and propane, but not to heavier hydrocarbons. The Dräger Polytest does not respond to methane or ethane, but gives a green stain in the presence of propane. In this case, the Natural Gas tube could be used to confirm suspicions regarding the identity of the gas. The Natural Gas tube cannot however, measure very low concentrations; the limit of sensitivity is approximately 0.5% or 5000 ppm for methane, and 0.05% or 500 ppm for ethane and propane.

Example 4:

A worker becomes unconscious while cleaning the interior of a stationary tank containing tetrahydrofuran (THF). The tank was washed with water and drained, then left open for several hours before the worker entered. Only a few inches of semi-solid residue is present on the bottom of the tank. The worker had started to shovel the residue into a 5-gallon bucket which was then to be hoisted to the top of the tank by another worker. Ambient conditions are 72°F, sunny, relative humidity 66%.

Chemical/physical data for THF are as follows:

Flash point: 6°F; LEL: 2%; UEL: 11.8%; Vapor Pressure: 132 mm Hg; Boiling Point: 151°F; Solubility: miscible in water; Specific Gravity: 0.89; Vapor Density: 2.5; OSHA TWA 200 ppm; NIOSH IDLH 20,000 ppm.

If the residue was composed entirely of THF, what would be the approximate headspace concentration of THF at the bottom of the tank?

Using the 1300-rule, the approximate maximum headspace concentration at the bottom of the tank is 170,000 ppm or 17% by volume.

The calculated headspace concentration is well above the UEL of 11.8% and the NIOSH IDLH of 20,000. Is it appropriate to consider the worker unsalvageable and begin recovery operations? Why or why not?

Definitely not! The tank was washed once with water; THF is miscible in water. The residue is undoubtedly not 100% THF; a significantly lower concentration may be present in the residue and therefore in the tank atmosphere. A %LEL CGI with an oxygen deficiency indicator should be lowered into the tank and readings taken as the sample inlet approaches the level of the worker. If possible, use a CGI with a response curve or conversion factor for THF.

Sample lines from two CGIs, an MSA 361 and a Scott Alert, are simultaneously lowered into the tank. The MSA gives readings between 50% and 53% LEL when the sample hose inlet is approximately even with the overcome worker; the Scott-Alert shows readings between 65% and 70% LEL at a similar sampling location. Are these readings consistent? What is the approximate actual concentration at worker level?

The MSA and Scott-Alert conversion factors for THF are 0.8 and 0.63, respectively. The actual concentration present based on the MSA reading is approximately 40% to 42% LEL, or about 8000 ppm; this value corresponds well to an actual concentration of 41% to 44% LEL, based upon the Scott-Alert reading.

Example 5:

A tractor trailer loaded with drums and plastic shrink-wrapped 5-gallon pails on pallets jackknifed and skidded to a stop on its side. No leaks or vapors are noted upon arrival; there is no diesel fuel leak from the tractor. The trailer is placarded as carrying flammable and corrosive materials. The manifest indicates the load includes roofing materials, paint stripper, and window wash concentrate. Responders in protective gear do not enter but take readings through the damaged portions of the sides and roof of the trailer. Readings from various areas within the cargo area are:

GasTechtor: 0-3% LEL, O_2 levels fluctuate between 20% and 22%
MSA 360: 0-2% LEL, O_2 levels between 19% and 21%, 2 ppm CO
Dräger Polytest tube: pale pink discoloration
PD (10.2 eV lamp): erratic readings, less than 0 to 5 units
FID: erratic readings, less than 1 to 7 units

Is there a hazard present?

An often overlooked hazard is the confined space hazard when responders must enter the interior of the box cargo trailer. There is no immediate flammability hazard. The CGIs indicate relatively low levels of contamination, however, these readings may represent high ppm concentrations which may pose a significant health hazard to unprotected personnel. When dealing with unknowns, a good rule of thumb to follow is that a 1% LEL meter reading probably represents at least 100 ppm of contaminant. The CGI gave readings up to 3% LEL, which may represent at least 300 ppm of one or more unknown contaminates. Higher concentrations may be present deeper in the truck interior. The lack of a significant response on both survey meters suggests the material eliciting the LEL readings is inorganic, or it is adversely affecting survey meter response. It is not possible to reliably determine the hazard present. Personnel should remain in a high level of PPE until the hazard is characterized.

What can be determined from the Polytest detector tube results?

*Pink discoloration on a Dräger Polytest tube usually indicates the presence of high humidity; the lack of another color change suggests the material present does not produce a color change on the tube. Commonly encountered contaminants that the Polytest tube does **not** indicate include methane, ethane, carbon dioxide, and some organic and inorganic amines, including ammonia. The FID should have responded to methane and ethane and organic amines; the PID with a 10.2 eV UV lamp has poor efficiency for ammonia.*

Figure 11-13: The interior of a box cargo trailer represents a confined space hazard. (Author photo.)

What other measurements or instruments can be used to further assess the hazard?

Wetted pH paper or acid and alkaline detector tubes can be used to determine if the presence of corrosive vapors is affecting oxygen readings or other meter functions. A charcoal filter on the CGI can be used to distinguish between heavy hydrocarbons and light hydrocarbons or inorganic materials. An amine detector tube can be used to detect the presence of amines, including ammonia, which is often used in window wash concentrates.

An MSA Ammonia and Organic Amine detector tube and a Dräger Ammonia 5/a tube are used to sample the interior atmosphere of the trailer. A dark blue stain saturates the indicating layer on both tubes after one pump stroke. What do these results suggest?

The detector tube results suggest that the material present within the truck responds well to the amine tubes, and may be ammonia from window wash concentrate. The tube results indicate that high concentrations of material, probably ammonia are present. The CGI results obtained earlier also suggest the presence of high concentrations of ammonia.

Use the Scott-Alert conversion factor in Appendix C; based on the conversion factor, what approximate concentration of ammonia is present?

The conversion factor for ammonia is 0.53, a maximum reading of 2% LEL was obtained on the Scott-Alert meter. The actual concentration present is therefore approximately 1% LEL. The LEL of ammonia is 15% by volume or 150,000 ppm. An actual concentration of 1% of the LEL represents a minimum of 1500 ppm, well above the NIOSH IDLH of 500 ppm for ammonia.

Example 6:

Sacks of granular material have fallen off a flatbed truck; several have spilled open onto a busy freeway. Vehicles have spread the material over a 100 yard section of the roadway. The driver has set up flares directly next to a broken sack. Sacks are labeled HEXACHLOROPHENYL ETHER. Instrument readings directly over the granules are:

MSA 260: 0% LEL, 20.9% O_2
PID with 11.7 eV UV lamp: background readings
FID: background readings
Dräger Polytest detector tube: no color change
Matheson Leak Detector: very slow to no alarm

Can you define the potential hazards present from the meter readings?

Although a flammability hazard is not present, it is not possible to define other hazards based on meter readings; the absence of meter readings does not automatically indicate the lack of air contamination or absence of other hazards. Potentially toxic materials may elicit no meter reading because they have negligible vapor pressure or cannot be detected by available air monitoring equipment. Many corrosive solids will also not be indicated. There will be instances when air monitoring instruments will not be helpful in determining the potential hazard. This is when pH paper, test kits, or HAZCAT, if available, can be used to help determine the potential hazardous properties of unknown materials.

How would you proceed?

Move flares away from sacks to prevent inadvertent ignition. Direct vehicular traffic away from the spilled material to prevent further spreading. If material has the potential to blow away, cover loosely with plastic sheeting. Contact CHEMTREC for assistance on substance identification, and recommendations on containment and ultimate disposition.

Example 7:

At an isopropyl alcohol (IPA) spill at an indoor manufacturing plant, a Fire Department Emergency Response Team obtains readings between 15% and 20% LEL on an MSA 360. What is the actual %LEL present?

The MSA 360 conversion factor for IPA is 1.1; the actual approximate concentration present is essentially the same as the meter reading, or between 16% and 22% LEL.

Although the spill was relatively small, some of the material has entered a sanitary sewer by way of an illegal sump pump connection. The MSA gives a reading of 10% LEL when the sample line is dropped into the sewer. How can you verify that the reading is from IPA and not sewer gas?

A charcoal filter can be used to distinguish between methane gas and IPA vapors. An alcohol detector tube can also be used to detect and measure IPA concentrations in the sewer.

A large nursing home is located 500 yards from the spill area. Response Team members use the MSA 360 and get readings between 1% and 2% LEL in the downstairs social room, which is in the basement of the building and also has an illegal sump connection to the sanitary sewer. A charcoal filter indicates the material present is not methane. Upstairs readings in the residents' rooms, kitchen, nurses' station, and hallways are consistently 0% LEL. There are no complaints regarding the odor of IPA from nursing staff or residents. Assuming IPA is causing the reading, what is the approximate concentration in the social room? Is it appropriate to certify the remainder of the facility free of IPA contamination? The LEL of IPA is 2%.

Since the conversion factor is so close to 1, the meter reading can be used as the actual concentration present. The LEL of IPA is 2% or 20,000 ppm, 2% LEL is equivalent to 400 ppm (2.2% LEL is equivalent to 440 ppm). Since the reading is so close to the limit of detection of 1% CGI for the CGI, a two-fold safety factor should be used; the actual concentration present may therefore be twice the calculated value, or up to 800 or 900 ppm. It is not appropriate to certify the rest of the facility free of IPA contamination based on the lack of a CGI reading. Non-detectable concentrations of less than 200-400 ppm may be present.

The nursing home administrator asks if the elderly residents of the facility should be evacuated to prevent overexposure to IPA vapors. Is this necessary? The OSHA TWA for IPA is 400 ppm.

The only location which had CGI-detectable concentrations was in the social room; other areas of the facility had IPA levels that were not detectable with the CGI, i.e. were less than 200-400 ppm. Residents live at the facility 24 hours a day; many have pre-existing medical conditions which could be exacerbated by IPA exposure. In this case, it is inappropriate to use CGI readings to determine the relative safety of an atmosphere to unprotected personnel. While initial CGI readings suggest that evacuation is not necessary, health officials should be notified and other low-concentration meters should be used to verify the relative lack of IPA contamination in the upstairs areas.

What other detection devices can be used to detect IPA? Be specific.

Detector tubes, FID, PID with 10.6 (or greater) eV UV lamp, IR analyzer with IPA-specific filter. The ionization potential of IPA is 10.10; a 10.2 eV lamp would have very poor ionization efficiency and should not be used.

Which detector tube available (see Appendix A) can be used to directly measure IPA concentrations present?

The MSA Alcohol tube is a mm scale conversion tube calibrated to a number of different alcohols. When only one of these alcohols is present, the mm scale can be converted directly to the ppm concentration present. Unfortunately, the multi-calibrated MSA Alcohol tube is being replaced by the MSA Ethanol tube (see Appendix A). Like the Dräger Alcohol tube, it is calibrated only to ethanol and methanol. While these tubes can be used to indicate the presence of other alcohols, they cannot be used to estimate the concentration present.

Example 8:

A hazmat team responds to a report by a security guard of an unknown liquid spill. Upon arrival, a liquid is observed slowly dripping from a badly dented drum in an unused warehouse. Team members don a high level of PPE and begin control operations. Instrument readings obtained during plugging and diking are:

GasTechtor (hexane calibrated): 0% LEL, 20.9% LEL, 140 ppm-units
MicroTIP (10.6 eV lamp): peak reading of 402 units
OVA-128: peak reading of 520 units
Dräger Polytest tube: green/brown/purple stain after 1 ps
Is this sufficient information to determine the hazard to response personnel?

There is sufficient information to determine that there is no hazard from flammability or oxygen deficiency. Survey meter readings indicate a potential for toxic overexposure to unprotected personnel; supplied air and skin protection must continue to be utilized. The actual hazard, however, cannot be determined until the material is identified.

What else could be done to determine what type of material has been spilled?

A detector tube specific for chlorinated hydrocarbons, field test kits for chlorinated hydrocarbons could be used to determine if the unknown material is chlorinated. HAZCAT quick tests for water reactivity, water solubility, and specific gravity (i.e. floats or sinks in water) could assist responders in determining the physical/chemical properties of the material.

According to the security guard who found the drum, the spilled material smells like chloroform. The warehouse manager reports that drums of waste trichloroethylene (TCE) were recently removed from an outdoor storage pad adjacent to the warehouse. The LEL of TCE is 8%. If the material is TCE, what is the approximate concentration present based on the GasTechtor readings?

The TCE ppm scale conversion factor for the hexane-calibrated GasTechtor is 4.4. The meter read 140 units on the ppm scale. The actual concentration present is approximately 600 ppm (140 x 4.4 = 616).

The MSA Halogenated Hydrocarbon tube gives a reading of 10 mm stain after 1/2 ps. What does this result suggest?

The MSA Halogenated Hydrocarbon tube, when properly used, does not respond to non-halogenated hydrocarbons. This tube uses a reaction tube containing potassium permanganate and sulfuric acid to literally "chew up" organic hydrocarbons. When halogenated hydrocarbons are present, the chlorine, bromine, or fluorine released in the reaction tube produces a blue color change in the indicator tube. Non-halogenated hydrocarbons do not produce a color change. According to the mm conversion table for TCE, a 10 mm stain after one-half pump stroke indicates a concentration of only 100 ppm. Survey meter results suggest a much larger concentration is present; the tube results suggest that halogenated material present is not TCE, or a combination of halogenated and non-halogenated contaminates are present in the spilled material.

Review the response obtained with the Polytest tube; what does it suggest?

A review of the Dräger Polytest tube results indicates that there is more than one class of chemicals present; multiple classes are indicated by the presence of more than one color within the stain area. This suggests that the material contains multiple components, including one or more halogenated hydrocarbons. The drum contents and spilled material will have to be treated as an unknown hazardous waste until further analyses can be conducted.

REFERENCES

1. U.S. Department of Transportation. 1990. Emergency Response Guidebook.
2. National Response Team. 1987. Hazardous Materials Emergency Planning Guide. U.S. Environmental Protection Agency, Federal Emergency Management Agency, U.S. Department of Transportation.
3. Research and Test Department and Bureau of Explosives, Safety and Hazardous Materials Division. 1990. Emergency Action Guides. Washington, DC: Association of American Railroads.
4. U.S. Department of Health and Human Services, Public Health Service, Centers for Disease Control, National Institute for Occupational Safety and Health. 1990. NIOSH Pocket Guide to Chemical Hazards, DHHS (NIOSH) Publication No. 90-117.
5. National Fire Protection Association. 1985. NFPA 704M Standard System for the Identification of the Fire Hazards of Materials. Quincy, MA: National Fire Protection Association.
6. Hazardous Materials Identification System, Implementation Manual. 1985. Washington, DC: National Paint and Coatings Association.
7. Acute Hazardous Events Data Base. 1985. Cambridge, MA: Industrial Economics, Inc.
8. Turkington, R. 1988. HazCat® Chemical Identification System. San Francisco: Haztech Systems, Inc.

PROBLEM SET

1. Emergency team members respond to the scene of a toluene tanker rollover and rupture. The entire cargo has run into a a large drainage ditch. The methane-calibrated GasTechtor gives a reading of 100% LEL, 20.6% O_2. What is the actual concentration of toluene present? What information does the oxygen reading supply?

2. At an ethanol spill, a Dräger Alcohol 100/a tube gives a green stain to 20 mm after 10 ps. What concentration is present? What meter reading should be obtained on the Scott-Alert CGI if the reading was taken at the same time and location as the detector tube?

3. After a fire at an active research facility, a radiation meter with a GM detector (shield open) shows maximum response, followed immediately by a zero mR/hr reading. The user places the GM detector next to the check source; the meter shows no response. What should the responder do?

NOTES

217

Appendices

NOTES

* Detector tube instructions are provided for illustrative and instructional purposes only. Some information provided by the manufacturers has been deleted or modified to conserve space. When using detector tubes, always refer to the **complete** instructions provided by the manufacturer.

1. **General and Application**
   The DRÄGER acid test tube qualitatively detects traces of acid compounds in air (e.g. hydrochloric acid, nitric acid, acetic acid, formic acid). The tubes are to be used in conjunction with the DRÄGER gas detector pump. For use, see Section 4 of these Operation Instructions.
   Important: These Tubes must **not** be used with pumps supplied by other manufacturers, since this could lead to considerable errors in indication. Such a combination would offend against relevant regulations.

2. **Description:** See illustration. Opening time (duration of one pump stroke until the limit chain is completely taut): 1 to 4 seconds.

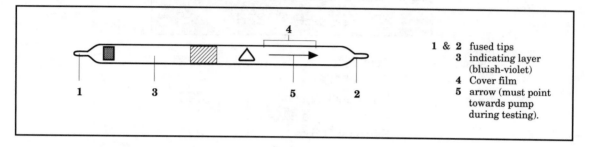

1 & 2	fused tips
3	indicating layer (bluish-violet)
4	Cover film
5	arrow (must point towards pump during testing).

3. **Sensitivity of indication**
   Some acid substances which give a discoloration of length app. 5 mm with n = 1 pump stroke are listed below:

Concentration	Substance	Colour of indication
5 ppm (mL/m^3)	Formic acid	Yellow
5 ppm (mL/m^3)	Acetic acid	Yellow
10 ppm (mL/m^3)	Nitric acid	Pinkish yellow
10 ppm (mL/m^3)	Hydrochloric acid	Pinkish yellow

4. **Test and evaluation of the result**
   4.1 Before each series of measurements, check the pump for leaks using an unopened DRÄGER tube.
   4.2 Break off the tips of the DRÄGER tube.
   4.3 Insert the DRÄGER tube tightly in the pump head (arrow points towards the pump).
   4.4 Suck the air sample through the tube with 1 pump stroke. Acid substances change the colour of the indicating layer from bluish violet to pink/yellow. Some examples of the sensitivity of indication are given in Section 3.

5. **Remarks**
   Following a negative test result, the Tube can be used again up to 3 times on the same day. Evaluate the indication immediately since the discoloration changes somewhat in the course of time.

6. **Influence of ambient conditions on the result of measurement**
   6.1 Temperature
   The DRÄGER Tubes can be used in a temperature range of from 0 to 40°C.
   6.2 Humidity
   Between 3 and 15 mg H$_2$0 per litre, humidity has no influence on the indication.

7. **Specificity (cross-sensitivity)**
   The indication is based on the colour reaction with an acid indicator.

8. **Shelf life**
   For expiration date and storage temperature, see data on package strip.

DRÄGER Tube Instructions (Modified)

1. General and Application

Determination of methanol and ethanol in air. The DRÄGER tube Alcohol 100/a must not be mistaken for the DRÄGER tube which is used for the determination of alcohol concentration in the breath (Alcotest). **Important:** It is not permissible to combine the tubes with pumps made by other manufacturers, since this may cause considerable errors in indication. Such a combination would offend against relevant regulations.

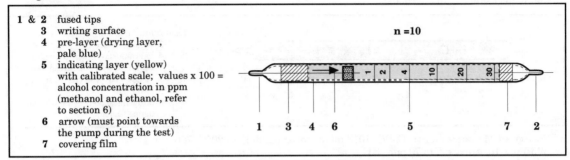

1 & 2   fused tips
3   writing surface
4   pre-layer (drying layer, pale blue)
5   indicating layer (yellow) with calibrated scale; values x 100 = alcohol concentration in ppm (methanol and ethanol, refer to section 6)
6   arrow (must point towards the pump during the test)
7   covering film

n =10

2. Description:
Refer to illustration. Opening period (time needed for the release of the compressed bellows, ending when the arrestor chain is fully tensioned): 6 to 12 seconds.

3. Measuring Range   (20°C, 1013 mbar)

Methanol: 100 to 3000 ppm

Ethanol: 100 to 3000 ppm

$$1 \text{ mg methanol per m}^3 \text{ of air} = 0.75 \text{ ppm}$$
$$1 \text{ ppm} = 1.33 \text{ mg/m}^3$$
$$1 \text{ mg ethanol per m}^3 \text{ of air} = 0.52 \text{ ppm}$$
$$1 \text{ ppm} = 1.92 \text{ mg/m}^3$$

$\left. \right\}$ 20°C, 1013 mbar

4. Test and Evaluation of the Result

Before each series of measurements check the pump for leaks using an unopened DRÄGER tube. Break off the tips of the tube and insert with arrow pointing toward pump. With 10 strokes of the pump, draw the air sample through the tube. The length of the pale green discolouration corresponds to the concentration. Scale values x 100 = ppm.

5. Remarks

Discolourations last for a considerable time, allowing tubes to be retained as evidence (seal the ends with rubber and caps protect the tubes against light). Even after a negative reaction, the tube cannot be reused.

6. Influence of the Ambient Conditions on the Measurement

The calibrated scale printed on the tube is valid for 10 strokes of the pump and for temperatures of 20°±5°C. If the ambient temperature deviates considerably from 20°C, the number of pump strokes is to be read from the following table. The calibrated scale on the tube is then valid again for both methanol and ethanol.

Ambient Temperature °C	Number of strokes for methanol and ethanol
5	16
10	12
15-25	10
30	9

7. Specificity (Cross - Sensitivity)

Indication is based on the oxidation of the alcohols with chromo-sulfuric acid. The reagent is so selective that only the alcohols with low boiling points are indicated. It is not possible to differentiate the different alcohols and it is therefore necessary to know which alcohols are present before the test. Aldehydes, ethers, ketones and esters react with the indicating layer only if present in concentrations above their Threshold Limit Values. The vapours from gasoline, benzene, carbon disulfide, acetic acid, chlorinated hydrocarbons, etc., give no indication.

8. Shelf Life:

See expiration date and storage temperature on label of the box.

Caution: Do not allow used tubes to fall into the hands of children. Contents are corrosive!

1. **General and Application:** Determination of anmonia (NH_3) in air and technical gases. The tubes are to be used in conjunction with the DRÄGER Gas Detector Pump. For use, see Section 4 of these Instructions.

   Important: It is not permissible to combine the tubes with pumps made by other manufactures, since this may cause considerable errors in indication. Such a combination would offend against relevant regulations.

3. **Description:** See illustration. Opening time (duration of one pump stroke until the limit chain is completely taut): 4 to 8 seconds.

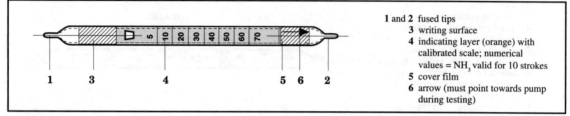

1 and 2 fused tips
3 writing surface
4 indicating layer (orange) with calibrated scale; numerical values = NH_3 valid for 10 strokes
5 cover film
6 arrow (must point towards pump during testing)

3. **Range of Measurement:** (20°C, 1013 mbar; corresponding to 20°C, 760 mm Hg)
   With n = 10 strokes: 5 to 70 ppm NH_3
   With less than ten strokes, up to a maximum of 700 ppm NH_3 (see section 4.5).
   1 ppm $NH_3 \approx 0.71$ mg/m^3 } 20°C, 1013 mbar
   1 mg NH_3/m$^3 \approx 1.41$ ppm

4. **Test and Evaluation of the Result**
   4.1 Before each series of measurement, check the pump for tightness using an unopened tube.
   4.2 Break-off the tips of the DRÄGER Tube.
   4.3 Insert the DRÄGER Tube tightly in the pump head (arrow points towards the pump).
   4.4 Measurement in the range of 5 to 70 ppm NH_3. Suck the air sample through the DRÄGER Tube with 10 pump strokes. NH_3 changes the colour of the indicating layer from orange to blue. The length of the discoloration depends on the concentration. Numerical values = ppm NH_3.
   4.5 Measurement in the concentration range above 70 ppm NH_3. Suck the air sample through the tube with less than 10 strokes. Evaluation is carried out from the length of the discoloration in accordance with the following table:

Numerical value read-off	No. Strokes n = 10	No. Strokes n = 5	No. Strokes n = 2
5	5 ppm	10 ppm	25 ppm
10	10 ppm	20 ppm	50 ppm
20	20 ppm	40 ppm	100 ppm
30	30 ppm	60 ppm	150 ppm
40	40 ppm	80 ppm	200 ppm
50	50 ppm	100 ppm	250 ppm
60	60 ppm	120 ppm	300 ppm
70	70 ppm	140 ppm	350 ppm

   Note: With n = 1 stroke, the range of measurement extends up to 700 ppm NH_3; but the scatter in the indication is higher than with other numbers of strokes.

5. **Remarks**: Evaluate indication immediately, since the lengths of the discolorations change somewhat in the course of time. After a negative test result, the Tubes can be used up to ten times on the same day.

6. **Influence of Ambient Conditions**
   6.1 Temperature: The DRÄGER Tubes can be used in a temperature range of from 10 to 50°C.
   6.2 Humidity: Between 5 and 12 mg H_2O per litre, humidity has no influence on the indication.
   6.3 Atmospheric pressure: For pressure correction, multiply the tube reading by the following factor:

$$\text{conversion factor} = \frac{1013}{\text{actual atmospheric presure (in mbar)}} \quad \text{or conversion factor} = \frac{760}{\text{actual atmospheric pressure (in Torr)}}$$

7. **Specificity (Cross Sensitivity)**: The NH_3 indication is based on the colour reaction of ammonia with bromophenol blue and acid. In addition to ammonia, other basic compounds are also indicated (e.g. amines).

8. **Shelf life:** For expiration date and storage temperature, see data on the label of the box.

Caution: Do not allow DRÄGER Tubes to fall into the hands of children. Contents are corrosive!

DRÄGER Tube Instructions (Modified)   Benzene 0.5/a

1. **General and Application:** Determination of benzene (C₆H₆) in air. The tubes are to be used in conjunction with the DRÄGER Gas Detector Pump. For use, see Section 4 of these Operating Instructions.
   Important: It is not permissible to combine the tubes with pumps made by other manufacturers, since this may cause considerable errors in indication. Such a combination would offend against relevant regulations.

2. **Description:** See illustration. Opening time (duration of one pump stroke until the limit chain is completely taut): 15 to 30 seconds.

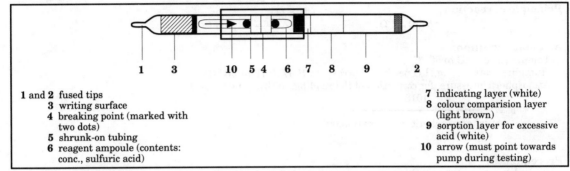

1 and 2	fused tips	7	indicating layer (white)
3	writing surface	8	colour comparision layer (light brown)
4	breaking point (marked with two dots)	9	sorption layer for excessive acid (white)
5	shrunk-on tubing	10	arrow (must point towards pump during testing)
6	reagent ampoule (contents: conc., sulfuric acid)		

3. **Range of Measurement:** (20°C, 1013 mbar) $1\ ppm\ C_6H_6 \approx 3.26\ mg/m^3$
   0.5 to 10 ppm benzene (C₆H₆) $1\ mg/m^3 \approx 0.31\ ppm\ C_6H_6$ $\}$ 20°C, 1013 mbar

4. **Test and Evaluation of the Result**
   4.1 Before each series of measurements, test the pump for leaks with an unopened DRÄGER Tube.
   4.2 Break-off the tips of the DRÄGER Tube.
   4.3 Bend the tube at the point marked with two black dots, so that the inner reagent ampoule (6) breaks. Be careful, contents of ampoule conc. sulfuric acid. Then shake the tube to propel the contents of the ampoule in the direction of the indicating layer; this layer (7) has to be wetted thoroughly by the sulfuric acid.
   4.4 Insert the DRÄGER Tube tightly in the pump head (arrow points towards the pump).
   4.5 Draw the air sample through the tube until the whole indicating layer (7) has turned from white to light-brown; the colour matches approximately the light-brown comparison layer (8). From the number of strokes, needed for colour matching, the benzene concentration is calculated according to the following table:

Number of Strokes	Benzene	
	ppm	mg/m³
40	0.5	1.6
30	0.7	2.3
25	0.8	2.6
20	1	3.3
10	2	6.5
7	3	9.8
4	5	16.3
3	7	22.8
2	10	32.6

5. **Remarks:** Even after a negative test result the DRÄGER tube benzene 0.5/a cannot be reused. The indication lasts for a short time only due to colour intensification.

6. **Influence of Ambient Conditions**
   6.1 Temperature: The DRÄGER Tubes can be used in a temperature range of from 10 to 40°C.
   6.2 Humidity: Between 3 and 15 mg H₂O per litre, humidity has no influence on the indication.
   6.3 Atmospheric pressure: For pressure correction multiply the tube reading by the following factor:

   $$\text{conversion factor} = \frac{1013}{\text{actual atmospheric presure (in mbar)}}$$

7. **Specificity (Cross Sensitivity):** The indication is based on the benzene reaction with formaldehyde in the presence of sulfuric acid, colour change from white to light-brown. Other aromatic compounds (e.g. toluene, xylene, ethylbenzene) are indicated with approximately the same sensitivity as benzene.

8. **Shelf life:** For expiration date and storage temperature, see data on the label of the box.

Caution: Do not allow DRÄGER tubes to fall into the hands of children. Contents are corrosive!

Application ranges
– Determination of CO in air
– Determination of CO contained in technical gases with less than 50 vol.% of hydrogen

Mode of operation
The tube contains two layers: an orange-coloured pre-layer and a white indicating layer. When air or a gas sample are sucked through the tube, interfering gases are retained in the pre-layer. The indicating layer changes to a brownish-green in the presence of carbon monoxide.

Principle of reaction

$$5\,CO + I_2O_5 \xrightarrow{\quad SeO_2 + H_2S_2O_7 \quad} I_2 + 5\,CO_2$$

Ambient conditions
– Temperature: 0 °C to 50 °C
– Humidity: max. 50 mg H_2O per litre (corresp. to 100% R.H. at 40 °C)
– Atmospheric pressure: for correction of the reading, multiply by factor F

$$F = \frac{1013}{\text{actual atmospheric pressure (mbar)}}$$

Prerequisites
– The tubes may only be used in conjuction with the DRÄGER Gas Detector Pump or the DRÄGER Quantimeter 1000, since otherwise, considerable measurement errors are liable to occur.
– Observe the Instructions for Use of the pump.
– Before each series of measurement, check the pump for leaks with an unopened tube.

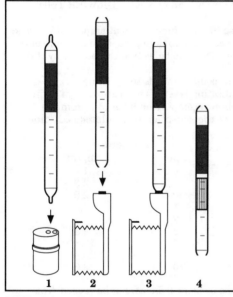

Measurement and Evaluation
1 Break off both tips of the tube in the tube opener.
2 Insert the tube tightly in the pump. Arrow points towards the pump.
• Select the measuring range.
    10 to   300 ppm (10 strokes, scale n = 10)
    20 to   600 ppm (5 strokes, scale n = 10)
    100 to 3000 ppm (1 stroke, scale n = 1)
3 Suck air or gas sample through the tube with the appropriate number of strokes.
One stroke takes 15 to 25 seconds, until the limit chain is completely taut.
4 Read the entire length of the brownish-green discoloration from the scale n = 1 (1 stroke) or n = 10 (5 to 10 strokes).
• Mulitiply the value with factor F for correction of the atmospheric pressure.
Enter the result in the measurement record.
Relative standard deviation: ± 10 to 15%
• Observe possible cross sensitivities.
• Flush the pump with air after operation.

Disposal
Avoid skin contact with the tube filling. Contents are corrosive. Keep out of the reach of children.

Cross sensitivities
– Acetylene reacts similar to carbon monoxide, however, with less sensitivity.
– Benzine, benzene, halogenated hydrocarbons and hydrogen sulfide are retained in the pre-layer. In the case of higher concentrations of interfering hydrocarbons and halogenated hydrocarbons, use should be made of a carbon pre-tube (Order No. CH 24101).
– Higher concentrations of easily cleavable halogenated hydrocarbons (e.g. trichloroethylene), are liable to form chromyl chloride in the pre-layer which changes the indicating layer to a yellowish-brown.
– Technical gases which contain more than 50 vol.% of hydrogen, change the entire indicating layer to green. These gases should be measured with the DRÄGER CO 8/a tubes.
– CO determination is not possible in the case of high olefine concentrations.

Caution: Do not allow DRÄGER tubes to fall into the hands of children. Contents are corrosive!

DRÄGER Tube Instructions (Modified)   Cyanide 2/a (KCN, NaCN)

1. **General and Application:** Determination of potassium cyanide and/or sodium cyanide in aerosol form in air. The tubes are to be used in conjunction with the DRÄGER Gas Detector Pump. For use, see Section 4 of these Instructions.
   Important: It is not permissible to combine the tubes with pumps made by other manufacturers, since this may cause considerable errors in indication. Such a combination would offend against relevant regulations.

2. **Description:** See Illustration. Opening time (duration of one pump stroke until the limit chain is completely taut): 10 to 20 seconds.

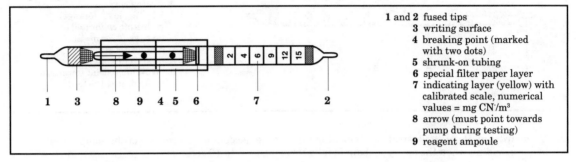

1 and 2 fused tips
3 writing surface
4 breaking point (marked with two dots)
5 shrunk-on tubing
6 special filter paper layer
7 indicating layer (yellow) with calibrated scale, numerical values = mg CN⁻/m³
8 arrow (must point towards pump during testing)
9 reagent ampoule

3. **Range of Measurement:** (20°C, 1013 mbar, corresponding to 20°C, 760 mm Hg)
   With n = 10 strokes: 2 to 15 mg CN⁻/m³

4. **Test and Evaluation of the Result**
   4.1 Before each series of measurements, check the pump for leaks using an unopened DRÄGER Tube.
   4.2 Break-off the tips of the DRÄGER Tube.
   4.3 Insert the DRÄGER Tube tightly in the pump head (arrow points towards the pump).
   4.4 Suck the air sample through the DRÄGER Tube with n = 10 strokes.
   4.5 Remove the tube from the pump head and crack it at the point marked with two black dots so that the inner reagent ampoule (9) breaks. The shrunk-on tubing holds the two parts of the tube together. Now propel the ampoule liquid in the direction of the arrow onto the special filter paper layer (6) by gently tapping the tube and make n = 2 activation strokes in air free from cyanide with the Gas Detector Pump. In the presence of cyanide, the yellow indicating layer turns red. The total length of the discoloration is a measure of the cyanide concentration. Numerical values = mg CN⁻/m³

5. **Remarks**
   The DRÄGER tube cannot be used again even after a negative result. Evaluate the indication immediately, since the length of the discoloration can change in the course of time.

6. **Influence of Ambient Conditions on the Result of Measurement**
   6.1 Temperature: The DRÄGER Tubes can be used within a temperature range of from 0 to 30°C.
   6.2 Humidity: Below 20 mg H_2O per litre, humidity has no influence on the indication.
   6.3 Atmospheric pressure: For pressure correction, multiply the Tube reading by the following conversion factor:

   $$\text{Conversion factor} = \frac{1013}{\text{actual atmospheric pressure (in mbar)}}$$

7. **Specificity (Cross Sensitivity)**
   Cyanide aerosols (KCN, NaCN) are precipitated on the filter layer. Gaseous hydrogen cyanide is produced by reaction with sulphuric acid (ampoule liquid). The hydrogen cyanide reacts with mercury chloride and methyl red in the yellow indication layer to give a colour reaction (colour change from yellow to red).

   If the air sample contains free hydrogen cyanide (HCN), a colour indication occurs even before the ampoule is broken. If KCN or NaCN are also present, this indication increases by the value of their concentration when continuing to proceed as in 4.5

8. **Shelf life:** For expiration date and storage temperature, see information on package strip.

Caution: Do not allow DRÄGER tubes to fall into the hands of children. Contents are corrosive!

1. **General and Application:** Determination of ethyl acetate (and other esters of acetic acid) in air. The tubes are to be used in conjunction with the DRÄGER Gas Detector Pump.

   Important information: It is not permissible to combine the tubes with pumps made by other manufacturers, since this may cause considerable errors in indication. Such a combination would offend against relevant regulations.

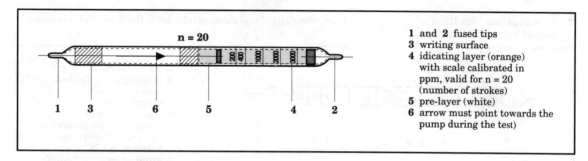

1 and 2 fused tips
3 writing surface
4 idicating layer (orange) with scale calibrated in ppm, valid for n = 20 (number of strokes)
5 pre-layer (white)
6 arrow must point towards the pump during the test)

2. **Description:** Refer to illustration. "Opening time" (time needed for the release of the compressed bellows, ending when the arrestor chain is fully tensioned): 8 to 16 seconds.

3. **Measuring Range:**   (20°C, 1013 mbar)
   With 20 strokes: 200 to 3000 ppm
   1 mg ethyl acetate per m³ of air = 0.27 ppm
   1 ppm = 3.68 mg/m³   } 20°C, 1013 mbar

4. **Test and Evaluation of the Result**
   With 20 strokes of the pump, draw the air sample through the tube. The length of the brownish-green discolouration corresponds to the concentration of the ester. Scale values = ppm.

5. **Remarks**
   Discolourations last for a limited time only, the lengths of the discolouration vary.

6. **Influence of Ambient Conditions on the Measurement**
   Humidity of the air has no influence. Water vapour is retained in the pre-layer of the tube. The calibrated scale printed on the tube is valid for 20 strokes of the pump at temperatures from 17°C to 40°C. If the ambient temperature is below 17°C, the calibrated scale is valid for the following number of strokes:

Ambient Temperature	Number of Strokes
0 to 9°C	30
10 to 16°C	25
17 to 40°C	20

7. **Specificity (Cross Sensitivity)**
   Indication is based on oxidation of ester with chromosulphuric acid. Colour change: orange to brownish-green. Other esters of acetic acid react similarly, but the sensitivity of indication decreases with increasing molecular weight and increasing concentration (e.g. 1000 ppm butyl acetate give approximately the same indication as 1000 ppm ethyl acetate; with 3000 ppm butyl acetate, the indication corresponds to only about 1800 ppm ethyl acetate). Ketones are also indicated (e.g. 1000 ppm acetone; approximately the same indication as 200 ppm ethyl acetate; methyl ethyl ketone reacts with approximately the same sensitivity of indication as ethyl acetate). Benzene, toluene and petroleum hydrocarbons turn the whole indicating layer pale green even in a concentration of 100 ppm.

8. **Predetermined Period of Use:** For expiration data and storage temperature, see date on package strip. Protect tubes from light.

Caution: Do not allow DRÄGER tubes to fall into the hands of children. Contents are corrosive!

1. **Important information:** This DRÄGER - Tube consists of two individual tubes joined together by combination of a sleeve and shrunk - on tubing. Each individual tube is fused at both ends. Consequently, before use of this DRÄGER - Tube, not only the two outer tips, but also the two tips covered by the sleeve are to be opened (see Section 5.3).

2. **General and Application:** Determination of formaldehyde (HCHO) vapour in air. The tubes are to be used in conjunction with the DRÄGER Gas Detector Pump. For use, see Selection 5 of these Operating Instructions.

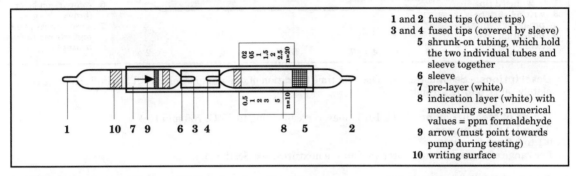

> **1** and **2** fused tips (outer tips)
> **3** and **4** fused tips (covered by sleeve)
> **5** shrunk-on tubing, which hold the two individual tubes and sleeve together
> **6** sleeve
> **7** pre-layer (white)
> **8** indication layer (white) with measuring scale; numerical values = ppm formaldehyde
> **9** arrow (must point towards pump during testing)
> **10** writing surface

3. **Description:** See illustration. Opening time (duration of one pump stroke until the limit chain is completely taut): 5 to 10 seconds.

4. **Measuring Range**
   (20°C, 1013 mbar; corresponding to 20°C, 760 mm Hg)
   with n = 20 strokes: 0.2 to 2.5 ppm HCHO
   with n = 10 strokes: 0.5 to 5 ppm HCHO

5. **Test and Evaluation of the Result**
   5.1 Before each series of measurement, check the pump for leaks with an unopened DRÄGER Tube.
   5.2 Break-off the outer tips (1 and 2) of the DRÄGER Tube.
   5.3 Break-off the tips (3 and 4) covered by the sleeve; to do this, hold the tube firmly at the sleeve and bend first one and then the other of the individual tubes until the tips can clearly be heard to break-off inside the sleeve. The shrunk-on tubing must not tear in the process.
   5.4 When all 4 tube tips are broken-off, insert the DRÄGER Tube tightly in the pump head (arrow points towards the pump).
   5.5 Suck the air sample through the DRÄGER Tube with the prescribed number of strokes. HCHO turns the white indicating layer reddish. The total length of the discoloration is a measure of the HCHO concentration. With concentrations between 0.5 and 5 ppm, the test should be carried out with 10 pump strokes. Evaluation is made on the 10-stroke scale. Numerical values = ppm HCHO. With concentrations between 0.2 and 2.5 ppm, the test should be carried out with 20 pump strokes. Evaluation is made on the 20-stroke scale. Numerical values = ppm HCHO.

6. **Remarks**
   The discolorations persist for few hours if the tubes are sealed with rubber caps. The tubes cannot be used again even after a negative test result.

7. **Influence of Ambient Conditions**
   7.1 Temperature: The DRÄGER Tubes can be used in a temperature range of from 10 to 40°C.
   7.2 Humidity: Between 3 and 15 mg H_2O per litre, humidity has no influence on the indication.

8. **Specificity (Cross Sensitivity)**
   The indication is based on the reaction of formaldehyde with xylene in the presence of sulphuric acid. Colour change: white → pink. Acetaldehyde, acrolein, diesel fuel, furfuryl alcohol and styrene turns the indicating layer yellow to brown.

9. **Shelf life:** For expiration date and storage temperature, see data on package strip.

Caution: Do not allow DRÄGER tubes to fall into the hands of children. Contents are corrosive!

1. General and Application

Measurement of the vapours of hydrocarbons, preferably of liquid gases (propane, butane) in air. The tubes are to be used in conjunction with the DRÄGER Gas Detector pump. For use, see Section 4.

Important: It is not permissible to combine the tubes with pumps made by other manufacturers, since this may cause considerable errors in indication. Such a combination would offend against relevant regulations. For measurement of hydrocarbons of higher boiling range, use DRÄGER tubes Petroleum Hydrocarbons 100/a or Hydrocarbons 2.

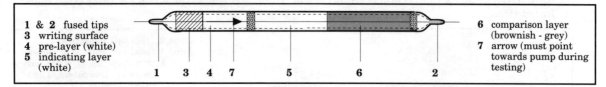

1 & 2 fused tips	6 comparison layer (brownish - grey)
3 writing surface	
4 pre-layer (white)	7 arrow (must point towards pump during testing)
5 indicating layer (white)	

1　3　4　7　　5　　6　　2

2. Descriptions:

See illustration. Opening time (duration of one pump stroke until the limit chain is completely taut): 6 to 12 seconds.

3. Range of Measurement (20°C, 1013 mbar, corresponding to 20°C, 760 mm Hg)

0.5 to 1.3 vol. % propane
0.1 to 0.8 vol. % butane
For range of measurement of propane/butane mixtures, see Section 4.4.

4 Test and Evaluation of the Result

4.1 Before each series of measurement, check the pump for leaks using an unopened tube.

4.2 Break-off the tips of the tube.

4.3 Insert the tube tightly in the pump head.

4.4 Suck the air sample through the tube until the white indicating layer (5) has turned brownish-grey over its whole length and the discoloration corresponds approximately to that of the brownish-grey comparison layer (6). The concentration follows from the table below, from the number of pump strokes necessary for colour comparison:

Number of suction strokes for colour comparison	Propane acc. DIN 51622 Volume %	Butane acc. DIN 51622 Volume %	Liquid gas acc. DIN 51621 Volume %
n = 3	-	0.8	-
n = 4	-	0.5	-
n = 5	-	0.4	0.8
n = 6	-	0.3	0.6
n = 7	1.3	0.25	0.5
n = 8	1.1	0.2	0.4
n = 9	0.9	-	0.3
n = 11	0.8	0.15	0.2
n = 13	0.7	-	0.15
n = 14	0.6	-	-
n = 15	0.5	0.10	-

5. Remarks

The tubes cannot be used again, even after a negative test result. The discoloration lasts for only a short time, since the colour deepens.

6. Influence of Ambient Conditions on the Result of Measurement

6.1 Temperature: The tubes can be used in a temperature range of from 0 to 40°C.

6.2 Humidity: Between 3 and 15mg H_2O per litre, humidity has no influence on the indication.

7. Specificity (Cross-Sensitivity)

The indication is based on the reaction of the hydrocarbon with iodine pentoxide and fuming sulphuric acid. The reagent changes colour from white to brownish-grey. In addition to the liquid gases, hydrocarbons of higher boiling range and unsaturated hydrocarbons, and also carbon monoxide are indicated. (CO turns the indicating layer green.) However, the sensitivity of indication is different.

8. Shelf Life: (See Label of the box.)

Caution: Do not allow used tubes to fall into the hands of children. Contents are corrosive!

1. General and Application: Determination of hydrocyanic acid (HCN) in air, combustible and flue gases, etc. For the use of DRÄGER tubes and Gas Detector, see Section 4 of this instructions.

Important: It is not permissible to combine the tubes with pumps made by other manufacturers, since this may cause considerable errors in indication. Such a combination would offend against relevant regulations.

2. Description: See illustration. Opening time (duration of one pump stroke until the limit chain is fully taut): 7 to 12 seconds.

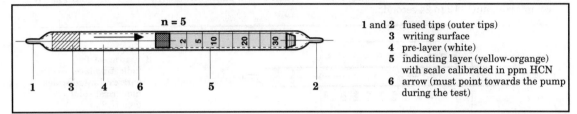

1 and 2	fused tips (outer tips)
3	writing surface
4	pre-layer (white)
5	indicating layer (yellow-organge) with scale calibrated in ppm HCN
6	arrow (must point towards the pump during the test)

3. Range of Measurements
(20°C, 1013 mbar; corresponding to 20°C, 760 mm Hg)
with n = 5 strokes: 2 to   30 ppm HCN
with n = 2 strokes: 5 to   75 ppm HCN
with n = 1 strokes: 10 to 150 ppm HCN

1ppm HCN ≈ 1.13 mg/m^3
1ppm HCN/m^3 ≈ 0.89 ppm   } 20^0 C, 1013 mbar

4. Test and Evaluation of the Result
  4.1 Before each series of measurements, test the pump for tightness using an unopened tube.
  4.2 Break-off the tips of the tube.
  4.3 Insert the tube tightly in the pump head (arrow points towards the pump).
  4.4 Measurement within the range of 2 to 30 ppm HCN.
  Suck the air sample through the tube with 5 strokes. HCN changes the colour of the indicating layer to red. The total length of the discoloration is a measure of the concentration. Numerical values = ppm HCN.
  4.5 Measurement in the range above 30 ppm HCN.
  The test is to be performed with 2 strokes. The indication is evaluated from the entire length of the discoloration. Numerical values multiplied by the factor 2.5 result in ppm HCN. In case of concentration above 75 ppm the test should be performed with 1 stroke. Numerical values mutiplied by the factor 5 result in ppm HCN.

5. Remarks
After a negative test result, the tube can be used up to five times on the same day. After a positive result, the discolorations persists for only a limited time, that evaluation should be carried out immediately after the measurement.

6. Influence of Ambient Conditions
  6.1 Temperature: The tubes can be used in a temperature range of from 0 to 30°C.
  6.2 Humidity: Below 20 mg H$_2$O per litre, humidity has no influence on the indication.
  6.3 Atmospheric pressure: For pressure correction, multiply the tube reading by the following factor:

$$\text{Conversion factor} = \frac{1013}{\text{actual atmospheric pressure (in mbar)}} \quad \text{or}$$

$$\text{Conversion factor} = \frac{760}{\text{actual atmospheric pressure (in Torr)}}$$

7. Specificity
The indication is based on the reaction of hydrocyanic acid with mercury salts, liberating an acid which is which is determined by a colour indicator.

8. Shelf life: For shelf life and storage temperature, see data on package strip.

Caution: Do not allow tubes to fall into the hands of children. Contents are corrosive!

1. General and Application:

Determination of hydrogen sulphide (H_2S) in air and in technical gases. The tubes are to be used in conjunction with the DRÄGER Gas Detector Pump. For use, see Section 4 of these Instructions.
Important information: It is not permissible to combine the tubes with pumps made by other manufacturers, since this may cause considerable errors in indication. Such a combination would offend against relevant regulations.

2. Description:

See illustration. Opening time (duration of one pump stroke until the limit chain is completely taut): 12 to 24 seconds.

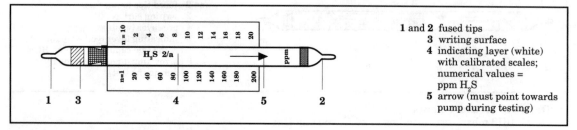

1 and **2**	fused tips
3	writing surface
4	indicating layer (white) with calibrated scales; numerical values = ppm H_2S
5	arrow (must point towards pump during testing)

3. Range of Measurement

(20°C, 1013 mbar; corresponding to 20°C, 760 mm Hg)
With 10 strokes: 2 to 20 ppm H_2S
With 1 stroke: 20 to 200 ppm H_2S
$$\left. \begin{array}{l} 1 \text{ ppm } H_2S \approx 1.42 \text{ mg/m}^3 \\ 1 \text{ mg } H_2S/m^3 \approx 0.71 \text{ ppm} \end{array} \right\} \quad 20°C, 1013 \text{ mbar;}$$

4. Test and Evaluation of the Result

4.1 Before each series of measurements, check the pump for leaks with an unopened DRÄGER tube.
4.2 Break-off the tips of the tube.
4.3 Insert the tube tightly in the pump head (arrow points towards the pump).
4.4 Suck the air sample through the tube with the prescribed number of strikes. H_2S turns the white indicating layer pale brown. The total length of the discoloration is a measure of the H_2S concentration. With concentrations between 2 and 20 ppm, the test should be carried out with 10 pump strokes. Evaluation is made on the 10-stroke scale. Numerical values = ppm H_2S. With concentrations between 20 and 200 ppm, the test should be carried out with 1 pump strokes. Evaluation is made on the 1-stroke scale. Numerical values = ppm H_2S.

5. Remarks

The discolorations last for several days if the tubes are sealed with rubber caps. The tubes can be use up to 10 times on the same day after a negative test result.

6. Influence of Ambient Conditions on the Result of Measurement

6.1 Temperature: The DRÄGER Tubes can be used in a temperature range of from 0 to 40°C.
6.2 Humidity: Between 3 and 30 mg H_2O per litre, humidity has no influence on the indication.
6.3 Atmospheric Pressure: For pressure correction, multiply the tube reading by the following conversion factor:

$$\text{Conversion factor} = \frac{1013}{\text{actual atmospheric pressure (in mbar)}}$$

7. Specificity (Cross Sensitivity)

The indication is based on the reaction of mercury complex, producing pale brown metal sulphide.
No interference with the H_2S indication by up to 200 ppm SO_2, 100 ppm HCl, 100 ppm ethyl mercaptan.

8. Shelf life:

For expiration date and storage temperature, see data on package strip. Store tubes away from light.

Caution: Do not allow DRÄGER tubes to fall into the hands of children. Contents are corrosive!

1. **General and Application:**
   Determination of hydrogen sulphide (H_2S) in air and in technical gases. For use of DRÄGER Tubes and Gas Detector, see Section 4 of these Operating Instructions.
   Important information: It is not permissible to combine the tubes with pumps made by other manufacturers, since this may cause considerable errors in indication. Such a combination would offend against relevant regulations.

2. **Description:** See illustration. Opening time (duration of one pump stroke until the limit chain is completely taut): 15 to 30 seconds.

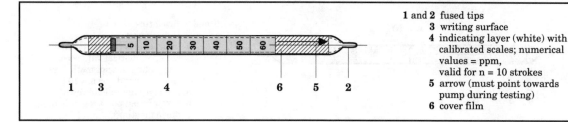

         1 and **2** fused tips
            3 writing surface
            4 indicating layer (white) with calibrated scales; numerical values = ppm, valid for n = 10 strokes
            5 arrow (must point towards pump during testing)
            6 cover film

3. **Range of Measurement**
   (20°C, 1013 mbar; corresponding to 20°C, 760 mm Hg)
   with n = 10 strokes: 5 to 60 ppm H_2S
   with n = 1 stroke:　50 to 600 ppm H_2S
   1 ppm H_2S ~ 1.42 mg/m³　}　20°C, 1013 mbar (760 mm Hg)
   1 mg H_2S/m³ ~ 0.71 ppm

4. **Test and Evaluation of the Result**
   4.1 Before each series of measurements, check the pump for leaks with an unopened DRÄGER Tube.
   4.2 Break-off the tips of the DRÄGER Tube.
   4.3 Insert the DRÄGER Tube tightly in the pump head (arrow points towards the pump).
   4.4 Measurement in the range from 5 to 60 ppm. Suck the air sample through the Tube with 10 pump strokes. The length of the zone turned brown by H_2S is a measure of the H_2S concentration in ppm (numerical values = ppm H_2S).
   4.5 Measurement in the range above 60 ppm. Here the air sample should be sucked through the tube with less than 10 strokes. With 1 stroke, the numerical value read-off should be multiplied by 10. Result = ppm H_2S.

5. **Remarks**
   After a negative test result, the DRÄGER Tube H_2S 5/b can be used again up to five times on the same day. After a positive result, it cannot be used again. Discolorations last for some time, provided that the tips are sealed with rubber caps.

6. **Influence of Ambient Conditions**
   6.1 Temperature: The DRÄGER Tubes can be used in a temperature range of from 0 to 60°C.
   6.2 Humidity: Between 3 and 30 mg H_2O per litre, humidity has no influence on the indication.
   6.3 Atmospheric Pressure: For correction of the atmospheric pressure, the tube indication should be multiplied by the following conversion factor:

$$\text{Conversion factor} = \frac{1013}{\text{actual atmospheric pressure (in mbar)}}$$

7. **Specificity (Cross Sensitivity)**
   The indication is based on the colour reaction with a lead compound. Brown lead sulphide is produced. In the presence of fairly high SO_2 concentrations, the H_2S indication will be a little too high (e.g. mixture of 10 ppm H_2S and 10 ppm SO_2 gives an indication of about 15 ppm H_2S): SO_2 alone does not change the colour of the indicating layer. Other gases and vapours do not affect the H_2S indication.

8. **Shelf life:** For expiration data and storage temperature, see date on package strip.

Caution: Do not allow DRÄGER Tubes to fall into the hands of children. Contents are corrosive!

1. **General and Application:** Determination of methylene chloride CH_2Cl_2 in air. The tubes are to be used together with the DRÄGER Gas Detector Pump. For use, refer to Section 4 of these Instructions.

   Important: It is not permissible to combine the tubes with pumps made by other manufacturers, since this may cause considerable errors in indication. Such a combination would offend against relevant regulations.

2. **Description:** Refer to illustration. "Opening period" (time needed for the release of the compressed bellows, ending when the arrestor chain is fully tensioned): 10 to 20 seconds.

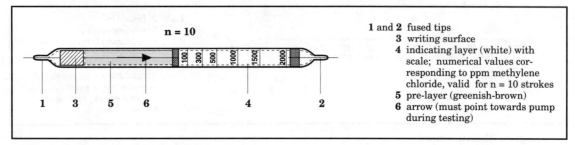

	1 and 2 fused tips
	3 writing surface
	4 indicating layer (white) with scale; numerical values corresponding to ppm methylene chloride, valid for n = 10 strokes
	5 pre-layer (greenish-brown)
	6 arrow (must point towards pump during testing)

3. **Measuring Range:** (20°C, 1013 mbar).
   With 10 strokes: 100 to 2000 ppm methylene chloride.
   1 ppm methylene chloride = 3.53 mg/m^3 } 20°C, 1013 mbar
   1 mg/m^3 = 0.28 ppm

4. **Test and Evaluation of the Result**
   4.1 Before each test, check the pump for leaks with the sealed tube.
   4.2 Break-off the fused tips of the DRÄGER Tube.
   4.3 Insert the DRÄGER Tube tightly into the pump head (arrow points towards the pump).
   4.4 Draw the air sample with 10 strokes through the tube. If methylene chloride is present, the white indicating layer discolours to a brownish-green. The length of the discoloration depends on the concentration. Numerical value = ppm methylene chloride.

5. **Remarks**
   The discolourations may change after some time. Even after a negative result, the tubes cannot be reused.

6. **Influence of Ambient Conditions**
   The tubes can be used in a temperature range of 10 to 30°C. The moisture does not influence the measurement in a range of 3 to 15 mg H_2O/litre.

7. **Specificity (Cross Sensitivity)**
   The indication is based on a decomposition of the methylene chloride, through a chromate compound (pre-layer preparation) and the reaction of the decomposition products with iodine pentoxide, selenium dioxide and fuming sulphuric acid (colour change from white to a brownish-green). Besides methylene chloride, hydrocarbon halides and organic compounds (e.g. alcohols, benzine hydrocarbons) and carbon monoxide will also be indicated.

8. **Shelf life**
   For expiration date and storage temperature, see data on the label of the box.

Caution: Do not allow DRÄGER tubes to fall into the hands of children. Contents are corrosive!

DRÄGER Tube Instructions (Modified)   Natural Gas (Erdgastest)

1. **General and Application:**
    The DRÄGER Natural Gas Test is used for qualitative indication of natural gas. For use of DRÄGER tubes and Gas Detector Pump, refer to Section 5 of these Operating Instructions.
    Important information: It is not permissible to combine the tubes with pumps made by other manufacturers, since this may cause considerable errors in indication. Such a combination would offend against relevant regulations.

2. **Description:**
    The DRÄGER Natural Gas Test consists of two tubes, the pre-tube and the indicating tube. For measurement they are connected together by a short rubber tube (enclosed in the packet). The free end of the indicating tube is inserted in the mouthpiece of the pump with the arrow pointing towards the pump.

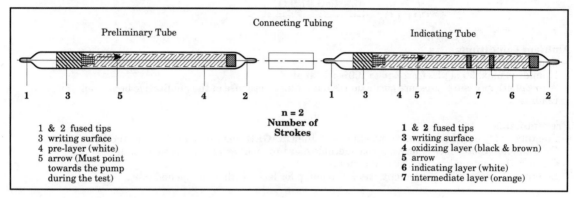

Preliminary Tube	Connecting Tubing	Indicating Tube
1 & 2 fused tips 3 writing surface 4 pre-layer (white) 5 arrow (Must point towards the pump during the test)	n = 2 **Number of Strokes**	1 & 2 fused tips 3 writing surface 4 oxidizing layer (black & brown) 5 arrow 6 indicating layer (white) 7 intermediate layer (orange)

3. **Shelf life:**
    1 year at storage temperatures not exceeding 20°C. The lower the temperature at which the tubes are stored, the smaller the change in their chemical properties.

4. **Sensitivity:**
    Approximately 5 mm of the white indicating layer is discoloured brownish-green to greyish-violet with concentrations of 0.5 % by volume CH_4 (methane)
    or 0.05% by volume C_2H_6 (ethane)
    or 0.05% by volume C_3H_8 (propane)

    The length of the discoloration does not increase linearly with increasing concentration but to a lesser extent. Therefore it is not possible to determine the actual concentration from the length of the discoloration.

5. **Test and Evaluation of the Result**
    With 2 strokes of the pump, draw the air sample through the DRÄGER Natural Gas Test (refer to section 2). Natural gas is present if the white indicating layer (or a part of it) is discolored brownish-green to greyish violet. Direction: Do not value colorations appearing in the pre-layers.

6. **Remarks:** Even after a negative reaction, the DRÄGER Natural Gas Test can not be reused.

7. **Influence of Ambient Conditions on the result of Measurement**
    Temperatures between 0 and + 50°C have no influence on the result. Sensitivity is reduced at lower temperatures. Moisture in the air sample has no effect on the measurement.

8. **Stability of Discoloration**
    Discolorations last for some time providing that the tubes are air-tightly sealed (with rubber caps) after the test. The indicating tube may then be kept as evidence.

9. **Reaction Principle and Specificity**
    Methane is oxidized to carbon monoxide (CO) by fuming sulphuric acid (in the pre-tube) and by potassium permanganate (black-brown oxidizing layer in the indicating tube). The CO developed is indicated by iodine pentoxide + sulphuric acid (white indicating layer); discoloration: white - brownish-green to greyish-violet. Ethane, propane and carbon monoxide are also indicated. Other hydrocarbons (e.g. gasoline, benzene, alcohol) are retained n the pre-tube and have no effect on the indication of natural gas.

Caution: Do not allow tubes to fall into the hands of children. Contents are corrosive!

Application Ranges
Qualitative determination of easily oxidizable substances in air or technical gases, e.g. for a pre-test.

Mode of Operation
The tube contains a white indicating layer. When air or a gas sample are sucked through the tube, the indicating layer changes colour as a function of the substance in question to brown, green or violet in the presence of e.g. ethylene, carbon monoxide, perchloroethylene, benzene or the like. If there is no reading, this does not always indicate that easily oxidizable substances are not present. In the individual case, the use of the DRÄGER Polytest should be qualified by independent methods, particularly when combustible gases and vapours close to the LEL, or toxic substances are suspected.

Principle of Reaction   (example)

$$CO + I_2O_5 \xrightarrow{\quad SeO_2/H_2S_2O_7 \quad} I_2 + CO_2$$

Ambient Conditions
– Temperature: 0 °C to 50 °C
– Humidity: max. 50 mg/L (corresp. to 100% r.h. at 40 °C)
– Atmospheric pressure: pressure correction is not required because of the qualitative behaviour of the reading.

Prerequisites
– The tubes may only be used in conjunction with the DRÄGER Gas Detector Pump or the DRÄGER Quantimeter 1000, since otherwise, considerable measurement errors are liable to occur.
– Observe the Instructions for Use of the pump.
– Before each series of measurement, check the pump for leaks with an unopened tube.

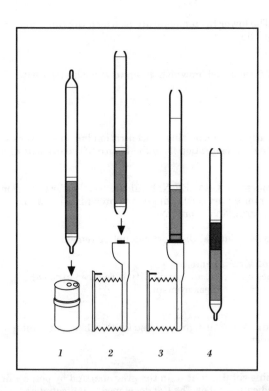

Measurement and Evaluation
1 Break off both tips of the tube in the tube opener.
2 Insert the tube tightly in the pump. The direction is irrelevant.
3 Suck air or gas sample through the tube with 5 strokes. One stroke takes 10 to 20 seconds.
4 Verify the discoloration of the indicating layer. It is not possible to make a quantitative statement with respect to the length of discoloration Should a distinct reading appear after less than 5 strokes, the concentration in question is well above the stated threshold values.
• Behaviour of the reading, examples: each 2000 ppm acetone, 50 ppm ethane, 10 ppm octane, 100 ppm propane, 5 ppm carbon monoxide, 20 ppm perchloroethylene, 2 ppm hydrogen sulfide, 10 ppm acetylene, 1 ppm arsine, 50 ppm benzene, 100 ppm butane, 10 ppm styrene, 1 ppm carbon disulfide, 10 ppm toluene or xylene result in a distinct reading.
    Methane, ethane and carbon dioxide for instance, are not indicated.
• Enter the result in the measurement record. Relative standard deviation: + 50% related to the values of concentration threshold.
• Please pay attention to possible limits of application and cross sensitivities.
• Flush the pump with air after operation.

Cross Sensitivities: A greater number of easily oxidizable compounds (but not all of them) are also indicated due to the principle of reaction.

Disposal
Avoid skin contact with the tube filling. Contents are corrosive. Keep out of the reach of children.

1. **General and Application:** Determination of toluene ($C_6H_5CH_3$) in air. The tubes are to be used in conjunction with the DRÄGER Gas Detector Pump. For use, see Section 4 of these Operating Instructions.

   Important: It is not permissible to combine the tubes with pumps made by other manufacturers, since this may cause considerable errors in indication. Such a combination would offend against relevant regulations.

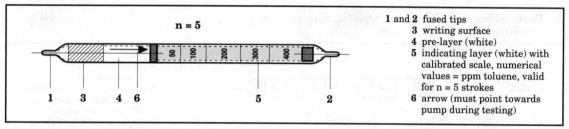

n = 5

1 and 2 fused tips
3 writing surface
4 pre-layer (white)
5 indicating layer (white) with calibrated scale, numerical values = ppm toluene, valid for n = 5 strokes
6 arrow (must point towards pump during testing)

2. **Description:** See illustration. Opening time (duration of one pump stroke until the limit chain is completely taut): 10 to 20 seconds.

3. **Range of Measurement:** (20°C, 1013 mbar, corresponding to 20°C, 760 mm Hg)
   With n = 5 strokes: 5 to 400 ppm toluene (The first scale division corresponds to 50 ppm, smaller concentrations can only be estimated).
   With n = 4 strokes: Up to a maximum of 500 ppm toluene (see Section 4.5)
   1 ppm toluene ≈ 3.84 mg/m³
   1 mg toluene/m³ ≈ 0.26 ppm   }   20°C, 1013 mbar

4. **Test and Evaluation of the Result**
   4.1 Before each series of measurements, test the pump for tightness using an unopened DRÄGER Tube.
   4.2 Break-off the tips of the DRÄGER Tube.
   4.3 Insert the DRÄGER Tube tightly in the pump head (arrow points towards the pump).
   4.4 Suck the air sample through the DRÄGER Tube with 5 pump strokes. If toluene is present, the white indicating layer turns brown. The length of the discoloration depends upon the concentration. Numerical values = ppm toluene.
   4.5 Measurements in the concentration range above 400 ppm. Suck the air sample through the tube with 4 pump strokes. The indication (numerical values on the scale) multiplied by a factor of 1.25 gives ppm toluene.

5. **Remarks**
   The discoloration lasts for a few days if the tubes are sealed with rubber caps. The DRÄGER tube cannot be used again even after a negative result.

6. **Influence of Ambient Conditions**
   6.1 Temperature: The DRÄGER Tubes can be used within a temperature range of from +10°C to +30°C.
   6.2 Humidity: Between 5 and 12 mg H_2O per litre, humidity has no influence on the indication.
   6.3 Atmospheric pressure: For pressure correction, multiply the Tube reading by the following conversion factor:

$$\text{Conversion factor} = \frac{1013}{\text{actual atmospheric pressure (in mbar)}} \quad \text{or}$$

$$\text{Conversion factor} = \frac{760}{\text{actual atmospheric pressure (in Torr)}}$$

7. **Specificity (Cross Sensitivity)**
   The indication is based on the toluene reacting with iodic and sulphuric acid (colour change from white to brown). Petroleum hydrocarbons change the colour of the indicating layer to reddish-brown. Benzene affects the indication only in concentrations above 150 ppm. Benzene changes the colour of the indicating layer to a diffuse yellow. Xylenes are indicated with a lower sensitivity than toluene.

8. **Shelf life:** For expiration date and storage temperature see label on the box.

Caution: Do not allow DRÄGER tubes to fall into the hands of children. Contents are corrosive!

1. **General and Application:** Determination of o-xylene $C_6H_4(CH_3)_2$ (1,2-dimethylbenzene) in air. The tubes are to be used in conjunction with the DRÄGER Gas Detector Pump, model 31. For use, see Section 4 of these Instructions.
   Important: It is not permissible to combine the tubes with pumps made by other manufactures, since this may cause considerable errors in indication. Such a combination would offend against relevant regulations.

2. **Description:** See illustration. Opening time (duration of one pump stroke until the limit chain is completely taut): 6 to 12 seconds.

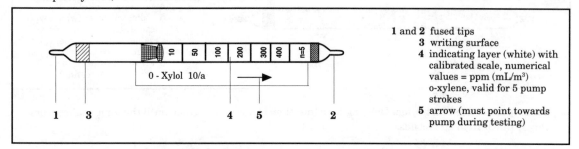

1 and 2 fused tips
3 writing surface
4 indicating layer (white) with calibrated scale, numerical values = ppm (mL/m³) o-xylene, valid for 5 pump strokes
5 arrow (must point towards pump during testing)

3. **Range of Measurement:** (20°C, 1013 mbar, corresponding to 20°C, 760 mm Hg)
   With n = 5 strokes 10 to 400 ppm (mL/m³) o-xylene
   1 ppm (mL/m³) o-xylene ≈ 4.44 mg/m³   }   20°C, 1013 mbar
   1mg o-xylene/m³ ≈ 0.23 ppm

4. **Test and Evaluation of the Result**
   4.1 Before each series of measurements, check the pump for leaks using an unopened DRÄGER Tube.
   4.2 Break-off the tips of the DRÄGER Tube.
   4.3 Insert the DRÄGER Tube tightly in the pump head (arrow points towards the pump).
   4.4 Suck the air sample through the DRÄGER Tube with 5 pump strokes. o-Xylene turns the white indicating layer reddish brown. The total length of the discoloration is a measure of the o-xylene concentration. Numerical values = ppm (mL/m³) o-xylene.

5. **Remarks**
   The DRÄGER Tube cannot be used again even after a negative result. Discoloration lasts for a few days if the Tube is sealed with rubber caps.

6. **Influence of Ambient Conditions on the result of measurement**
   6.1 Temperature: The DRÄGER Tubes can be used within a temperature range of from 0 to 40°C.
   6.2 Humidity: Between 3 and 15 mg H_2O per litre, humidity has no influence on the indication.
   6.3 Atmospheric pressure: For pressure correction, multiply the Tube reading by the following conversion factor:

   $$\text{Conversion factor} = \frac{1013}{\text{actual atmospheric pressure (in mbar)}}$$

7. **Specificity (Cross Sensitivity)**
   The o-xylene indication is based on the colour reaction with formaldehyde and sulphuric acid. m-xylene and p-xylene are indicated with about the same sensitivity as o-xylene. In addition to o-xylene, other organic compounds are indicated.
   Some examples of the sensitivity of indication are:
   100 ppm monostyrene give a reddish brown indication of app. 40
   100 ppm toluene give a dark brown indication of app. 200
   200 ppm ethyl benzene give a brown indication of app. 350
   1,000 ppm butadiene give a brown indication of app. 100.
   No interference with the indication by, for example: 200 ppm methanol, 500 ppm n-octane, 400 ppm ethyl acetate.

8. **Shelf life**
   For expiration date and storage temperature, see data on package strip.

Caution: Do not allow DRÄGER tubes to fall into the hands of children. Contents are corrosive!

1. **WARNING**: *These instructions are applicable for use with the Universal Tester™ Pump or the Samplair® Pump only. When using these tubes with any other MSA® sampling device, use the instructions and calibration values provided with that device. These tubes are not calibrated for use with other than MSA sampling devices.*

2. **Storage**
    Do not store at temperatures above 90°F. Protect from prolonged exposure to light. Either of these conditions could cause a change in the sensitivity and accuracy of the tubes.

3. **Shelf Life**
    If stored properly, these tubes can be used up to 30 months from date of manufacture. They must be used prior to the expiration date stamped on the box.

4. **Chemical Reaction and Color Change**
    This detector tube provides a quantitative method for estimating the concentration of alcohols in air. Indication is based on the reduction of potassium dichromate by the alcohols. The color change is from yellow to green. Furfuryl alcohol generally results in a partly dark brown, partly green stain.

5. **Test Procedure**
    5.1 Check pump for leaks in accordance with the instruction manual for the pump.
    5.2 When using the Universal Tester Pump, set the rotating pump head on #2 index mark. With the Samplair Pump, no indexing is necessary.
    5.3 Remove one detector tube from the box and break off both tips using the tube breaker hole in the head of the pump.
    5.4 Insert the detector tube into the rubber tube holder of the pump, with the arrow on the tube pointing toward the pump.
    5.5 Refer to the calibration chart and decide on the number of pump strokes to be taken. This is dependent on the compound of interest and the expected concentration.
    5.6 Align the index marks on the handle and back plate of the pump.
    5.7 Face the mounted tube into the atmosphere to be tested. Take an appropriate volume of air to be sampled by pulling the pump handle out the required number of strokes. Wait a mininum of 50 seconds for each full pump stroke for the evacuated pump to fill. If alcohol is present, a greenish stain develops in the detector chemical. The stain may have a light end point when first formed, but in a few minutes intensity increases producing a distinct end point. As mentioned, furfuryl alcohol gives a partly brown, partly green stain.
    5.8 To determine the concentration, read the length of stain in millimeters directly from the tube. If the end of stain should be uneven, read at the position of the average length of stain. Then refer to the Calibration Chart. Read opposite the compound of interest and the number of pump strokes used, the concentration in ppm corresponding to the measured length of stain in millimeters.

    NOTE: If the stain length is too short, additional pump strokes may be taken immediately and the concentration read opposite the compound of interest and the total number of pump strokes taken.

6. **Limitations and Corrections**
    6.1 Interferences: Other common gases in the range of their allowable limits do not interfere with accurate measurements, although strong reducing agents could interfere by increasing the length of stain.
    6.2 Temperature: Where required, temperature correction factors are given on the Calibration Chart. In order to avoid correction factors at test conditions colder than 60°F, it is recommended that the tube be kept in an inner pocket prior to use and warmed with the hand during test.
    6.3 Relative Humidity: Moist air causes an increase in stain length. At 80% relative humidity, multiply the stain length by 0.8 and then convert this corrected stain length to concentrations in ppm.
    6.4 Pressure: Calibrations are made at 740 mm mercury. A pressure correction must be used for pressures greatly different from this. To correct for pressure, use the following formula:

    $$\text{corrected reading, ppm} = \text{actual reading, ppm} \times \frac{740}{\text{test pressure in mm mercury}}$$

7. **Measurement Range**
    This tube measures the alcohols listed in the Calibration Chart over the concentration ranges.

8. **Calibration and Accuracy**
    Each lot of tubes is separately calibrated. This instruction sheet should be used only with the designated lot of detector tubes.

CALIBRATION CHART

For the various temperatures, where a correction is required, multiply the test stain length by the correction factor and then *convert this corrected length of stain* to ppm concentration.

Light Figures Below Indicate Length of Stain in Millimeters — **Bold** Figures Indicate Concentration in Parts Per Million

Compound	Number of Strokes										°F Temperature Correction Factors
n-Amyl Alcohol	**5**	0 mm / **0**	3.5 / **50**	6 / **100**	8.5 / **200**	11 / **400**	12.5 / **600**	15 / **1000**	18.5 / **2000**		@ 110° x 0.9
	10	0 mm / **0**	4 / **25**	6 / **50**	9 / **100**	12.5 / **200**	17 / **400**	19.5 / **600**	22 / **1000**	27 / **2000**	
Iso-Amyl Alcohol	**5**	0 mm / **0**	5 / **50**	7 / **100**	9 / **200**	11 / **300**	12.5 / **400**	14 / **500**	17 / **750**	18.5 / **1000**	
	10	0 mm / **0**	7.5 / **50**	10 / **100**	14 / **200**	18 / **300**	21 / **400**	24.5 / **500**	29.5 / **750**	34 / **1000**	
sec-Amyl Alcohol	**5**	0 mm / **0**	3.5 / **25**	5 / **50**	7 / **100**	9 / **200**	13.5 / **400**	17 / **600**	23 / **1000**	30 / **2000**	
	10	0 mm / **0**	5 / **25**	7 / **50**	10 / **100**	14 / **200**	18 / **300**	21 / **400**	27 / **600**	37 / **1000**	
tert-Amyl Alcohol	**5**	0 mm / **0**	3.5 / **50**	6 / **100**	9 / **200**	12 / **300**	13.5 / **400**	18 / **600**	24.5 / **1000**	38 / **2000**	
	10	0 mm / **0**	5 / **25**	7 / **50**	10 / **100**	14 / **200**	18.5 / **300**	22 / **400**	27 / **600**	38 / **1000**	
2-Butoxy Ethanol (Butyl Cellosolve)	**10**	0 mm / **0**	3.5 / **30**	4 / **50**	5 / **100**	6 / **175**	7.5 / **375**	9 / **650**	10 / **900**		
n-Butyl Alcohol	**3**	0 mm / **0**	4 / **50**	5 / **100**	7.5 / **200**	10 / **400**	12.5 / **600**	15 / **1000**	20 / **2000**	27 / **4000**	@ 110° x 0.9
Iso-Butyl Alcohol	**3**	0 mm / **0**	4 / **50**	6 / **100**	8.5 / **200**	12 / **400**	14 / **600**	18.5 / **1000**	26 / **2000**	34 / **4000**	@ 110° x 0.8
sec-Butyl Alcohol	**3**	0 mm / **0**	3.5 / **50**	5 / **100**	7.5 / **200**	12 / **400**	15 / **700**	18.5 / **1000**	25 / **2000**	34 / **4000**	
tert-Butyl Alcohol	**10**	0 mm / **0**	12 / **100**	18 / **200**	22 / **300**	25 / **400**	31 / **600**	37 / **800**	42 / **1000**		@ 110° x 0.9
Cyclohexanol	**10**	0 mm / **0**	3.5 / **25**	5 / **50**	7 / **100**	9 / **200**	12 / **300**	14 / **500**	17 / **750**	18.5 / **1000**	@ 110° x 0.9
2-Ethoxyethanol (Cellosolve)	**10**	0 mm / **0**	1 / **50**	2.5 / **100**	4 / **200**	5 / **400**	6 / **700**	6.5 / **1000**			@ 110° x 0.8
Ethyl Alcohol	**3**	0 mm / **0**	8 / **200**	10.5 / **500**	13 / **1000**	17 / **2000**	22.5 / **5000**	28.5 / **10,000**			@ 110° x 0.9
Furfuryl Alcohol	**10**	0 mm / **0**	3.5 / **25**	4 / **50**	6 / **100**	7.5 / **200**	8.5 / **300**	10 / **400**	11 / **500**		
Methyl Alcohol	**3**	0 mm / **0**	4 / **100**	7.5 / **200**	10 / **400**	13.5 / **1000**	18 / **2000**	25 / **5000**	34 / **10,000**		@ 110° x 0.8
2-Methyl Cyclohexanol	**5**	0 mm / **0**	2 / **25**	3 / **50**	5 / **100**	7.5 / **200**	11 / **400**	12.5 / **600**	13.5 / **800**		
	10	0 mm / **0**	2.5 / **25**	5.5 / **50**	9 / **100**	13.5 / **200**	18 / **400**	19.5 / **600**	20 / **800**		
Methyl Isobutyl Carbinol (methyl amyl alcohol)	**5**	0 mm / **0**	3.5 / **25**	4 / **50**	7 / **100**	8.5 / **200**	10 / **300**	13.5 / **500**	18 / **700**	22 / **1000**	
	10	0 mm / **0**	5 / **25**	7 / **50**	10 / **100**	13.5 / **200**	17 / **300**	23.5 / **500**	30 / **700**	40 / **1000**	
n-Propyl Alcohol	**3**	0 mm / **0**	6 / **100**	9.5 / **200**	12 / **400**	15 / **700**	17 / **1000**	21 / **2000**	30 / **5000**	37 / **10,000**	@ 110° x 0.9
Iso-Propyl Alcohol	**3**	0 mm / **0**	6 / **100**	7 / **200**	10 / **400**	13.5 / **700**	15 / **1000**	20 / **2000**	30 / **5000**	36 / **10,000**	

1. **WARNING**: *These instructions are applicable for use with the Universal Tester™ Pump or the Samplair® Pump only. When using these tubes with any other MSA® sampling device, use the instructions and calibration values provided with that device. These tubes are not calibrated for use with other than MSA sampling devices.*

2. **Storage**
   Do not store at temperatures above 90°F. Protect from prolonged exposure to light. Either of these conditions could cause a change in the sensitivity and accuracy of the tubes.

3. **Shelf Life**
   If stored properly, these tubes can be used up to 30 months from date of manufacture. They must be used prior to the expiration date stamped on the box.

4. **Chemical Reaction and Color Change**
   This detector tube provides a quantitative method for estimating the concentration of ammonia and organic amines in air. Indication is based on the reaction of ammonia or organic amines with Bromophenol blue causing the indicator to change color. The color change is from orange to blue.

5. **Test Procedure**
   5.1 Check pump for leaks in accordance with the instruction manual for the pump.
   5.2 When using the Universal Tester Pump, set the rotating pump head on #2 index mark. With the Samplair Pump, no indexing is necessary.
   5.3 Remove one detector tube from the box and break off both tips using the tube breaker hole in the head of the pump.
   5.4 Insert the detector tube into the rubber holder of the pump, with the arrow on the tube pointing toward the pump.
   5.5 Refer to the calibration chart and decide on the number of pump strokes to be taken. This is dependent on the compound of interest and the expected concentration.
   5.6 Align the index marks on the handle and back plate of the pump.
   5.7 Face the mounted tube into the atmosphere to be tested. Take an appropriate volume of air to be sampled by pulling the pump handle out the required number of strokes. Wait a minimum of 100 seconds for each full pump stroke, 50 seconds for each 1/2 pump stroke, 25 seconds after each 1/4 pump stroke for the evacuated pump to fill. If ammonia or an organic amine is present, a blue stain developes in the detector chemical.
   5.8 To determine the concentration, read the length of stain in millimeters directly from the tube. If the end of stain should be uneven, read at the position of the average length of stain. Then refer to the Calibration Chart. Read opposite the compound of interest and the number of pump strokes used the concentration in ppm corresponding to the measured length of stain in millimeters.

   NOTE: If the stain length is too short, additional pump strokes may be taken immediately and the concentration read opposite the compound of interest and the total number of pump strokes taken.

6. **Limitations and Corrections**
   6.1 Interferences: Since indication is based on an acid-base reaction, acid gases may interfere by decreasing stain length and other basic gases would increase stain length.
   6.2 Temperature: Accuracy of indication is not affected by temperatures between 40°F and 120°F. Test conditions colder than 40°F. will result in an indication of ammonia or organic amines below that actually present, therefore, the tube must be kept at a temperature between 40°F. and 120°F. as by keeping it in an inner pocket before use and keeping it warm with the hand during the test. Do not use at temperatures above 120°F.
   6.3 Relative Humidity: Accuracy of indication is not affected by relative humidity between 20% and 90%. Relative humidity greater than 90% will result in an indication of ammonia or organic amines below that actually present.
   6.4 Pressure: Calibrations are made at 740 mm mercury. A pressure correction must be used for pressures greatly different from this. To correct for pressure, use the following formula:

   $$\text{corrected reading, ppm} = \text{actual reading, ppm} \times \frac{740}{\text{test pressure in mm mercury}}$$

7. **Measurement Range**
   This tube measures the compounds listed in the Calibration Chart over the concentration ranges.

8. **Calibration**
   Each lot of tubes is separately calibrated. This instruction sheet should be used only with the designated lot of detector tubes.

CALIBRATION CHART

For the various temperatures, where a correction is required, multiply the test stain length by the correction factor and then *convert this corrected length of stain* to ppm concentration.

Light Figures Below Indicate Length of Stain in Millimeters · **Bold** Figures Indicate Concentration in Parts Per Million

Each cell is shown as *length in mm* / **concentration in ppm**.

Compound	Number of Strokes										
Ammonia	1/2	0 / 0	7 / 25	12 / 50	16 / 75	19 / 100	24.5 / 150	29 / 200	37.5 / 300	45 / 400	
Ammonia	1	0 / 0	8.5 / 20	15 / 40	21 / 60	30.5 / 100	42 / 150	52 / 200			
Ammonia	2	0 / 0	7.5 / 10	17.5 / 25	32.5 / 50	47 / 75	60 / 100				
n-Butylamine	1	0 / 0	3 / 3	4 / 5	9 / 10	21 / 25	36 / 50	60 / 100			
n-Butylamine	3	0 / 0	6 / 2	8 / 3	12 / 5	21 / 10	29 / 15	43 / 25	60 / 40		
Cyclohexylamine	1/2	0 / 0	5 / 10	10 / 25	13.5 / 35	17.5 / 50	23 / 75	28 / 100			
Cyclohexylamine	2	0 / 0	7 / 5	13 / 10	25 / 20	36 / 30	47 / 40	57.5 / 50			
Diethylamine	1/2	0 / 0	10 / 20	18 / 40	31 / 75	38.5 / 100	51 / 140	57 / 160	68.5 / 200		
Diethylamine	1	0 / 0	9.5 / 10	17 / 20	35 / 50	48 / 75	60 / 100	69 / 120			
Dimethylamine	1	0 / 0	10.5 / 10	21 / 25	26 / 35	33 / 50	43.5 / 75	53 / 100			
Dimethylamine	3	0 / 0	14 / 5	24 / 10	32 / 15	47 / 25	54 / 30				
Di-n-Propylamine	1	0 / 0	7 / 10	17 / 25	23 / 35	32 / 50	46 / 75	59 / 100			
Di-n-Propylamine	2	0 / 0	6 / 5	12 / 10	17 / 15	27.5 / 25	37.5 / 35	51.5 / 50	65 / 65		
Ethylamine	1/2	0 / 0	4.5 / 15	8 / 30	15 / 60	23 / 100	32 / 150	41 / 200	49 / 250	57.5 / 300	
Ethylamine	1	0 / 0	5 / 10	10 / 20	13.5 / 30	25 / 60	39 / 100	55 / 150	65 / 180		
Ethylene imine	1	0 / 0	1 / 2	2 / 5	4 / 10	7 / 20	14 / 50	20 / 75	26 / 100		
Ethylene imine	4	0 / 0	5 / 2	8 / 5	12 / 10	17 / 15	21 / 20	25 / 25	33 / 35	43 / 50	
n-Ethylmorpholine	1/4	0 / 0	7.5 / 10	14 / 20	19.5 / 30	29 / 50	40 / 75	50 / 100			
n-Ethylmorpholine	1	0 / 0	13 / 5	25 / 10	35 / 15	45 / 20	64 / 30				
Isopropylamine	1	0 / 0	8.5 / 10	19 / 25	26 / 35	35 / 50	50 / 75	64 / 100			
Isopropylamine	3	0 / 0	11.5 / 5	21 / 10	29 / 15	45 / 25	66 / 40				
Methylamine	1	0 / 0	6.5 / 10	15 / 25	20 / 35	27 / 50	38.5 / 75	50 / 100			
Methylamine	2	0 / 0	6 / 5	11.5 / 10	26 / 25	31 / 30	49 / 50	66 / 70			
Propylene imine	1/4	0 / 0	8 / 25	14 / 50	19 / 75	24 / 100	34 / 150	44 / 200	54 / 250	64 / 300	
Propylene imine	1	0 / 0	2 / 2	5 / 5	9 / 10	13 / 15	17 / 20	20 / 25	36 / 50	51 / 75	66 / 100
Triethylamine	1/2	0 / 0	5 / 10	12 / 25	22 / 50	31 / 75	40 / 100				
Triethylamine	1	0 / 0	6 / 5	11 / 10	23 / 25	40 / 50	55 / 75	69 / 100			
Trimethylamine	1/4	0 / 0	3.5 / 10	6.5 / 20	10 / 35	13.5 / 50	22 / 100	28.5 / 150	34 / 200		
Trimethylamine	1	0 / 0	6 / 5	10 / 10	17 / 20	28.5 / 40	38 / 60	47 / 80	55 / 100		

1. **WARNING**: *These instructions are applicable for use with the Universal Tester™ Pump or the Samplair®
Pump only. When using these tubes with any other MSA® sampling device, use the instructions and
calibration values provided with that device. These tubes are not calibrated for use with other than MSA
sampling devices.*

2. **Storage**
Do not store at temperatures above 90°F. Protect from prolonged exposure to light. Either of these conditions could cause a change in the sensitivity and accuracy of the tubes.

3. **Shelf Life**
If stored properly, these tubes can be used up to 30 months from date of manufacture. They must be used
prior to the expiration date stamped on the box.

4. **Chemical Reaction and Color Change**
This detector tube provides a quantitative method for estimating the concentration in air of the aromatic
hydrocarbons listed in the Calibration Chart. Indication is based on the reaction of these aromatic
hydrocarbons with iodine pentoxide and sulfuric acid. The color change is from white to brown.

5. **Test Procedure**
   5.1 Check pump for leaks in accordance with the instruction manual for the pump.
   5.2 When using the Universal Tester Pump, set the rotating pump head on #2 index mark. With the
   Samplair Pump, no indexing is necessary.
   5.3 Remove one detector tube from the box and break off both tips using the tube breaker hole in the
   head of the pump.
   5.4 Insert the detector tube into the rubber tube holder of the pump, with the arrow on the tube pointing
   toward the pump.
   5.5 Refer to the calibration chart and decide on the number of pump strokes to be taken. This is dependent on the compound of interest and the expected concentration.
   5.6 Align the index marks on the handle and back plate of the pump.
   5.7 Face the mounted tube into the atmosphere to be tested. Take an appropriate volume of air to be
   sampled by pulling the pump handle out the required number of strokes. Wait a minimum of 40
   seconds for each full pump stroke for the evacuated pump to fill. If aromatic hydrocarbons are
   present, a brownish stain develops in the detector chemical.
   5.8 To determine the concentration, read the length of stain in millimeters directly from the tube. If the
   end of stain should be uneven, read at the position of the average length of stain. Then refer to the
   Calibration Chart. Read opposite the compound of interest and the number of pump strokes used the
   concentration in ppm corresponding to the measured length of stain in millimeters.

   NOTE: If the stain length is too short, additional pump strokes may be taken immediately and the
   concentration read opposite the compound of interest and the total number of pump strokes
   taken.
   NOTE: Moist air (80% R.H.) causes a slight fading (1 to 2 mm per pump stroke) at the beginning of the
   stain; therefore, when reading the stain length, the moisture faded area must be included in the
   length of stain.

6. **Limitations and Corrections**
   6.1 Interferences: Some aliphatic hydrocarbons and other aromatic hydrocarbons may interfere with
   accurate measurements. Carbon monoxide in high concentration (above 200 ppm) forms a light stain
   which disappears in a few minutes.
   6.2 Temperature: Where required, use temperature correction factors given on the Calibration Chart.
   At test conditions below 32°F., the tube should be kept at a temperature where no correction is
   required, as by keeping it in an inner pocket before use and keeping it warm with the hand during
   test. The detector tubes should not be used at temperatures above 120°F.
   6.3 Relative Humidity: The note at the end of paragraph 5.8 indicates that moisture at 80% relative
   humidity or higher causes a fading at the beginning of stain. This faded section must be included in
   the overall stain length otherwise relative humidity does not affect this tube.
   6.4 Pressure: Calibrations are made at 740 mm mercury. A pressure correction must be used for
   pressures greatly different from this. To correct for pressure, use the following formula:

   $$\text{corrected reading, ppm} = \text{actual reading, ppm} \times \frac{740}{\text{test pressure in mm mercury}}$$

   6.5 Gasoline: As gasoline is a mixture, response will vary with manufacture grade geographic area and
   season.

7. **Measurement Range:** This tube measures the compounds listed in the Calibration Chart over the
concentration ranges.

8. **Calibration:** Each lot of tubes is separately calibrated. This instruction sheet should be used only with
the designated lot of tubes.

CALIBRATION CHART

For the various temperatures, where a correction is required, multiply the test stain length by the correction factor and then **convert this corrected length of stain** to ppm concentration.

Light Figures Below Indicate Length of Stain in Millimeters. **Bold** Figures Indicate Concentration in Parts Per Million.

Detectable Compound	Number of Strokes									°F Temperature	Correction Factors
Benzene (benzol)	**1**	0 mm **0**	10.0 **25**	14.0 **50**	16.0 **75**	18.0 **100**	21.0 **150**	23.0 **200**		32-40	1.3
	3	0 mm **0**	11.0 **5**	13.5 **10**	16.0 **25**	21.0 **50**	27.0 **100**	32.0 **150**	35.0 **200**	40-50 / 50-110 / 110-120	1.2 / None / 1.2
Bromobenzene (monobromobenzene)	**1**	0 mm **0**	6.0 **40**	9.0 **75**	12.0 **100**	15.0 **150**	18.0 **200**	27.0 **400**	39.0 **800**	32-40	1.3
	3	0 mm **0**	8.0 **10**	14.0 **50**	18.0 **75**	22.0 **100**	31.0 **200**	44.0 **400**	55.0 **600**	40-50 / 50-120	1.2 / None
Chlorobenzene (monochlorobenzene)	**1**	0 mm **0**	6.0 **40**	9.0 **75**	12.0 **100**	15.0 **150**	18.0 **200**	27.0 **400**	39.0 **800**	32-40	1.3
	3	0 mm **0**	8.0 **10**	14.0 **50**	18.0 **75**	22.0 **100**	31.0 **200**	44.0 **400**	55.0 **600**	40-50 / 50-120	1.2 / None
Gasoline	**1**	0 mm **0**	8.0 **100**	13.0 **200**	17.0 **300**	20.0 **400**	23.0 **500**	31.0 **750**	37.0 **1000**	See Section 6.5	
	3	0 mm **0**	4.0 **10**	8.0 **25**	10.0 **50**	17.0 **100**	26.0 **200**	33.0 **300**	40.0 **400**		
Toluene (toluol)	**1**	0 mm **0**	6.0 **25**	9.0 **50**	13.0 **100**	19.0 **200**	24.0 **300**	28.0 **400**	36.0 **800**	32-40 / 40-100	1.1 / None
	3	0 mm **0**	8.0 **10**	13.0 **25**	17.0 **50**	23.0 **100**	31.0 **200**	42.0 **400**	57.0 **800**	100-110 / 110-120	1.2 / 1.3
Xylene (xylol)	**1**	0 mm **0**	5.0 **25**	6.0 **50**	9.0 **100**	13.0 **200**	15.0 **300**	18.0 **400**	24.0 **800**	32-40	0.9
	3	0 mm **0**	5.0 **10**	9.0 **25**	12.0 **50**	14.0 **100**	19.0 **200**	27.0 **400**	37.0 **800**	40-120	None

* READ AND UNDERSTAND THESE INSTRUCTIONS BEFORE USE.

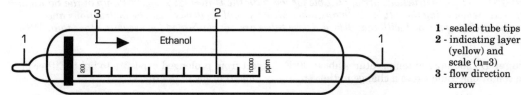

1 - sealed tube tips
2 - indicating layer
   (yellow) and
   scale (n=3)
3 - flow direction
   arrow

1. **Sampling Devices:** Ethanol Detector Tubes are calibrated for use with the following MSA devices. The Kwik-Draw® Pump or the Samplair® Pump. **WARNING:** *Do Not Use These Detector Tubes with any other Sampling Devices. Use with other devices may result in improper readings.*

2. **Storage:** Do not store at temperatures above 90°F (32°C). Protect from prolonged exposure to light. Either of these conditions could cause a change in the sensitivity and accuracy of the tubes.

3. **Shelf Life:** These detector tubes must not be used after the end of the month stamped on the box.

4. **Chemical Reaction and Color Change:** This detector tube quantitatively estimates the concentration of ethanol in air. Indication is based on the reduction of potassium dichromate by ethanol. The color change is from yellow to green.

5. **Test Procedure**
    5.1 Check pump for leaks in accordance with the instruction manual for the pump.
    5.2 When using the Kwik-Draw Deluxe Pump or the Kwik-Draw Pump, set the stroke counter to zero.
    5.3 When using the Samplair Pump, align the index marks on the handle with those located on the backplate of the pump.
    5.4 Remove one detector tube from the box and break off tips using the tube breaker hole in the pump.
    5.5 Insert the detector tube into the tube holder of the pump, with the direction-of-flow arrow pointing toward the pump.
    5.6 Place the mounted detector tube into the atmosphere to be tested and take the appropriate number of pump strokes. For 200 to 10,000 ppm use 3 pump strokes and read the scale directly.
    NOTE:　• When using the Kwik-Draw Deluxe Pump, wait until the end-of-stroke indicator changes.
    　　　　　• When using the Kwik-Draw Pump, wait until the bellows inflates to the stop plus an additional 10 seconds.
    　　　　　• When using the Samplair wait a minimum of 100 seconds after each full pump stroke for the pump to fill.
    5.7 To determine the concentration, read to the end of the stain and note the scale reading. If the end of the stain is uneven, use the average length of stain.

6. **Limitations and Corrections**
    6.1 Interferences: This tube is calibrated using ethanol. Strong reducing agents may interfere by increasing stain length. Some other alcohols will also indicate.
    　　　　　　200 ppm methyl alcohol indicates 200 ppm ethanol.
    　　　　　　200 ppm n-propyl alcohol indicates 200 ppm ethanol.
    　　　　　　400 ppm isopropyl alcohol indicates 400 ppm ethanol.
    Several other alcohols may react but with less sensitivity at their Threshold Limit Values.
    The following give an indication of less than 200 ppm ethanol

100 ppm n-amyl alcohol	100 ppm iso-amyl alcohol
100 ppm sec-amyl alcohol	100 ppm tert-amyl alcohol
25 ppm 2-butoxy ethanol	50 ppm n-butyl alcohol
50 ppm iso-butyl alcohol	100 ppm sec-butyl alcohol
100 ppm tert-butyl alcohol	50 ppm cyclohexanol
5 ppm 2-ethoxyethanol	10 ppm furfuryl acohol
50 ppm 2-methyl cyclohexanol	25 ppm methyl isobutyl carbinol

    6.2 Temperature: Accuracy of indication is not affected by temperatures between 60°F (15°C.) and 100°F (38°C.). Test conditions colder than 60°F. will result in an indication of ethanol below that actually present; therefore, the tube should be kept above 60°F. by keeping it in an inner pocket before use and warming it with the hand during test. Do not use at temperatures over 100°F.
    6.3 Relative Humidity: Moist air causes an increase in stain length. At 80% relative humidity, multiply reading by 0.8.
    6.4 Pressure: Calibrations are made at 740 mm of mercury (986 mbar). A correction must be used for pressures greatly different from this. To correct for pressure, multiply the apparent ppm reading by the compensation factor F:

$$F = \frac{740}{\text{atm. press. (mm Hg)}} = \frac{986}{\text{atm. press. (mbar)}}$$

7. **Measurement Range and Calibration:** This tube measures and calibrates ethanol over the range from 200 to 10,000 ppm. Each lot of tubes is separately calibrated using ethanol.

1. **WARNING**: *These instructions are applicable for use with the Universal Tester™ Pump or the Samplair® Pump only. When using these tubes with any other MSA® sampling device, use the instructions and calibration values provided with that device. These tubes are not calibrated for use with other than MSA sampling devices.*

2. **Storage:** Do not store at temperatures above 90°F. Protect from prolonged exposure to light. Either of these conditions could cause a change in the sensitivity and accuracy of the tubes.

3. **Shelf Life:** If stored properly, these tubes can be used up to 30 months from date of manufacture. They must be used prior to the expiration date stamped on the box.

4. **Chemical Reaction and Color Change**
   This detector tube provides a quantitative method for estimating the concentration of certain halogenated hydrocarbons, listed in the calibration chart, in air. Indication is based on (1) the halogenated hydrocarbon reacting with potassium permanganate and sulfuric acid the reactor tube to liberate the halogen (chlorine or bromine); and (2) the color reaction of the halogen with tetraphenylbenzidine to give a quinoidimonium halide salt in the detector tube. The color change is white to blue.

5. **Test Procedure**
   5.1 Check pump for leaks in accordance with the instruction manual for the pump.
   5.2 When using the Universal Tester Pump, set the rotating pump head on the #2 index mark. With the Samplair Pump, no indexing is necessary.
   5.3 Remove from the box: 1 detector tube, 1 reactor tube and 1 charcoal tube. Break off both tips of the detector tube and the charcoal tube using the tube breaker hole in the pump.
   NOTE: The charcoal tube is reusable for up to 12 tests.
   5.4 Insert the charcoal tube into the reactor tube. The end of the charcoal tube with the fiberglass plug (white) goes into the reduced diameter end of the reactor tube. Using a gentle rocking motion securely seat the charcoal tube in the reactor tube.
   5.5 Completely crush each of the two glass ampoules in the reactor tube by placing clamp, Pt. No. 58238, around the plastic tubing and squeezing the jaws of the pinch clamp. It is necessary to completely crush the glass ampoules in order to get good mixing and so that all of the reactants drop to the bottom of the tube in part 5.6. Remove clamp and, putting your index finger over the filter (white) end of the reactor tube, shake vigorously the contents of the reactor tube for $1^1/_2$ minutes in order to thoroughly mix the reagents.
   WARNING: Avoid contact with contents of the reactor tube as they are highly corrosive!
   NOTE: Once the ampoules are crushed and mixed, the remainder of the test should be conducted as quickly as possible.
   5.6 Hold the reactor tube and attached charcoal tube in a vertical position, charcoal tube down, and tap the mixed reagents to the bottom of the reactor tube.
   5.7 Again holding the tubes vertically with the charcoal tube down, place the pinch clamp around the tubing at the top of the mixed reagents in the reactor tube such that it restricts the reagents in that portion of the tube, but does not seal off air flow through the tubes.
   5.8 Insert the free end of the charcoal tube into the tube holder on the pump.
   5.9 Align the index marks on the handle and back plate of the pump.
   5.10 Facing the tube into the atmosphere to be tested, take two pump strokes to flush the system in the reactor tube. Wait 60 seconds after each full pump stroke to allow the evacuated pump to fill.
   NOTE: After the second pump stroke do not push in the pump handle until the tube assembly is removed. This procedure prevents forcing chlorine, if present, from the reactor tube.
   5.11 After waiting 60 seconds after the second pump stroke, remove the charcoal and reactor tube from the pump, and then push in the pump handle.
   5.12 Replace the charcoal tube with the detector tube. Be sure the arrow on the detector tube points toward the pump, i.e. in the direction of air flow.
   5.13 With the reactor tube, detector tube and pump assembled together, face the free end of the reactor tube into the atmosphere to be sampled. Take an appropriate volume of air to be sampled by pulling the pump handle out the required number of strokes. Wait a minimum of 40 seconds for each pump stroke and 20 seconds for each $1/2$ pump stroke for the evacuated pump to fill. If one of the halogenated hydrocarbons listed in the table is present, a bluish stain develops in the detector chemical.
   NOTE: A short (about 2 mm) brown or bluish stain may appear when sampling uncontaminated air samples due to the reagents used in the reactor tube; however, this has no effect on the overall performance of the detector tube.
   5.14 To determine the concentration, read the length of stain in millimeters directly from the tube. If the end of stain should be uneven, read at the position of average length of stain. Then refer to the Calibration Chart. Read opposite the compound of interest and the number of pump strokes used the concentration in ppm corresponding to the measured length of stain in millimeters.
   NOTE: If the stain length is too short, additional pump strokes may be taken immediately and the concentraion read opposite the compound of interest and the **total** number of pump strokes taken.

6. Limitations and Corrections

6.1 Interferences: Other halides and halogens in high concentrations may interfere with accurate measurements.

6.2 Temperature: Where required, temperature correction factors are given on the Calibration Chart. Multipy the test stain length by the correction factor and then convert this corrected length of stain to ppm concentration. At test conditions colder than 40°F., it is recommended that the tube be kept in an inner pocket before use and warmed with the hand during test.

6.3 Relative Humidity: Calibration of these tubes is based on air at 50% relative humidity at 75°F. Relative humidity between 30 and 80% have little effect on the performance of this tube, but at relative humidities outside of this range the tube will read slightly low (6-8%).

6.4 Pressure: Calibrations are made at 740 mm mercury. A pressure correction must be used for pressures greatly different from this. To correct for pressures use the following formula:

$$\text{corrected reading, ppm} = \text{actual reading, ppm} \times \frac{740}{\text{test pressure in mm mercury}}$$

7. Measurement Range:
This tube measures the halogenated hydrocarbons listed in the Calibration Chart over the ranges, indicated in the chart.

8. Calibration:
Each lot of tubes is separately calibrated. This instruction sheet should be used only with indicated lot numbers of detector and reactor tubes.

CALIBRATION CHART

For the various temperatures, where a correction is required, multiply the test stain length by the correction factor and then *convert this corrected length of stain* to ppm concentration.

Light Figures Below Indicate Length of Stain in Millimeters. **Bold** Figures Indicate Concentration in Parts Per Million

"Group B" Halogenated Hydrocarbons	Number of Strokes								°F Temperature Correction Factors
Chlorobenzene (MONO)	1½	2 mm / **0**	12.5 mm / **30**	18 mm / **50**	28 mm / **75**	30 mm / **100**	39 mm / **150**	44 mm / **200**	@ 100° x 0.7
Chlorobromomethane	1	2 mm / **0**	7.5 mm / **50**	11 mm / **100**	20 mm / **200**	29 mm / **300**	38 mm / **400**		
1,2-Dibromoethane	1	2 mm / **0**	6 mm / **5**	10 mm / **10**	18 mm / **25**	26 mm / **50**	32 mm / **75**	35 mm / **100**	@ 40° x 1.5 @ 50° x 1.2
Dichlorobenzene (ORTHO)	1	2 mm / **0**	4 mm / **25**	7.5 mm / **50**	10 mm / **75**	14 mm / **100**	23 mm / **150**	30 mm / **200**	
	2	2 mm / **0**	6 mm / **25**	14 mm / **50**	24 mm / **75**	30 mm / **100**	43 mm / **150**		
1,1-Dichloroethane	½	2 mm / **0**	5 mm / **25**	7.5 mm / **50**	12.5 mm / **100**	20 mm / **200**	24 mm / **300**	26.5 mm / **400**	@ 40° x 1.5 @ 50° x 1.2
	1	2 mm / **0**	7.5 mm / **25**	15 mm / **50**	25 mm / **100**	40 mm / **200**			
1,2-Dichloroethane	1	2 mm / **0**	5 mm / **25**	10 mm / **50**	18 mm / **100**	30 mm / **200**	39 mm / **300**	48 mm / **400**	@ 40° x 1.3 @ 50° x 1.2
Dichloroethyl Ether	2	2 mm / **0**	5 mm / **5**	6 mm / **10**	7.5 mm / **15**	10 mm / **25**	23 mm / **50**	38 mm / **75**	
	3	2 mm / **0**	6 mm / **5**	9 mm / **10**	11 mm / **15**	18 mm / **25**	36.5 mm / **50**		
Dichloroethylene CIS.-1,2)	¼	2 mm / **0**	10 mm / **25**	15 mm / **50**	21.5 mm / **100**	33 mm / **200**	43 mm / **250**	50.5 mm / **300**	
	½	2 mm / **0**	15 mm / **10**	23 mm / **25**	31.5 mm / **50**	41.5 mm / **100**	47 mm / **150**	50.5 mm / **200**	
Dichloroethylene (TRANS.-1,2)	½	2 mm / **0**	5 mm / **50**	7.5 mm / **100**	14 mm / **200**	24 mm / **300**	33 mm / **400**		@ 40° x 1.2
	1	2 mm / **0**	9 mm / **50**	16 mm / **100**	24 mm / **150**	33 mm / **200**	40 mm / **250**		
Ethyl Bromide	1	2 mm / **0**	10 mm / **50**	15 mm / **100**	20 mm / **200**	25 mm / **400**			@ 40° x 1.3
	2	2 mm / **0**	15 mm / **25**	24 mm / **50**	33 mm / **100**	40 mm / **200**			
Ethyl Chloride	¼	2 mm / **0**	4 mm / **100**	7.5 mm / **300**	10 mm / **500**	15 mm / **1000**	24 mm / **1500**	31.5 mm / **2000**	@ 100° x 0.7
	½	2 mm / **0**	5 mm / **100**	12.5 mm / **300**	15 mm / **400**	19 mm / **500**	28 mm / **750**	36.5 mm / **1000**	
Methyl Bromide	2½	2 mm / **0**	5 mm / **10**	10 mm / **20**	15 mm / **30**	25 mm / **50**	39 mm / **75**	53 mm / **100**	
Methylene Chloride	½	2 mm / **0**	9 mm / **100**	12 mm / **250**	16 mm / **500**	20 mm / **750**	25 mm / **1000**	41.5 mm / **2000**	@ 40° x 1.3

CALIBRATION CHART

For the various temperatures, where a correction is required, multiply the test stain length by the correction factor and then ***convert this corrected length of stain*** to ppm concentration.

"Group B" Halogenated Hydrocarbons	Number of Strokes	Light Figures Below Indicate Length of Stain in Millimeters / **Bold** Figures Indicate Concentration in Parts Per Million							°F Temperature Correction Factors
Perchloroethylene	1/2	2 mm **0**	6 mm **25**	9 mm **50**	12.5mm **100**	20 mm **200**	29 mm **300**	35 mm **350**	@ 40° x 1.4 @ 50° x 1.2 @100° x 1.2
Propylene Dichloride	1/2	2 mm **0**	5 mm **25**	9 mm **50**	14 mm **75**	19 mm **100**	29 mm **150**	47 mm **250**	
1,1,2,2-Tetrabromoethane	2	2 mm **0**	4 mm **2**	7.5 mm **5**	10 mm **10**	18 mm **20**	24 mm **30**	35 mm **50**	
1,2,2,2-Tetrachloroethane	2	2 mm **0**	5 mm **5**	10 mm **10**	15 mm **15**	20 mm **20**	25 mm **25**	45 mm **50**	@ 40° x 1.2
1,2,2,2-Tetrachloroethane	4	2 mm **0**	5 mm **2**	11 mm **5**	21.5mm **10**	29 mm **15**	35 mm **20**	41.5mm **25**	
1,1,3,3-Tetrachloropropene	1	2 mm **0**	5 mm **10**	7.5 mm **20**	12.5mm **30**	21.5mm **50**	33 mm **75**	43 mm **100**	
1,1,3,3-Tetrachloropropene	2	2 mm **0**	5 mm **5**	7.5 mm **10**	16 mm **20**	25 mm **30**	43 mm **50**		
Trichloroethane (BETA-1,1,2)	1/2	2 mm **0**	9 mm **100**	15 mm **200**	21.5mm **300**	31.5mm **500**	38 mm **700**		
Trichloroethane (BETA-1,1,2)	1	2 mm **0**	12.5mm **50**	23 mm **100**	34 mm **150**	45 mm **200**			
Trichloroethylene	1/2	2 mm **0**	5 mm **50**	10 mm **100**	18 mm **200**	21.5 mm **300**	25 mm **400**	33 mm **600**	@ 40° x 1.3 @ 50° x 1.2
1,2,3-Trichloropropane	1/4	2 mm **0**	5 mm **50**	7.5 mm **75**	9 mm **100**	16mm **200**	25 mm **300**	35 mm **500**	@ 40° x 1.3 @ 50° x 1.2
1,2,3-Trichloropropane	1 1/2	2 mm **0**	7.5 mm **10**	11 mm **25**	19 mm **50**	25 mm **75**	33 mm **100**	36.5mm **125**	
Vinyl Chloride	1	2mm **0**	4 mm **25**	10 mm **50**	19 mm **100**	25 mm **150**	33 mm **200**	38 mm **250**	@ 40° x 1.2
Vinylidene Chloride	1/4	2 mm **0**	6 mm **20**	12.5 mm **50**	21.5mm **100**	29 mm **150**	35 mm **200**	48 mm **300**	
Vinylidene Chloride	1	2 mm **0**	5 mm **10**	11 mm **20**	18 mm **35**	25 mm **50**	44 mm **100**	48 mm **125**	

NOTE: This Calibration Chart should be used only with the designated lot of Tubes.

1. **WARNING:** *These instructions are applicable for use with the Universal Tester™ Pump or the Samplair® Pump only. These tubes are not calibrated for use with other than MSA® sampling devices.*

2. **Storage:** Do not store at temperatures above 90°F. Protect from prolonged exposure to light. Either of these conditions could cause a change in the sensitivity and accuracy of the tubes.

3. **Shelf Life:** If stored properly, these tubes can be used up to 30 months from date of manufacture. They must be used prior to the expiration date stamped on the box.

4. **Chemical Reaction and Color Change:** This detector tube provides a quantitative method for estimating the concentration of hexane in air. Indication is based on the color reaction of hexane with chromium trioxide in the presence of sulfuric acid. The color change is from orange to greenish brown.

5. **Test Procedure**
   5.1 Check pump for leaks in accordance with the instruction manual for the pump.
   5.2 When using the Universal Tester Pump, set the rotating pump head on #2 index mark. With the Samplair Pump, no indexing is necessary.
   5.3 Remove one detector tube from the box and break off tips using the tube breaker hole in the head of the pump.
   5.4 Insert the detector tube into the rubber tube holder of the pump, with the arrow on the tube pointing toward the pump.
   5.5 Align the index marks on the handle and back plate of the pump.
   5.6 Face the mounted detector tube into the atmosphere to be tested. Take an appropriate volume of air to be sampled by pulling the pump handle outward, the required number of strokes: Use 1 pump stroke (100 cc) for 25 to 5000 ppm.

   NOTE: Wait a minimum of 60 seconds after each stroke for the evacuated pump to fill. If hexane is present a greenish brown stain develops in the detector chemical.

   5.7 To determine the concentration, read the appropriate scale printed on the tube in ppm for the number of pump strokes used. Read at the end of the stain. If the end of stain should be uneven, read at the position of average length of stain.

6. **Limitations and Corrections**
   6.1 **Interferences:** Other aromatic and aliphatic hydrocarbons may interfere with accurate measurements. Also, some other reducing gases may interfere, for example carbon monoxide. These interferents generally increase the length of stain. Some examples of interferents are:

2000 ppm Benzene	740 ppm as Hexane
2000 ppm Toluene	225 ppm as Hexane
500 ppm n-Pentane	940 ppm as Hexane
50 ppm Carbon Monoxide	10 ppm as Hexane
500 ppm Hexanes (as n-hexane)	520 ppm as Hexane

   It should be noted that Benzene and Toluene at their TLV's would not interfere.

   6.2 **Temperature:** The Calibration Scale is based on air at about 77°F. To correct for temperature, multiply the ppm indicated by the correction factor in the table. Test conditions colder than 40°F. will result in an indication of hexane below that actually present; therefore, the tube should be kept above 40°F. as by keeping it in an inner pocket before use and warming it with the hand during test. Do not use at temperatures over 100°F.

   TEMPERATURE CORRECTION FACTOR

Temperature, °F	40	50	60	70	80	90	100
Correction Factor	1.53	1.34	1.19	1.06	0.95	0.85	0.77

   6.3 **Relative Humidity:** Relative humidity causes little effect on the accuracy of these tubes.
   6.4 **Pressure:** Calibrations are made at 740 mm mercury. A pressure correction must be used for pressures greatly different from this. To correct for pressure, use the following formula:

   $$\text{corrected reading, ppm} = \text{actual reading, ppm} \times \frac{740}{\text{test pressure in mm mercury}}$$

7. **Measurement Range:** This tube measures hexane over the range from 25 to 5000 ppm.

8. **Calibration and Accuracy:** Each lot of tubes is separately calibrated. Up to 250 ppm the tube is accurate to about ± 35% and above this concentration it is accurate to about ± 25%.

1. **WARNING:** *These instructions are applicable for use with the MSA Universal Tester™ Pump or the Samplair® Pump only. When using these tubes with any other MSA® sampling device, use the instructions and calibration values provided with that device. These tubes are not calibrated for use with other than MSA sampling devices.*

2. **Storage**
   Do not store at temperatures above 90°F. Protect from prolonged exposure to light. Either of these conditions could cause a change in the sensitivity and accuracy of the tubes.

3. **Shelf Life**
   If stored properly, these tubes can be used up to 30 months from date of manufacture. They must be used prior to the expiration date stamped on the box.

4. **Chemical Reaction and Color Change**
   This detector tube provides a quantitative method for estimating the concentration of mercury in air. Indication is based on the reaction of mercury with cuprous iodide. The color change is from light gray to light red.

5. **Test Procedure**
   5.1 Check pump for leaks in accordance with the instruction manual for the pump.
   5.2 When using the Universal Tester Pump, set the rotating pump head on #2 index mark. With the Samplair Pump, no indexing is necessary.
   5.3 Remove one detector tube from the box and break off both tips using the tube breaker hole in the head of the pump.
   5.4 Insert the detector tube into the rubber tube holder of the pump, with the arrow on the tube pointing toward the pump.
   5.5 Align the index marks on the handle and back plate of the pump.
   5.6 Face the mounted detector tube into the atmosphere to be tested. Take an appropriate volume of air to be sampled by pulling the pump handle outward to the selected position.
   Use 35 pump strokes (3500 cc) for 0 to 1.0 mg/m^3
   Use 18 pump strokes (1800 cc) for 1.0 to 2.0 mg/m^3

   NOTE: Wait a minimum of 40 seconds after each full stroke for the evacuated pump to fill. If mercury vapor is present, a light reddish stain develops in the detector chemical.

   5.7 To determine the concentration, read the appropriate scale printed on the tube in mg/m^3 for the number of pump strokes used. Read at the end of stain. If the end of stain should be uneven, read at the position of the average length of stain.

6. **Limitations and Corrections**
   6.1 Interferences: Other common gases in the range of their allowable limits do not interfere with accurate measurements.
   6.2 Temperature: Accuracy of indication is not affected by temperatures between 30° and 120°F. Test conditions colder than 30°F will result in an indication of mercury vapor below that actually present; therefore, the tube must be kept at a temperature between 30° and 120°F as by keeping it in an inner pocket before use and keeping it warm with the hand during the test. Do not use at temperatures above 120°F.
   6.3 Relative Humidity: Accuracy of indication is not affected by relative humidity between 10% and 90%. Relative humidity less than 10% or greater than 90% will result in an indication of mercury vapor below that actually present.
   6.4 Pressure: Calibrations are made at 740 mm mercury. A pressure correction must be used for pressures greatly different from this. To correct for pressure, use the following formula:

   $$\text{corrected reading, ppm} = \text{actual reading, ppm} \times \frac{740}{\text{test pressure in mm mercury}}$$

7. **Measurement Range:** This tube measures mercury over the range from 0 mg/m^3 to 2.0 mg/m^3.

8. **Calibration and Accuracy:** Each lot of tubes is separately calibrated.

1. **WARNING**: *These instructions are applicable for use with the Universal Tester™ Pump or the Samplair® Pump only. These tubes are not calibrated for use with other than MSA® sampling devices.*

2. **Storage**
   Do not store at temperatures above 90°F. Protect from prolonged exposure to light. Either of these conditions could cause a change in the sensitivity and accuracy of the tubes.

3. **Shelf Life**
   If stored properly, these tubes can be used up to 30 months from date of manufacture. They must be used prior to the expiration date stamped on the box.

4. **Chemical Reaction and Color Change**
   This detector tube provides a quantitative method for estimating the concentration of petroleum hydro-carbons in air. Indication is based on the color reaction of hydrocarbon compounds with chromium trioxide in the presence of sulfuric acid. The color change is from orange to green-brown.

5. **Test Procedure**
   5.1 Check pump for leaks in accordance with the instruction manual for the pump.
   5.2 When using the Universal Tester Pump, set the rotating pump head on #2 index mark. With the Samplair Pump, no indexing is necessary.
   5.3 Remove one detector tube from the box and break off tips using the tube breaker hole in the head of the pump.
   5.4 Insert the detector tube into the rubber tube holder of the pump, with the arrow on the tube pointing toward the pump.
   5.5 Align the index marks on the handle and back plate of the pump.
   5.6 Face the mounted tube into the atmosphere to be tested. Take an appropriate volume of air to be sampled by pulling the pump handle outward, the required number of strokes. Use 2 pump strokes (200 cc) for 200 to 2000 ppm.

   NOTE: Wait a minimum of 40 seconds after each stroke for the evacuated pump to fill. If any petroleum hydrocarbon is present a green-brown stain develops in the detector chemical.

   5.7 To determine the concentration, read the appropriate scale printed on the tube in mm for the number of pump strokes the end of stain. If the end of stain should be uneven, read at the position of the average length of stain. Refer to the Calibration Chart for the actual concentration.

6. **Limitations and Corrections**
   6.1 Interferences: Other aromatic and aliphatic hydrocarbons may give an indication.
   6.2 Temperature: The Calibration Scale is based on air at about 77°F. Test conditions colder than 40°F. will result in an indication of petroleum hydrocarbon below that actually present, therefore, the tube should be kept above 40°F. as by keeping it in an inner pocket before use and warming it with the hand during test. Do not use at temperatures over 100°F.
   6.3 Relative Humidity: Relative humidity causes little effect on the accuracy of these tubes.
   6.4 Pressure: Calibrations are made at 740 mm mercury. A pressure correction must be used for pressures greatly different from this. To correct for pressure, use the following formula:

$$\text{actual reading, ppm} \times \frac{740}{\text{test pressure in mm mercury}}$$

7. **Measurement Range**
   This tube measures the Petroleum Hydrocarbons over the range from 200 to 2000 ppm.

   Extension of the Range of Measurement: Concentration ranges of the compounds can be expanded by using more or less pump strokes. The gas concentration in ppm is then obtained by using the 2 pump stroke ppm value on the chart and the pump strokes actually used to determine the value.
   Example: 1 pump stroke reads 1500 ppm Pentane:

$$\text{Calculation:}\qquad 1500 \text{ ppm} \times \frac{2 \text{ P.S. scale}}{1 \text{ P.S. actually used}} = 3000 \text{ ppm}$$

   Extension of Usage
   For added usage the detector tube may be reversed and sampled from the opposite end, the same day. This enables one to get two tests from the same tube. In this procedure one must determine the length of the stain in mm by reading the mm scale at the start of the stain and at the finish of the stain and subtracting the two munerical readings. Or one can use their own mm scale to measure the stain length and then refer to the calibration chart.

8. **Calibration and Accuracy**
   Each lot of tubes is separately calibrated. Up to 200 ppm the tube is accurate to about 35% and above this concentration it is accurate to about ±25%.

PETROLEUM HYDROCARBONS CALIBRATION CHART

2 pump stroke sample of samplair pump

	200 ppm	500 ppm	1000 ppm	1500 ppm	2000 ppm
1. Propane	24.5 mm	59 mm			
2. n-Butane	35 mm	59 mm			
3. n-Pentane	22 mm	40.5 mm	50 mm	65.5 mm	
4. n-Hexane	15.5 mm	25 mm	36 mm	46 mm	53 mm
5. n-Heptane	11 mm	15.5 mm	22 mm	26.5 mm	31 mm
6. n-Octane	10 mm	14 mm	18 mm	21 mm	24 mm
7. iso-Octane	10.5 mm	17 mm	25 mm	29 mm	33.5 mm
8. n-Nonane	8.5 mm	12.5 mm	18 mm	22 mm	26 mm
9. Benzene	3 mm	5.5 mm	10 mm	14 mm	15.5 mm
10. Toluene	3 mm	4 mm	8.5 mm	11 mm	14 mm
11. Xylene	3 mm	4 mm	5.5 mm	7 mm	8.5 mm
12. Cyclohexane	12.5 mm	21.5 mm	32 mm	39 mm	45 mm
13. Carbon Monoxide	0 mm	0 mm	0 mm	0 mm	0 mm
14. Hydrogen Sulfide	3 mm	5.5 mm	10 mm	12.5 mm	15.5 mm
15. Gasoline *	7 mm	14 mm	22 mm	28 mm	33.5 mm
16. Kerosene * (grade 1)	1.5 mm	2 mm	3 mm	4 mm	5.5 mm

* Gasoline and Kerosene are mixtures so response varies with manufacture and grade.

1. **WARNING:** *These instructions are applicable for use with the Universal Tester™ Pump or the Samplair® Pump only. When using these tubes with any other MSA® sampling device, use the instructions and calibrated values provided with that device. These tubes are not calibrated for use with other than MSA sampling devices, and are not certified except when used with the Universal Tester or the Samplair Pump.*

2. **Storage**: Do not store at temperatures above 90°F. Protect from prolonged exposure to light. Either of these conditions could cause a change in the sensitivity and accuracy of the tubes.

3. **Shelf Life:** If stored properly, these tubes can be used up to 30 months from date of manufacture. They must be used prior to the expiration date stamped on the box.

4. **Chemical Reaction and Color Change:** This detector tube provides a quantitative method for estimating the concentration of toluene in air. Indication is based on the color reaction of iodine pentoxide and sulfuric acid impregnated silica gel with toluene. The color change is from white to brown.

5. **Test Procedure**
   5.1 Check pump for leaks in accordance with the instruction manual for the pump.
   5.2 When using the Universal Tester Pump, set the rotating pump head on #2 index mark. With the Samplair Pump no indexing is necessary.
   5.3 Remove one detector tube from the box and break off both tips using the tube breaker hole in the head of the pump.
   5.4 Insert the detector tube into the rubber tube holder of the pump, with the arrow on the tube pointing toward the pump.
   5.5 Align the index marks on the handle and back plate of the pump.
   5.6 Face the mounted detector tube into the atmosphere to be tested. Take an appropriate volume of air to be sampled by pulling the pump handle outward the required number of strokes. Use 4 pump strokes (400 cc) for 12.5 to 700 ppm.
      NOTE: Wait a minimum of 60 seconds after each full stroke for the evacuated pump to fill. If toluene is present, a brownish stain develops in the detector chemical. Measure the stain within 4 minutes after sampling.
   5.7 To determine the concentration, read the ppm PS scale on the tube at the end of the stain. Divide this number by the number of pump strokes taken to obtain the concentration in ppm. For example, if a reading of 400 ppm PS is obtained & 4PS's are used: 400 ppm PS ÷ 4 PS=100 ppm. If the end of stain is uneven, read at the position of average length of stain.

6. **Limitations and Corrections**
   6.1 **Interferences:** Other aromatic and aliphatic hydrocarbons interfere with accurate measurements. Also some other reducing gases may interfere for example, carbon monoxide and hydrogen sulfide. Some examples of interferents are:

10	ppm benzene indicates 10 ppm toluene
100	ppm xylene indicates 50 ppm toluene
100	ppm hexane indicates 15 ppm toluene
100	ppm carbon monoxide indicates 50 ppm toluene (very light stain)*
100	ppm carbon monoxide in 50 ppm toluene indicates 55 ppm toluene (10% high)
100	ppm carbon monoxide in 100 ppm toluene indicates 110 ppm toluene (10% high)
10	ppm hydrogen sulfide indicates 5 ppm toluene (very light stain)
10	ppm hydrogen sulfide in 50 ppm toluene indicates 55 ppm toluene (10% high)
0.3	ppm phosphine indicates 0 ppm toluene (no interference)
5	ppm hydrogen chloride indicates 0 ppm toluene (no interference)

   *Stain fades and disappears in 2 min. and would not constitute a serious interference with an experienced user.
   6.2 **Temperature and Relative Humidity:** The calibration scales were obtained at 75°F and 50 % relative humidity. At other temperatures and relative humidities, the concentration should be adjusted by multiplying by the correction factor given in the chart for the actual sampling temperature and relative humidity.

TEMPERATURE - HUMIDITY CORRECTION FACTORS							
%\°F	40	50	60	70	75	80	90	95
10	1.25	1.20	1.18	1.14	1.13	1.11	1.09	1.06
30	1.23	1.18	1.14	1.09	1.07	1.05	1.00	0.99
50	1.20	1.16	1.10	1.04	1.00	0.98	0.94	0.92
70	1.18	1.12	1.06	1.00	0.98	0.96	0.92	0.89
90	1.14	1.08	1.01	0.96	0.94	0.93	0.87	0.85

   6.3 **Pressure:** Calibrations are made at 740 mm mercury. A pressure correction must be used for pressures greatly different from this. To correct for pressure, use the following formula:

   $$\text{corrected reading, ppm} = \text{actual reading, ppm} \times \frac{740}{\text{test pressure in mm mercury}}$$

7. **Measurement Range:** This tube measures toluene over the range from 12.5 ppm to 700 ppm.

8. **Calibration and Accuracy:** Each lot of tubes is separately calibrated. Over the range from 100 to 500 ppm, the accuracy is ± 25%, while at 50 ppm it is ± 35%.

B

RESPONSE CURVES

NAME OF GAS	CURVE No.	PAGE No.	NAME OF GAS	CURVE No.	PAGE No.
Acetaldehyde	2	258	n-Heptane	19	258
Acetone	1	256	n-Hexane	20	260
Acetylene	45	259	Hydrogen	47	260
Acrylonitrile	3	259	Isobutane	49	259
Allyl Alcohol	4	257	Isobutyl Alcohol	34	257
Ammonia	52	260	Isopropyl Acetate	22	257
Amyl Alcohol (1-Pentanol)	5	257	Isopropyl Alcohol	23	256
Amylene (1-Pentene)	44	258	JP-4 (Use Heptane Curve)	19	256
Benzene	6	258	LPG (Use Propane Curve)	39	256
1,3-Butadiene	60	258	Methane	37	259
n-Butane	40	256	Methyl Acetate	25	258
n-Butyl Acetate	7	257	Methyl Alcohol	26	260
n-Butyl Alcohol	8	258	Methyl Ethyl Ketone	27	256
Butylene (1-Butene)	43	256	Methyl Isobutyl Ketone	24	256
Carbon Monoxide	10	260	Octane	28	256
Coke Oven Gas	74	260	Pentane	29	259
Cyclohexane	11	258	Propane	39	256
1,4-Dioxane	55	259	n-Propyl Acetate	73	257
Ethane	38	259	n-Propyl Alcohol	50	257
Ethyl Acetate	13	259	Propylene	42	258
Ethyl Alcohol	14	260	Styrene	30	257
Ethyl Chloride	72	259	Tetrahydrofuran	31	258
Ethyl Ether (Diethyl Ether)	15	258	Toluene	32	260
Ethylene	41	257	V.M. & P. Naphtha	46	256
Ethylene Oxide	17	260	Xylene	33	259
Gasoline, Unleaded	75	260			

WARNING! These response curves are provided for instructional use only and are not valid for all MSA Model 260 instruments. Always use the response curves or conversion factors provided with the instrument by the manufacturer.

Response Curve

Model 260

No.	Name	Formula	LEL
39	Propane	CH₃CH₂CH₃	2.2
43	Butylene	CH₃CH₂CH:CH₂	1.6
23	Isopropyl Alcohol	(CH₃)₂CHOH	2.0
19	n-Heptane	CH₃(CH₂)₅CH₃	1.05
28	Octane	CH₃(CH₂)₆CH₃	1.0

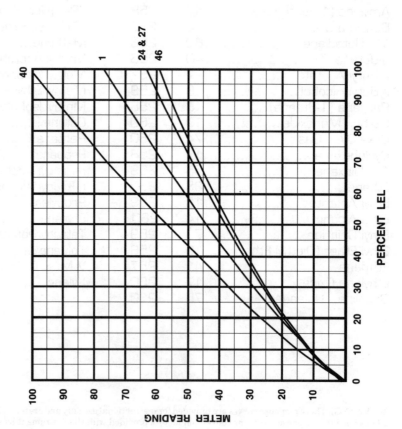

No.	Name	Formula	LEL
1	Acetone	CH₃COCH₃	2.15
24	Methyl Isobuty Ketone	CH₃CHOCH₂CH(CH₃)₂	1.2
27	Methyl Ethyl Ketone	C₂H₅COCH₃	1.7
46	V.M. & P. Naphtha		0.9
40	n-Butane	CH₃CH₂CH₂CH₃	1.6

Response Curve

No.	Name	Formula	LEL
41	Ethylene	H₂C:H₂	2.7
4	Allyl Alcohol	CH₂:CHCH₂OH	2.5
34	Isobutyl Alcohol	(CH₃)CHCH₂OH	1.7
5	Amyl Alcohol	CH₃(CH₂)₃CH₂OH	1.2
73	n-Propyl Acetate	C₃H₇OOCCH₃	1.7

Model 260

No.	Name	Formula	LEL
50	n-Propyl Alcohol	CH₂CH₂OH	2.1
22	Isopropyl Acetate	(CH₃)₂CHOOCH₃	1.8
30	Styrene	C₆H₅CH:CH₂	1.1
7	n-Butyl Acetate	CH₃COOC₄H₉	1.7

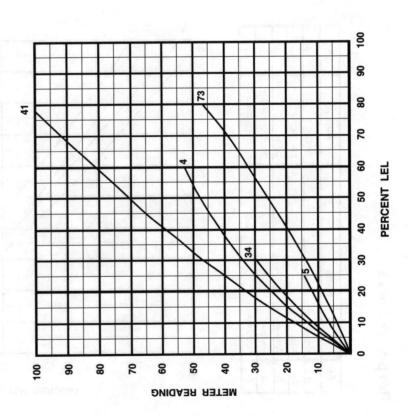

257

Response Curve

Model 260

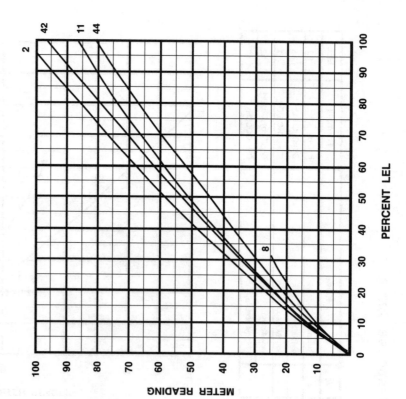

No.	Name	Formula	LEL
60	1,3-Butadiene	$CH_2{:}CHCH{:}CH_2$	2.0
25	Methyl Acetate	CH_3COOCH_3	3.1
15	Ethyl Ether	$C_2H_5OC_2H_5$	1.9
6	Benzene	C_6H_6	1.3
31	Tetrahydrofuran	$OCH_2CH_2CH_2CH_2$	2.0

No.	Name	Formula	LEL
2	Acetaldehyde	CH_3CHO	4.0
42	Propylene	$CH_2{:}CHCH_3$	2.0
11	Cyclohexane	C_6H_{12}	1.3
44	Amylene	$CH_3(CH_2)_2CH{:}CH_2$	1.5
8	n-Butyl Alcohol	$CH_3(CH_2)_2CH_2OH$	1.4

258

Model 260

No.	Name	Formula	LEL
37	Methane	CH₄	5.0
45	Acetylene	CH:CH	2.5
72	Ethyl Chloride	C2H5Cl	3.8
55	1, 4-Dioxane	OCH2CH2OCH2CH2	2.0
33	Xylene	C6H4(CH3)2	1.1

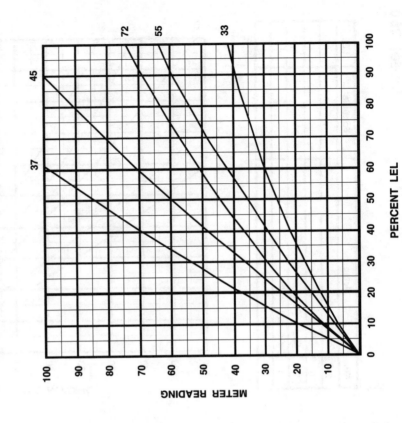

Response Curve

No.	Name	Formula	LEL
38	Ethane	CH3CH3	3.0
49	Isobutane	(CH3)3CH	1.8
29	Pentane	CH3(CH2)3CH3	1.5
3	Acrylonitrile	CH2:CHCN	3.0
13	Ethyl Acetate	CH3COOC2H5	2.0

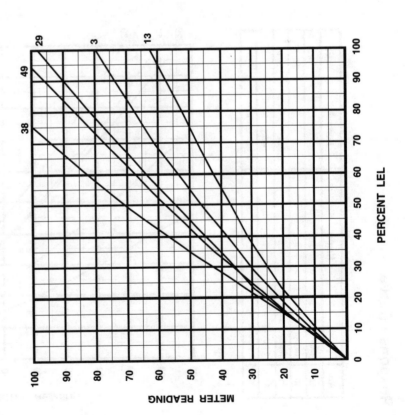

Response Curve

Model 260

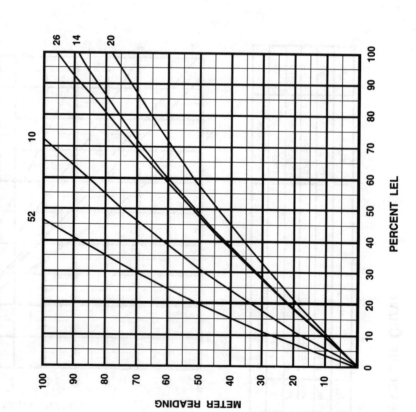

No.	Name	Formula	LEL
52	Ammonia	NH₃	16.0
10	Carbon Monoxide	CO	12.5
26	Methyl Alcohol	CH₃OH	6.0
14	Ethyl Alcohol	C₂H₅OH	3.3
20	n-Hexane	CH₃(CH₂)₄CH₃	1.1

No.	Name	Formula	LEL
74	Coke Oven Gas		4.8
47	Hydrogen	H₂	4.0
17	Ethylene Oxide	CH₂OCH₂	3.0
75	Gasoline, Unleaded		1.4
32	Toluene	C₆H₅CH₃	1.2

CONVERSION FACTORS FOR THE SCOTT ALERT MSA 360/361, GASTECHTOR

SCOTT ALERT
CONVERSION FACTORS FOR INSTRUMENT CALIBRATED TO HEXANE

GAS OR VAPOR	FACTOR	GAS OR VAPOR	FACTOR
ACETALDEHYDE	0.52	HEXANE	0.68
ACETONE	0.86	HYDROGEN	0.39*
ACETYLENE	0.56	ISOPROPYL ALCOHOL	0.73
AMMONIA	0.52	METHYL ETHYL KETONE	0.90
BENZENE	0.86	METHYL ISOBUTYL KETONE	0.95
1,3 - BUTADIENE	0.56	METHANE	0.38**
N-BUTANE	0.66	METHANOL	0.58
ISO-BUTANE	0.68	MINERAL SPIRITS	1.58
ISO-BUTYLENE	0.70	NAPHTHA, V.M. & P.	1.19
ISO-OCTANE	1.06	NITRO PROPANE	0.95
N-BUTYL ALCOHOL	1.06	N-OCTANE	1.46
CYCLOHEXANE	0.95	OCTENE	1.36
CYCLOHEXANONE	1.19	PENTENE	0.86
DIMETHYL FORMAMIDE	0.86	ISO-PENTANE	0.86
DIETHYL ETHER	0.95	ISOPRENE	0.58
N-DECANE	1.46	PROPANE	0.56
ETHANE	0.52	PROPYLENE	0.63
ETHYL ACETATE	0.90	STYRENE	1.27
ETHYL ALCOHOL	0.63	TETRAHYDROFURAN	0.63
ETHYL BENZENE	1.06	TOLUENE	1.00
ETHYLENE	0.53	VINYL ACETATE	0.70
HEPTANE	1.00	VINYL CHLORIDE	1.06
ETHYLENE OXIDE	0.61	O-XYLENE	1.36
N-HEXANE	1.00		

* above 39% LEL is off scale -- ** above 38% LEL is off scale

The actual % LEL of the gas/vapor concentration present is equal to the instrument indication multiplied by the factor listed.

EXAMPLE: Diethyl ether
-Instrument indicates 60
60 x 0.95 = 57% of LEL of diethyl ether

CAUTION: Correlation data is accurate to within ±10% full scale

WARNING! These conversion factors are provided for instructional use only and are not valid for all Scott Alert instruments. Always use the conversion factors provided with the instrument by the manufacturer.

MSA 360/361
CONVERSION FACTORS FOR INSTRUMENT CALIBRATED TO PENTANE

GAS OR VAPOR	FACTOR
ACETONE	1.0
ACETYLENE	0.8
BENZENE	1.1
ISO-BUTANE	1.0
N-BUTANE	1.0
BUTADIENE	0.8
BUTENE-1	1.0
CARBON MONOXIDE	0.6
ETHANE	0.7
ETHYL ACETATE	1.1
ETHYL ALCOHOL	1.0
ETHYLENE	0.8
ETHYLENE OXIDE	0.9
ETHYL ETHER	1.1
GASOLINE	1.0
HEPTANE	1.1
HEXANE	1.3
HYDROGEN	0.6
METHANE	0.6
METHYL ALCOHOL	0.7
METHYL ETHYL KETONE	1.6
PENTANE	1.0
PROPANE	0.9
PROPYLENE	1.0
PROPYLENE OXIDE	1.0
ISOPROPYL ALCOHOL	1.1
TETRAHYDROFURAN	0.8
TOLUENE	1.2

EXAMPLE:

When a Model 360 or Model 361, calibrated on Pentane, is used to test for propane, the conversion factor is 0.9. If the meter reads 50% LEL, the true propane concentration is:

(50% x 0.9) or 45% LEL

The response of a particular Model 360 or Model 361 may be higher or lower than the stated response. For this reason an accuracy tolerance of ± 25% should be applied in the interpretation of any meter response.

WARNING! These conversion factors are provided for instructional use only and are not valid for all MSA Model 360/361 instruments. Always use the conversion factors provided with the instrument by the manufacturer.

GAS OR VAPOR	METHANE		HEXANE	
	LEL FACTOR	PPM FACTOR	LEL FACTOR	PPM FACTOR
ACETONE	1.5	0.65	0.65	1.55
ACRYLONITRILE	-	0.54	-	1.31
BENZENE	2.1	0.46	0.87	1.11
BUTADIENE	2.1	0.82	0.88	2.0
CHLOROFORM	-	3.3	-	8.0
CARBON MONOXIDE	1.4	1.8	0.60	4.4
ETHYL ACETATE	2.0	0.65	0.81	1.61
ETHYL ALCOHOL	1.4	0.82	0.60	2.0
FORMALDEHYDE *	4.6	3.1	1.96	7.4
HEPTANE	2.4	0.40	1.03	0.98
HEXANE	2.3	0.41	1.00	1.00
HYDROGEN	1.1	0.59	0.47	1.43
METHYL CHLOROFORM	-	1.5	-	3.7
METHYL ETHYL KETONE	2.0	0.54	0.84	1.3
METHYLENE CHLORIDE	6.6	1.8	2.78	4.4
METHANE	1.0	1.0	0.42	2.4
PENTANE	1.6	0.76	0.71	1.81
PERCHLOROETHYLENE	-	1.9	-	4.5
N-PROPYL ACETATE	1.9	0.56	0.77	1.36
N-PROPYL ALCOHOL	1.9	0.65	0.80	1.52
STYRENE	3.0	0.54	1.24	1.31
TOLUENE	2.4	0.49	1.03	1.19
TRICHLOROETHYLENE	1.4	1.8	0.59	4.4

Please note: *that response varies from one sensor to another and the relative response of a sensor can change with age, so this data should be used for estimation purposes only.*

Assuming an instrument calibrated directly for methane or hexane, but used to observe a different gas, the equivalent response in %LEL (or ppm) for that gas is obtained by multiplying the observed %LEL (or ppm) reading by the LEL (or ppm) scale conversion factor.

* Methanol Free   *Relative Response / GasTechtor*

WARNING! These conversion factors are provided for instructional use only and are not valid for all GasTech instruments. Always use the conversion factors provided with the instrument by the manufacturer.

D

COMPOUND	IP (ev)	COMPOUND	IP (ev)
Acetaldehyde	10.21	2,3-Butadione	9.23
Acetic Acid	**10.37**	**n-Butanal**	**9.83**
Acetone	9.69	2-Butanal	9.73
Acetylene	**11.41**	**n-Butane**	**10.63**
Acetylene dichloride	9.80	1-Butanethiol	9.14
Acrolein	**10.10**	**2-Butanone**	**9.53**
Acrylonitrile	10.91	iso-Butanol	10.47
Allene	**9.83**	**s-Butanol**	**10.23**
Allyl alcohol	9.67	t-Butanol	10.25
Allyl chloride	**10.20**	**2-Butanol**	**10.1**
Aminoethanol	9.87	1-Butene	9.58
Ammonia	**10.15**	**cis-2-Butene**	**9.13**
Aniline	7.70	trans-2-Butene	9.13
Anisole	**8.22**	**3-Butene nitrile**	**10.39**
Arsine	9.89	n-Butyl acetate	10.01
Benzaldehyde	**9.53**	**s-Butyl acetate**	**9.91**
Benzene	9.25	n-Butyl alcohol	10.04
Benzenethiol	**8.33**	**n-Butyl amine**	**8.71**
Benzyl chloride	10.16	iso-Butyl amine	8.70
Benzonitrile	**9.71**	**s-Butyl amine**	**8.70**
Benzotrifluoride	9.68	t-Butyl amine	8.64
Bromobenzene	**8.98**	**n-Butyl benzene**	**8.69**
1-Bromobutane	10.13	iso-Butyl benzene	8.68
2-Bromobutane	**9.98**	**t-Butyl benzene**	**8.68**
1-Bromobutanone	9.54	Butyl Cellosolve	8.68
1-Bromo-2-chloroethane	**10.63**	**n-Butyl mercaptan**	**9.15**
Bromochloromethane	10.77	iso-Butyl ethanoate	9.95
Bromoethane	**10.28**	**iso-Butyl mercaptan**	**9.12**
Bromoethene	9.80	iso-Butyl methanoate	10.46
Bromoform	**10.48**	**1-Butyne**	**10.18**
1-Bromo-3-hexanone	9.26	2-Butyne	9.85
Bromoethyl ethyl ether	**10.08**	**n-Butyraldehyde**	**9.86**
Bromomethane	10.53	Carbon disulfide	10.13
1-Bromo-2-methylpropane	**10.09**	**Carbon tetrachloride**	**11.28**
2-Bromo-2-methylpropane	9.89	Chlorobenzene	9.07
1-Bromopentane	**10.10**	**Chlorobromomethane**	**10.77**
1-Bromopropane	10.18	1-Choro-2-bromoethane	10.63
2-Bromopropane	**10.08**	**1-Chlorobutane**	**10.67**
1-Bromopropene	9.30	2-Chlorobutane	10.65
2-Bromopropene	**10.06**	**1-Chlorobutanone**	**9.54**
3-Bromopropene	9.70	1-Chloro-2,3-epoxy propane	10.60
2-Bromothiophene	**8.63**	**Chloroethane (ethyl chloride)**	**10.97**
o-Bromotoluene	8.79	Chlorethene	10.00
m-Bromotoluene	**8.81**	**2-Chloroethoxyethene**	**10.61**
p-Bromotoluene	8.67	1-Chloro-2-fluorobenzene	9.16
1,3-Butadiene	**9.07**	**1-Chloro-3-fluorobenzene**	**9.21**

1-Chloro-2-fluoroethene (cis)	9.87	**1,2-Difluorobenzene**	**9.31**
1-Chloro-2-fluoroethene (trans)	**9.87**	1,4-Difluorobenzene	9.15
Chloroform	11.37	**Difluorodibromomethane**	**11.18**
o-Chloroiodobenzene	**8.35**	Difluorormethylbenzene	9.45
1-Chloro-2-methylbenzene	8.72	**Di-iodomethane**	**9.34**
1-Chloro-3-methylbenzene	**8.61**	Di-iosobutyl ketone	9.04
1-Chloro-4-methylbenzene	8.78	**Di-isopropylamine**	**7.73**
Chlorobutadiene	**8.79**	1,1-Dimethoxyethane	9.65
Chloromethylethyl ether	10.08	**Dimethoxymethane**	**10.00**
Chloromethymethyl ether	**10.25**	Dimethyl amine	8.24
1-Chloro-2-methylpropane	10.66	**2,3-Dimethylbutadiene**	**8.72**
Chloroprene	**8.80**	2,2-Dimethyl butane	10.06
1-Chloropropane	10.82	**2,2-Dimethyl butane-3-one**	**9.18**
2-Chloropropane	**10.78**	2,3-Dimethyl butane	10.02
3-Chloropropene	10.04	**2,3-Dimethyl-2-butene**	**8.30**
2-Chlorothiophene	**8.68**	3,3-Dimethyl butanone	9.17
o-Chlorotoluene	8.83	**Dimethyl disulfide**	**8.46**
m-Chlorotoluene	**8.83**	Dimethyl ether	10.00
p-Chlorotoluene	8.70	**Dimethylformamide**	**9.45**
Cumene (isopropyl benzene)	**8.75**	3,5-Dimethyl-4-heptanone	9.04
Crotonaldehyde	9.73	**2,2-Dimethyl-3-pentanone**	**8.98**
Cyanoethene	**10.91**	2,2-Dimethyl-propane	10.35
Cyanogen bromide	10.91	**Dimethyl sulfide**	**8.69**
3-Cyanopropene	**10.39**	Di-n-propyl disulfide	8.27
Cyclobutane	10.50	**Di-n-propyl ether**	**9.27**
Cyclohexane	**9.98**	Di-isopropyl ether	9.20
Cyclohexanone	9.14	**Di-n-propyl amine**	**7.84**
Cyclohexene	**8.95**	Di-n-propyl sulfide	8.30
Cyclo-octatetraene	7.99	**p-Dioxane**	**9.13**
Cyclopentadiene	**8.55**	Epichlorohydrin	10.60
Cyclopentane	10.52	**Ethane**	**11.65**
Cyclopentanone	**9.26**	Ethanal	10.21
Cyclopentene	9.01	**Ethanol**	**10.62**
Cyclopropane	**10.06**	Ethene (ethylene)	10.52
2-Decanone	9.40	**Ethyl acetate**	**10.11**
Diborane	**11.00**	Ethyl amine	8.86
Dibromochloromethane	10.59	**Ethyl amyl ketone**	**9.10**
1,1-Dibromoethane	**10.19**	Ethyl benzene	8.76
Dibromomethane	10.49	**Ethyl bromide**	**10.29**
1,2-Dibromopropane	**10.26**	Ethyl butyl ketone	9.02
1,2-Dichlorobenzene	9.07	**Ethyl chloride (chloroethane)**	**10.98**
1,3-Dichlorobenzene	**9.12**	Ethyl chloroacetate	10.20
1,4-Dichlorobenzene	8.94	**Ethyl ethanoate**	**10.10**
Dichlorodifluoromethane	**11.75**	Ethyl disulfide	8.27
1,1-Dichloroethane	11.06	**Ethylene chlorohydrin**	**10.90**
1,2-Dichloroethane	**11.04**	Ethylene dibromide (EDB)	10.37
cis-Dichloroethene	9.65	**Ethyleneimine**	**9.94**
trans-Dichloroethene	**9.66**	Ethylene oxide	10.56
Dichloromethane	11.35	**Ethyl formate**	**10.61**
1,2-Dichloropropane	**10.87**	Ethyl iodide	9.33
1,3-Dichloropropane	10.85	**Ethyl mercaptan**	**9.29**
1,1-Dichloropropanone	**9.71**	Ethyl methanoate	10.61
2,3-Dichloropropene	9.82	**Ethyl isothiocyanate**	**9.14**
Dicyclopentadiene	**7.74**	Ethyl methyl sulfide	8.55
Dibutyl amine	7.69	**Ethyl propanoate**	**10.00**
Diethoxymethane	**9.70**	Ethyl trichloroacetate	10.44
Diethyl amine	8.01	**Mono-fluorobenzene**	**9.20**
Diethyl ether	**9.53**	Mono-fluoroethene	10.37
n-Diethyl formamide	8.89	**Mono-fluoromethanal**	**11.4**
Diethyl ketone	**9.32**	Fluorotribromomethane	10.67
Diethyl sulfide	8.43	**o-Fluorotoluene**	**8.92**

m-Fluorotoluene	8.92	**2-Methylbutanal**	**9.71**
p-Fluorotoluene	**8.79**	2-Methylbutane	10.31
Freon 11	11.77	**2-Methyl-1-butene**	**9.12**
Freon 12	**12.31**	3-Methyl-1-butene	9.51
Freon 13	12.91	**3-Methyl-2-butene**	**8.67**
Freon 14	**16.25**	Methyl-n-butyl ketone	9.34
Freon 22	12.45	**Methyl butyrate**	**10.07**
Freon 113	**11.78**	Methyl Cellosolve	9.6
2-Furaldehyde	9.21	**Methyl chloroacetate**	**10.35**
Furan	**8.89**	Methyl chloride	11.28
Furfural	9.21	**Methyl chloroform**	**11.25**
Hexachloroethane	**11.22**	Methylcyclohexane	9.85
n-Hexane	10.18	**4-Methylcyclohexene**	**8.91**
n-Heptane	**10.07**	Methylcyclopropane	9.52
2-Heptanone	9.33	**Methyl dichloroacetate**	**10.44**
4-Heptanone	**9.12**	Methyl ethanoate	10.27
1-Hexene	9.46	**Methyl ethyl ketone**	**9.53**
Hexanone	**9.34**	Methyl ethyl sulfide	8.55
Hexamethylbenzene	7.85	**2-Methyl furan**	**8.39**
Hydrazine	**8.93**	Methyl iodine	9.54
Hydrogen cyanide	13.91	**Methyl isobutyl ketone**	**9.30**
Hydrogen selenide	**9.88**	Methyl isobutyrate	9.98
Hydrogen sulfide	10.46	**Methyl isopropyl ketone**	**9.32**
Hydrogen telluride	**9.14**	Methyl isothiocyanate	9.25
Iodine	9.28	**Methyl methacrylate**	**9.74**
Iodobenzene	**8.73**	Methyl methanoate	10.82
1-Iodobutane	9.21	**Methyl mercaptan**	**9.44**
2-Iodobutane	**9.09**	2-Methylpentane	10.12
Iodoethane (ethyl iodide)	9.33	**3-Methylpentane**	**10.08**
Iodomethane (methyl iodide)	**9.54**	2-Methylpropane	10.56
1-Iodo-2-methylpropane	9.18	**2-Methylpropanal**	**9.74**
1-Iodopentane	**9.19**	2-Methyl-2-propanol	9.70
1-Iodopropane	9.26	**2-Methylpropene**	**9.23**
2-Iodopropane	**9.17**	Methyl-n-propyl ketone	9.39
o-Iodotoluene	8.62	**Methyl styrene**	**8.35**
m-Iodotoluene	**8.61**	Morpholine	8.88
p-Iodotoluene	8.50	**Naphthalene**	**8.10**
Isoamyl acetate	**9.90**	Nitric oxide	9.25
Isoamyl alcohol	10.16	**Nitrobenzene**	**9.92**
Isobutane	**10.57**	Nitrotoluene	9.43
Isobutyl amine	8.70	**n-Nonane**	**10.21**
Isobutyl acetate	**9.97**	5-Nonanone	9.10
Isobutyl alcohol	10.47	**n-Octane**	**10.24**
Isobutyl formate	**10.46**	3-Octanone	9.19
Isobutyraldehyde	9.74	**4-Octanone**	**9.10**
Isopentane	**10.32**	1-Octene	9.52
Isoprene	8.85	**n-Pentane**	**10.35**
Isopropyl acetate	**9.99**	Pentachloroethane	11.28
Isopropyl alcohol	10.16	**1,3-Pentadiene (cis)**	**8.59**
Isopropyl amine	**8.72**	1,3-Pentadiene (trans)	8.56
Isopropyl benzene	8.75	**Pentafluorobenzene**	**9.84**
Isopropyl ether	**9.20**	Pentamethylbenzene	7.92
Isovaleraldehyde	9.71	**n-Pentanal**	**9.82**
Mesitylene	**8.40**	2,4-Pentanedione	8.87
Mesityl oxide	9.08	**2-Pentanone**	**9.39**
Methanol	**10.85**	3-Pentanone	9.32
Methyl acetate	10.27	**1-Pentene**	**9.50**
Methyl acrylate	**10.72**	Perchloroethylene	9.32
Methyl amine	8.97	**Perfluoro-2-butene**	**11.25**
Methyl bromide	**10.53**	Perfluoro-1-heptene	10.48
2-Methyl-1,3-butadiene	8.85	**n-Perfluoropropyl iodide**	**10.36**

n-Perfluoropropyl iodomethane	9.96	**1,2,3-Trimethylbenzene**	**8.48**
n-Perfluoropropyl methyl ketone	**10.58**	1,2,4-Trimethylbenzene	8.27
Phenol	8.69	**1,3,5-Trimethylbenzene**	**8.39**
Phenyl ether	**8.09**	2,2,4-Trimethyl pentane	9.86
Phenyl isocyanate	8.77	**2,2,4-Trimethyl-3-pentanone**	**8.82**
Phosphine	**9.96**	n-Valeraldehyde	9.82
Pinene	8.07	**Valeric acid**	**10.12**
Propadiene	**10.19**	Vinyl acetate	9.19
n-Propanal	9.95	**Vinyl benzene (styrene)**	**8.47**
Propane	**11.07**	Vinyl bromide	9.80
1-Propanethiol	9.20	**Vinyl chloride**	**10.00**
n-Propanol	**10.51**	4-Vinylcyclohexene	8.93
Propanone	9.69	**Vinyl ethanoate**	**9.19**
Propenal (acrolein)	**10.10**	Vinyl fluoride	10.37
Propene	9.73	**Vinyl methyl ether**	**8.93**
Prop-1-ene-2-ol	**8.2**	Water	12.58
Prop-2-ene-1-ol	9.67	**Xenon difluoride**	**11.5**
Propionaldehyde	**9.98**	Xenon tetrafluoride	12.9
n-Propyl acetate	10.04	**o-Xylene**	**8.56**
n-Propyl alcohol	**10.20**	m-Xylene	8.56
n-Propyl amine	8.78	**p-Xylene**	**8.45**
n-Propyl benzene	**8.72**		
Propylene	9.73		
Propylene dichloride	**10.87**		
Propylene oxide	10.22		
n-Propyl ether	**9.27**		
n-Propyl formate	10.54		
Propyne	**10.36**		
Pyridine	9.32		
Styrene	**8.47**		
Tetrachloroethene	9.32		
1,1,2,2-Tetrachloroethane	**11.10**		
1,2,3,4-Tetrafluorobenzene	9.61		
1,2,3,5-Tetrafluorobenzene	**9.55**		
1,2,4,5-Tetrafluorobenzene	9.39		
Tetrafluoroethene	**10.12**		
Tetrahydrofuran	9.54		
Tetrahydropyran	**9.26**		
1,2,4,5-Tetramethylbenzene	8.03		
2,2,4,4-Tetramethyl-3-pentanone	**8.65**		
Thioethanol	9.29		
Thiomethanol	**9.44**		
Thiophene	8.86		
1-Thiopropanol	**9.20**		
Toluene	8.82		
Tribromoethene	**9.27**		
1,1,1-Trichlorobutanone	9.54		
1,1,1-Trichloroethane	**11.25**		
1,1,2-Trichloroethane	11.0		
Trichloroethene	**9.45**		
Trichloromethyl ethyl ether	10.08		
Triethylamine	**7.50**		
1,2,4-Trifluorobenzene	9.37		
1,3,5-Trifluorobenzene	**9.32**		
Trifluoroethene	10.14		
1,1,1-Trifluoro-2-iodoethane	**10.10**		
Trifluoroiodomethane	10.40		
Trifluoromethylbenzene	**9.68**		
Trifluoromethylcyclohexane	10.46		
1,1,1-Trifluoropropene	**10.9**		
Trimethyl amine	7.82		

SOURCES:

Photovac Technical Bulletin No. 11
USEPA Superfund Compendium
NIOSH Pocket Guide to Chemical Hazards
HNU Systems instruction manuals

E

INSTRUMENT PROCEDURES AND CHECKLISTS

OPERATING PROCEDURE SUMMARIES

HNU Photoionization Detectors

Photvac MicroTIP

Foxboro Model 128 OVA (Survey Mode)

MSA Model 360/361

MSA Model 260

Matheson 8057 Hazardous Gas Leak Detector

Ludlum Model 14C Radiation Meter

INSTRUMENT CHECK LISTS

_____ HNU Photoionization Detectors

_____ Photovac MicroTIP

_____ Foxboro Model 128 OVA (Survey Mode)

_____ MSA Model 360/361

_____ MSA Model 260

_____ Matheson 8057 Hazardous Gas Leak Detector

_____ Ludlum Model 14C Radiation Meter

HNU PHOTOIONIZATION DETECTOR (ANALOG DISPLAY*)

1. Examine cable plug to ensure that 12 pins inside socket are intact and not bent or missing.

2. With instrument OFF, attach probe assembly to box by matching alignment slot in the plug to the connector. If present, small red button must be firmly depressed by cable plug collar.

3. Turn switch to BATT for battery check, needle display should go to OK zone. If red LED lights-turn instrument off immediately and recharge. If there is insufficient battery power, the ISPI-101 will shut itself off.

4. Turn to STANDBY; unit is electronically on but UV lamp is still off; use ZERO knob to bring needle to zero. Allow warm up for a few minutes.

5. Instrument display should be stable; turn to most sensitive range, 0 - 20; lamp is now on. Slight needle movement should be noted; record initial background reading (usually between 0.2 and 1.0 unit) and span setting (usually around 9.8 for 10.2 probe, 5.0 for 11.7 probe). Allow to warm up for 5-10 minutes.

6. Calibrate with manufacture's span gas; unlock SPAN knob and adjust setting if required. Record new SPAN setting (whole number in window, decimal on dial). Lock span knob.

7. After calibration, check background readings on 0-20 scale; there should be no significant change. If still less than 1, set background reading to 1.0 with ZERO adjust knob. Adjusting background with zero knob does not affect calibration.

8. HNU now ready to use.

9. Remember that compounds with IPs greater than UV lamp eV intensity will NOT be detected.

* includes P1-101, HW 101, GP-101 and ISPI-101

PHOTOVAC MICROTIP

1. Press button on handle to turn instrument on. Pump should be activated, screen reads: WARMING UP NOW, PLEASE WAIT.

2. Press BATT (#1 button) to determine battery voltage level. Normal operating voltage range is 6.0-8.5. LoBat message indicates battery pack is low, shut off instrument and replace with fresh battery pack or all stored data will be lost. Recharge depleted battery pack

3. When warmed up, screen will read UC, and indicate time and date.

4. When meter indicates READY, press light switch (#3) to determine UV lamp intensity; backlighting will also be activated. Normal lamp intensity is between 1000 and 4000. Record lamp intensity. Hold light switch down to decrease backlight intensity, then release. Press and hold switch down again to turn backlight off. Press CLEAR or EXIT to return to READY mode.

5. Press MAX switch; the maximum concentration previously encountered will be displayed for 15 seconds, time and date it was recorded will also be displayed. Press MAX switch twice to clear maximum reading; message will read PRESS CLEAR TO RESET MAX. Press CLEAR switch, message will read MAX CLEARED.

6. Press EVENT switch, use up/down arrows until DELETE EVENTS is displayed. Push EVENT switch, display will ask DELETE WHAT?. Use arrows to find ALL EVENTS, then push ENTER again. EVENT data-logger now clear to accept up to 255 recording events.

7. Fill a sample bag with zero (hydrocarbon free) air and a second bag with a known concentration of calibration gas (isobutylene). Have bags filled and ready before starting calibration sequence. Do not use ambient air to zero the instrument; always use hydrocarbon free air as specified by the manufacturer.

8. Press CAL switch (#8), display will read RESPONSE FACTOR 1.00? If using isobutylene, press ENTER as a yes answer. Screen will read CONNECT ZERO GAS AND PRESS ENTER. Do not press another key until zero gas is attached to meter inlet. Wait a few seconds, then press ENTER. Screen will briefly read CALIBRATING, then it will quickly switch to CONNECT SPAN GAS AND PRESS ENTER. Immediately attach sample bag with isobutylene, then press ENTER. Screen will indicate the concentration used when last calibration was performed; if current span gas concentration is the same, press ENTER, if not, enter new concentration, then press ENTER. When calibration is complete, screen will indicate READY, and continue to give concentration readings as long as the sample bag is attached. MicroTIP readings should be the same as the calibration gas. If readings are significantly different repeat the entire calibration procedure.

9. Disconnect sample bag. Meter is now calibrated and ready for use. Record initial background readings.

10. It is possible to incorrectly calibrate the MicroTIP to ambient air instead of span gas if the ENTER key is pressed before the sample bag containing isobutylene is attached, or if the operator delays in attaching the bag to the meter inlet. If this occurs, the instrument will consistently give a reading of >2500; the entire calibration sequence must be repeated.

11. Previous EVENT data will remain stored until recorded over or deleted. To view previous data, press PLAY (#7), then press * SETUP. Display message will ask PLAY FROM LAST EVENT? and a number will be displayed. The number, e.g. 18, indicates data from 18 events are available for reveiw. To view all previous EVENT data, enter the number 1 on the keypad then press ENTER. Screen will read MIN AVE MAX? use arrows to select minimum, average or maximum readings, then press ENTER. The time, date and reading for each event will be displayed, the sequence will then be repeated until the EXIT switch is pressed. To increase or decrease speed of playback, press up or down arrows. To freeze playback press ENTER, then press up or down arrow to go forward or backward until desired event data is located.

12. To record data, press EVENT key, use arrows until screen reads ADVANCE EVENT, then press ENTER to record event data. Bottom left of display screen will indicate event number, e.g. 003. To record another reading, press EVENT, then ENTER.

FOXBORO MODEL 128 OVA (SURVEY MODE)

Start-Up Procedure:

1. Attach readout assembly sample line and readout jack to appropriate connectors on instrument box (sidepack). Use 1/2" wrench to gently tighten sample connector.

2. Make sure teflon collar is inside probe handle; insert probe extension into handle and hand tighten. (Note: the probe extension contains a particulate filter and should always be in place when the pump is on.)

3. Toggle INSTRUMENT switch to battery test position BATT (pull toggle switch away from sidepack, then move up or down); hold in position and check readout; needle should deflect into OK range.

4. Toggle INSTRUMENT switch to ON position, allow to warm up at least five minutes.

5. Toggle PUMP switch to ON.

6. Adjust ALARM VOLUME knob to lowest setting by turning all the way counter-clockwise.

7. Unit is now electronically on but flame has not been ignited yet. Allow to warm up a few minutes.

8. Set range switch to x1, adjust needle display to read 1.0. Turn ALARM VOLUME knob all the way clockwise. Using ADJUST knob, set needle to desired alarm level; turn rheostat knob on back of readout assembly until alarm comes on. The alarm is activated by needle position, not actual reading, e.g. if alarm is set at 5 units on the x1 range, readings of 50, and 500 will also elicit an alarm response.

9. Turn ALARM VOLUME knob counter-clockwise (off). Keep range switch set at x1: use ADJUST knob to bring needle to zero. Allow warm up to continue for a few more minutes.

10. Open H2 TANK valve at least one full turn; reading on high pressure indicator should be at least 300 psi (OVA burns about 150 psi per hour).

11. Open H2 SUPPLY valve at least one full turn; reading on low pressure indicator should be approximately 10 psi.

12. SWITCH TO X10 RANGE TO PREVENT DAMAGE TO NEEDLE DISPLAY

13. WAIT A FULL 60 SECONDS AFTER TURNING ON HYDROGEN SUPPLY

14 Press red igniter button located above readout connectors; hydrogen ignition will cause faint pop and needle deflection. DO NOT keep button depressed for more than 6 seconds. If ignition does not occur, wait 30-60 seconds and try again.

15. Switch back to x1 setting; record initial background reading (usually between 1 and 5 units) and span setting (usually around 3.00). Allow to warm up for 5-10 minutes. If needle appears to be drifting, wait another 5 minutes and then check for drift again.

16. If background appears inordinately high, use hydrocarbon-free air to zero out hydrogen fuel background.

17. Methane calibration gas must be placed in a sample bag and then drawn into the OVA under ambient pressure; DO NOT feed calibration gas from a pressurized cylinder directly into the OVA. Adjust range setting to accomodate calibration gas concentration (e.g. if methane concentration is 50 ppm, set range to x10-meter now reads in the 10-100 unit range).

18. Open valve on sample bag, then attach hose or tubing to OVA probe extension. Unlock and adjust GAS SELECT if required and record setting (whole number in window, decimal on dial). Optimum setting is 3.00. When calibration is complete, disconnect tubing and then shut valve on sample bag. Relock SELECT knob.

19. The GAS SELECT knob is the span potentiometer; it DOES NOT "select" which gas will be detected.

20. After calibration, check background reading on x1 range; there should be no significant change. If less than 1.0, set background reading to at least 1.0 with ADJUST knob. Adjusting the background reading does not affect calibration.

21. Turn ALARM VOLUME knob clockwise to on position.

22. OVA now ready to use.

Shut Down Procedure:

1. Close H2 TANK valve, then H2 Supply valve.

2. When H2 SUPPLY low pressure indicator reading is zero, toggle INSTRUMENT switch to OFF (H2 TANK high pressure indicator does not go to zero).

3. Wait a few seconds, then toggle PUMP switch to OFF.

4. OVA is shut off.

Recommendations:

1. Check BATTery before turning Instrument switch to off, recharge as needed.

2. Check fuel high pressure indicator and record pressure. Needle indicator will gradually fade toward zero over time, giving an inaccurate reading of actual tank pressure. When H2 TANKvalve is opened again, actual pressure will again be correctly displayed on indicator.

MSA MODEL 361/360

1. Turn Function knob to SCAN. Pulsating audible alarm should sound, and alarm light should go on. Press RESET button, alarm light and audible alarm should go off; if it does not, turn Function knob to horn off (amber light should go on in horn-off mode).

2. If continuous alarm sounds, check readout for BATT; if BATT message present, turn instrument off and recharge before use.

3. Flow indicator should be visible and audible, indicating adequate air flow.

4. Press SELECT button until %LEL value is visible in readout; adjust LEL ZERO knob until reading is 0% LEL. (Lift up on knob and then turn to adjust.)

5. Press SELECT button to obtain % OXYgen value; adjust OXY CALIBRATE knob until reading is 20.9% oxygen.

6. Press SELECT button to obtain PPM TOX value; adjust TOX ZERO knob until reading is 0 ppm. (MSA 361 TOX value is for 0 - 50 ppm H2S, MSA 360 TOX value is for 0 - 500 ppm CO.)

7. Turn Function knob to SCAN for automatic scanning of all gases, or MANUAL for continuous readout of only one gas. Readings should remain stable.

8. Check calibration with MSA calibrant gases.

9. Instrument now ready for use.

First-time users should acquaint themselves with alarm levels for each sensor as follows:

Turn Function knob to HORN OFF; press SELECT button to view %LEL reading. Slowly turn %LEL Zero knob until alarm light comes on. Note % LEL alarm level and record. Turn knob back to 0%, press RESET button.

Repeat procedure for OXYgen and PPM TOX values, note and record alarm levels; return to normal readings and reset after each.

MSA MODEL 260

1. Turn middle knob to ON position; pulsating alarm should sound, alarm lights should go on. Press ALARM RESET button, audible alarm and lights should go off, if not, turn middle knob to HORN OFF position (green light blinks when in horn-off mode).

2. Flow indicator float should be visible and audible, indicating adequate air flow.

3. Press CHECK button and observe needle deflection in %LEL window display (needle should be in battery OK portion of scale (80-100%). If battery reading not in OK range, change or recharge battery before use.

4. Turn ZERO LEL knob until needle is at 0% LEL (lift up on knob and then turn to adjust).

5. Turn CALIBRATE O_2 knob until needle lies directly over offset line which indicates 20.9% oxygen (darkened line above N in oxygen display window).

6. Readings should remain stable.

7. Check calibration with MSA calibration gas.

8. Instrument now ready for use.

First-time users should acquaint themselves with alarm levels for both sensors as follows:

Slowly turn %LEL ZERO knob and watch needle display until audible alarm and light are activated, note and record alarm level. Turn knob back to 0%, press RESET button.

Repeat procedure for oxygen, return to normal setting and reset.

MATHESON 8057 HAZARDOUS GAS LEAK DETECTOR

1. Turn instrument on; pump should start and alarm sounds. Battery light should be on. If light not on, turn off and recharge or replace batteries.

2. Adjust wheel setting clockwise until alarm sounds; turn counter-clockwise until alarm just shuts off.

3. Allow unit to warm up 2-5 minutes; check alarm setting again as described above.

4. Test unit by briefly exposing to sample in Check Vial; alarm will sound. Allow unit to run in clean air until alarm shuts off (may take several minutes or more).

5. If meter is needed immediately, exhale toward probe instead of using Check Vial; unit will alarm briefly and quickly return to normal condition.

6. Unit is now ready for use.

LUDLUM MODEL 14C RADIATION METER

1. Open battery compartment lid (turn screw 1/4 turn to left) and insert two D size batteries. Close lid by turning screw 1/4 turn to right.

2. Turn instrument on by turning range switch to x1000; push BAT switch to check battery. Needle should deflect into BAT TEST portion of scale. If needed, replace batteries.

3. Use electrical connector cord to attach external detector to box.

4. Switch response switch to F (fast). Turn audio ON to activate audible chirper response.

5. Expose detector to check source on x1000 scale, then move to lower scales until meter reading is obtained. Record reading obtained.

6. Press RES button to reset meter to zero.

7. Meter is ready for use.

HNU PID (ANALOG DISPLAY) CHECK LIST:

_____ Probe attached to box. UV lamp eV capacity is _____.

_____ Battery in OK range, red LED not illuminated.

_____ Zero unit on STANDBY.

_____ Instrument readings stable.

_____ Initial background reading on 0 - 20 range is _____ units.

_____ Calibrate with recommended span gas

     HNU should read _____ with _____ eV probe.

     Actual HNU reading _____ with Span setting at _____.

_____ Background on 0 - 20 scale after calibration is _____.

_____ Adjust background to 1.0 units if less than 1.0 after calibration.

Shut Down:

_____ Check battery status, recharge or replace as needed.

PHOTOVAC MICROTIP CHECK LIST:

_____ Instrument on, pump turns on, display screen lit.

_____ Battery voltage _____. LoBat message not displayed.

_____ Screen reads READY, time and date displayed are correct.

_____ UV lamp intensity is _____.

_____ MAX cleared.

_____ EVENTS cleared.

_____ Sample bag with zero air and calibration gas ready.

Calibration gas concentration _____ ppm.

_____ Meter reads _____units while sampling calibration gas at end of calibration sequence.

_____ Background reading after calibration _____units.

_____ Press EVENT key, arrow key to ADVANCE EVENT, then press ENTER. EVENT number indicated is _____.

_____ Instrument now ready for use.

Shut Down:

_____ Check battery status, recharge or replace as needed; all stored data will be lost if battery becomes severely depleted.

FOXBORO MODEL 128 ORGANIC VAPOR ANALYZER (OVA) CHECK LIST:

Start Up Procedure:

_____ Readout assembly connected to sidepack; probe extension in place.

_____ Battery in OK range.

_____ Instrument turned on; allowed to warm up at least 5 minutes.

_____ Pump switch on, alarm volume off.

_____ Total warm-up time at least 5 minutes.

_____ Alarm level set, then alarm volume turned off.

_____ Instrument zeroed on x1 range.

_____ Total warm-up time at least 5 minutes.

_____ Hydrogen tank psi_____, hydrogen supply psi_____.

_____ Ignite hydrogen on x10 range.

_____ Initial background on x1 scale is _____units.

_____ Warm-up time with flame on at least 5 minutes, readings stable.

_____ Calibrate with _____ppm methane.

OVA reads _____ units with Gas Select at _____.

_____ Background on x1 scale is _____units.

_____ Adjusted to 1.0 if less than 1.0.

_____ Alarm volume turned on.

Shut Down:

_____ H2 TANK valve closed.

_____ H2 UV valve closed.

_____ Final H2 TANK pressure is _____ psi.

_____ H2 SUPPLY pressure reading is zero, toggle INSTRUMENT OFF.

_____ Toggle PUMP OFF.

_____ Check battery status, recharge or replace as needed.

MSA MODEL 360/361 INSTRUMENT AND CALIBRATION CHECK LIST:

_____ Instrument on, pump turns on.

_____ Audible alarm sounds upon turn-on.

_____ Alarm light on upon turn-on.

_____ Reset button resets alarms, audible alarm and light off.

_____ Horn off mode; amber light on.

_____ Check alarm levels on manual mode (reset after each)

Alarm level for oxygen sensor is _____ %.

Alarm level for %LEL is _____ %.

Alarm level for toxic sensor (H_2S or CO) is _____ ppm.

_____ Set normal ambient levels for each sensor.

_____ Instrument readings stable.

_____ Calibrate instrument using MSA calibrant gases

_____ % oxygen _____ % LEL _____ ppm CO or H_2S.

_____ Instrument response during calibration is

_____ % oxygen _____ % LEL _____ ppm CO or H_2S.

_____ Instrument set to manual or scan mode, horn on.

_____ Instrument ready for use.

MSA MODEL 260 INSTRUMENT AND CALIBRATION CHECK LIST:

_____ Instrument on, pump turns on.

_____ Alarm sounds upon turn-on.

_____ Alarm lights (2) activated upon turn-on.

_____ Reset button resets alarm horn, lights off.

_____ Horn off mode; blinking green light.

_____ Battery in OK range.

_____ Check alarm levels on manual mode (reset after each)

Alarm level for oxygen sensor is _____%.

Alarm level for %LEL is _____%.

_____ Set normal ambient levels for both sensors.

_____ Instrument readings stable.

_____ Calibrate instrument using MSA calibrant gas

_____ % oxygen   _____ % LEL.

_____ Instrument response during calibration is

_____ % oxygen   _____ % LEL.

_____ Horn on, green light on.

_____ Instrument ready for use.

MATHESON 8057 HAZARDOUS GAS LEAK DETECTOR CHECK LIST:

_____ Instrument on, pump turns on.

_____ Alarm sounds, alarm light on.

_____ Red battery light is on (if not on, replace or recharge batteries).

_____ Adjust wheel until alarm just shuts off.

Wheel setting is _____.

_____ Allow unit to warm up 2-5 minutes.

_____ Readjust alarm; wheel setting is _____.

_____ Test unit with Check Vial or exhale into unit.

_____ Unit returns to normal, now ready for use.

LUDLUM MODEL 14C RADIATION METER

_____ Batteries inserted, instrument on x1000 scale.

_____ Batteries in OK range.

_____ Audio chirper on.

_____ Switch to F for fast response.

_____ External detector connected, responds to check source on

      x_____ range; approximate reading is _____mR/hr.

_____ Reset to zero, range on x1 or x 0.1, meter ready for use.

Shut Down:

_____ Remove batteries.

AIR MONITORING INSTRUMENTATION DAILY CALILBRATION LOG							
Operator				Date			
Project				Project Number			
Location				Time			
Instrument Model	Initial Bkgrd	Calibrant Gas and Concentration	Zero Air or Ambient	Span	Meter Response	Battery Check	Time

Comments:

Operator Signature _____   Date _____
(Sign all sheets)   Sheet No. _____

AIR MONITORING INSTRUMENTATION
DAILY SUMMARY LOG

Operator		Date	
Project		Project Number	
Location		HSO	

Site Conditions

General Weather Conditions

Temp		Humidity		Wind Conditions	

Instrument Models	Location	Time	Background	Reading	Time	Background	Reading

Comments:

Operator Signature _____

(Sign all sheets)

Date _____

Sheet No. _____

AIR MONITORING INSTRUMENTATION
DAILY SUMMARY LOG CONTINUATION SHEET

Operator				Date			
Project				Project Number			
Location				HSO			

Instrument Models	Location	Time	Background	Reading	Time	Background	Reading

Comments:

Operator Signature _____   Date _____
(Sign all sheets)   Sheet No. _____

AIR MONITORING INSTRUMENTATION
DAILY REPORT LOG

Operator		Date	
Project		Project Number	
Location		HSO	

Site Conditions

General Weather Conditions

Temp	Humidity	Wind Conditions

INSTRUMENT	CGI	OXY	FID	PID	TOX	RAD	PART
MODEL NO.							
TIME							
BACKGROUND							
TIME							
READ							
TIME							
READ							
TIME							
READ							
TIME							
READ							
TIME							
READ							
TIME							
READ							

Comments:

Operator Signature _____
(Sign all sheets)

Date _____
Sheet No. _____

AIR MONITORING INSTRUMENTATION
DAILY REPORT LOG CONTINUATION SHEET

Operator		Date	
Project		Project Number	
Location		HSO	

INSTRUMENT	CGI	OXY	FID	PID	TOX	RAD	PART
MODEL NO.							
TIME							
BACKGROUND							
TIME							
READ							
TIME							
READ							
TIME							
READ							
TIME							
READ							
TIME							
READ							
TIME							
READ							
TIME							
READ							

Comments:

Operator Signature _____ Date _____
(Sign all sheets)   Sheet No. _____

AIR MONITORING INSTRUMENTATION
WEEKLY CALILBRATION LOG

Operator		Week of	
Project		Project Number	
Instrument		HSO	

Calibration Date	Initial Bkgrd	Calibrant Gas and Concentration	Zero Air or Ambient	Span	Meter Response	Battery Check

Required Field Maintenance/Adjustments	Date	Battery Charged	Date

Comments:

Operator Signature _____ Date _____

REVIEW QUESTIONS

CHAPTER 1

1. characterize hazards and risk, select protective equipment, identify areas of contamination, evaluate mitigation and cleanup, assess need for medical monitoring
2. flammable gases and vapors, oxygen deficiency, radiation
3. 1
4. A
5. hydrogen, butadiene, ethylene oxide
6. not in a Group; requirements exceed those of Group A
7. intended use, detection range, ease of use, cost, inherent safety, limitations such as operating temperature range and battery life.
8. box or case, readout display, function switches, sample intake, sensor, battery
9. sampling rate, length of sample hose, temperature, type of sensor
10. calibration or calibration check
11. false
12. 2, 1
13. true
14. recovery
15. advantages: no pump or moving parts, short lag time, provides almost immediate instrument reading
disadvantages: sensor more exposed, more affected by dust and temperature extremes, may be more difficult to access remote sampling locations
16. advantages: sensor protected within instrument, can sample remote locations, less susceptible to external contamination
disadvantages: requires more battery power, susceptible to sample dilution and pump malfunction, increased lag time and recovery time
17. true
18. before initial use, in a clean, non-contaminated area
19. no
20. calibration gas

CHAPTER 2

1. 760
2. 760, 14.7
3. vapor pressure
4. 1300 x 10 = 13,000 ppm
5. air
6. 1.29
7. 1
8. 1.29
9. 1
10. 29
11. 29
12. 2; settle
13. 100°F
14. water
15. 1
16. heavier
17. miscible
18. 1000 gm in 0.5 liter
1 gm in 0.5 ml
19. 100
20. flammable
21. ignitable
22. gases
23. flammable range

24. ingestion, inhalation, skin/eye contact and absorption, injection
25. air-purifying respirator
26. positive pressure supplied air
27. 10,000
28. 19,000
29. 200,000 ppm (1300 x 150 = 195,000)
30. yes

CHAPTER 3

1. color change or stain
2. 100
3. toward
4. false
5. piston pump
6. 0 to 40°C (32 to 104°F)
7. time (expiration date), light, humidity
8. interfering or cross specific
9. distal end
10. approximate middle value
11. 85 to 115
12. 1000
13. continue pump stroke protocol and monitor color change
14. dosimeter tubes
15. no color change; concentration of contaminate, if present, is below detection limit of tube

CHAPTER 4

1. catalytic filament
2. true
3. increase
4. decrease
5. false
6. true
7. metal oxide sensor (MOS) or semiconductor
8. false
9. calibrant
10. conversion factors
11. true
12. multiplied
13. identity of gas is known, only one gas is present, must have response curve/factor
14. charcoal filter
15. extension hose, filters, liquid trap, calibration kit

16. inhibitor filter
17. silicones and silicates, organic metals, sulfates
18. warm-up
19. calibration
20. false
21. false
22. true
23. heavier or lighter
24. methane, pentane, hexane

CHAPTER 5

1. electrolyte solution
2. hydrogen sulfide, sulfur dioxide, carbon monoxide, carbon dioxide ammonia, chlorine, ozone, freon (see Table 5-3 for partial list)
3. temperature, interfering gases, atmospheric pressure, membrane susceptible to damage
4. false
5. 20.9
6. true
7. true
8. false
9. ammonia, methane, hydrogen sulfide (see Table 5-4 for partial list)
10. gold foil

CHAPTER 6

1. ultraviolet (UV) light
2. 10.2, 10.6
3. true
4. lamp window absorbs moisture, lamp window degraded by UV light
5. ionization potential (IP)
6. carbon monoxide, oxygen, carbon tetrachloride
7. water vapor, dust, non-ionizable gases and vapors, heated atmospheres
8. water vapor, methane
9. false
10. false
11. no, background varies by location and should not be zeroed out
12. true
13. false
14. 0 to 2000
15. benzene or isobutylene, isobutylene

CHAPTER 7

1. a flame
2. false
3. requires oxygen, requires hydrogen fuel
4. false
5. carbon monoxide, carbon dioxide, carbon disulfide, nitrogen dioxide
6. false
7. methane
8. false
9. oxygen deficiency, low hydrogen pressure, kinking of sample line, presence of flammable vapors
10. organic hydrocarbons heavier than methane and ethane
11. false
12. contamination in hydrogen fuel, contamination in ambient air
13. 60
14. 99.999
15. false

CHAPTER 8

1. GC column, injection system, detector
2. carrier gas
3. type of column packing material, column length, column temperature, and carrier gas flow rate
4. precolumn is short and designed to trap heavy, slow-moving materials; longer analytical column separates faster, more volatile compounds.
5. 99.999% pure, less than 0.1 ppm total hydrocarbon contamination
6. PIDs and FIDs
7. IR radiation source, wavelength filter, sample cell, detector
8. true
9. true
10. water vapor or humidity

CHAPTER 9

1. alpha, beta, gamma
2. radio and TV waves, sound, light
3. they are heavy and travel only a short distance

4. decrease time of exposure, increase distance from source, use shielding
5. 0.01 to 0.05 mR/hr
6. gamma
7. beta, alpha
8. survey instrument indicates radiation present at time of sampling; dosimeter measures accumulated radiation dose over exposure interval
9. beta
10. 100
11. 1 mR/hr, 50 mrems over the entire course of pregnancy
12. 5 rem
13. 0.64 mR
14. true
15. false
16. lead, earthen berms, concrete, containers filled with water or soil

CHAPTER 3

1. The Dräger Hydrocarbon 0.1%/b and the MSA Petroleum Hydrocarbon tubes both detect butane. The Dräger tube has a detection range of 0.1 to 0.8%, or 1000 to 8000 ppm; the MSA tube has a detection range only 200 to 500 ppm.

3. The instructions (section 7, Specificity) for the Dräger Alcohol 100/a tube clearly state that the tube will not cross react with benzene (an aromatic hydrocarbon similar to toluene) and gasoline (which contains toluene). There is insufficient information to determine cross-reactivity for MSA Alcohol and Aromatic Hydrocarbon tubes.

5. 80-90 ppm; 50 seconds.

7. The anticipated color change for the ethyl acetate tube is orange to brownish green; petroleum hydrocarbons turn the entire tube pale green. A color change to pale green suggests that a significant concentration of ethyl acetate is not present, but one or more other materials, which may include petroleum hydrocarbons, are present. It is not possible to estimate the actual concentration of the interfering gases or vapors present.

9. The Dräger Hydrogen Sulfide 2/a tube is not affected by sulfur dioxide concentrations up to 200 ppm.

CHAPTER 4

1. 15%

3. If the tank contains acetone liquid, acetone vapors must also be present above the liquid. The lack of a meter response is probably due to insufficient oxygen concentrations within the tank, or the presence of acetone vapors in excess of the UEL.

5. The Scott-Alert conversion factor list in Appendix C indicates that the meter is calibrated to hexane, at a hexane spill the meter reading indicates the actual %LEL concentration present, or 22-24% LEL. The MSA 360 conversion factor for hexane is 1.3. An MSA 360 meter reading around 18% would be expected (22-24% LEL divided by 1.3 = 17-19% LEL).

7. The concentration of formaldehyde, if present, is less than the detection range of the instrument.

CHAPTER 5

1. No, entry should not be made unless the identity of the displacing gas or vapor is known. The oxygen value obtained indicates that there is approximately 4% by volume (0.8% x 5) or 40,000 ppm of unknown contaminants in the vault.

3.   The presence of steam can affect the function of electrolyte sensors when the hot water vapor/aerosol mixture condenses on the sensor membrane. Such condensation can result in erroneous, low readings.

5.   Mercury detectors which use a gold foil sensor are not affected by hydrocarbons and solvents. Detectors which utilize UV absorption can be expected to give a false-positive response in the presence of solvents and other materials which absorb UV light.

CHAPTER 6

1.   Tetrachloroethylene, with an IP of 9.32, will be ionized by a 10.2 eV lamp; carbon tetrachloride has an IP of 11.47 and will not be ionized. The presence of any non-ionizable material will decrease meter response to those materials which can be ionized.

3.   The concentration of MEK directly over the sorbent pads is greater than the meter can handle; instruments often give erratic readings in this situation.

5.   This procedure is incorrect. The user should zero the instrument electronically or with zero air, then calibrate the instrument. The background reading obtained after calibration represents ionizable materials in the air. Considering the location, a background of 6.2 is very plausible; background may be greater or less, depending on meter location. Zeroing out the background would result in an inability to measure background or on-site readings of less than 6.2 units.

CHAPTER 7

1.   No, the presence of water vapor in air does not affect FID readings. FIDs should not, however, be used in the presence of water aerosols or steam.

3.   Repeated flame-out suggests that there is insufficient hydrogen fuel pressure, kinking in the sample line, or insufficient oxygen to maintain the hydrogen-fed flame.

5.   The hydrogen fuel contamination background was removed when the instrument was zeroed with hydrocarbon-free air; this procedure is correct. The meter can thus reflect the actual ambient air background, independent of the fuel background caused by the hydrocarbon contaminates in the fuel supply.

CHAPTER 10

1.   A charcoal filter may be used to determine the contribution of methane to the OVA (FID) and CGI response. Detector tubes may be used to rule out the presence of significant concentrations of specific chemical classes (i.e. acids, amines, alcohols, petroleum hydrocarbons). Further characterization would require use of column chromotography to determine the approximate number of contaminants.

3.   Air monitoring instrumentation can be used to assess the relative extent of contamination around the drums. Unless the identity of the materials are known, instrumen-

tation cannot be used to determine if the drum contents are hazardous. For example, high meter readings are obtained from non-hazardous materials such as citrus juices, spices, and other food products; non-flammable materials in confined spaces can elicit significant CGI readings. The contents of each drum must be sampled and evaluated for specific hazard criteria such as corrosivity, flammability, and reactivity; chemical analysis to identify the material will be necessary before a definitive hazard determination can be made.

CHAPTER 11

1. The GasTechtor conversion factor for toluene is 2.4, a meter reading of 100% LEL indicates that over 100% of the LEL is actually present. The oxygen reading suggests that toluene has displaced 0.3% oxygen, or 1.5% by volume of air (15,000 ppm); this indicates that the concentration present is greater than the LEL of toluene, which is 12,000 ppm or 1.2% by volume in air.

3. GM detectors can be saturated when exposed to excessively high radiation levels. The meter's failure to respond to the check source suggests detector saturation. The responder should immediately leave the area and check his personal radiation dosimeter. A different radiation detector, or a GM detector designed for use at higher radiation levels, should be employed.

NOTES

Index